ANNUAL EDITIONS

Violence and Terrorism

07/08

Tenth Edition

EDITOR

Thomas J. Badey

Randolph-Macon College

Thomas J. Badey is an associate professor of Political Science and the director of the International Studies Program at Randolph-Macon College in Ashland, Virginia. He received a B.S. in Sociology from the University of Maryland (University College) in 1986 and an M.A. in Political Science from the University of South Florida in 1987. In 1993 he received a Ph.D. in Political Science from the *Institut für Politische Wissenschaft of the Ruprecht-Karls Universität* in Heidelberg, Germany. He served as a security policeman in the United States Air Force from 1979 to 1988 and was stationed in the United States, Asia, and the Middle East. Dr. Badey regularly teaches courses on international terrorism and has written a number of articles on the subject. He is also the editor of McGraw-Hill Contemporary Learning Series' *Annual Editions: Homeland Security*.

2460 Kerper Blvd., Dubuque, IA 52001

Visit us on the Internet

http://www.mhcls.com

Credits

1. **The Concept of Terrorism**
   Unit photo—U.S. Customs and Border Protection photo by James R. Tourtellotte
2. **Tactics of Terrorism**
   Unit photo—The McGraw-Hill Companies, Inc./Leroy Webster
3. **State-Sponsored Terrorism**
   Unit photo—U.S. AID Photo
4. **International Terrorism**
   Unit photo—Tomi/PhotoLink/Getty Images
5. **Terrorism in America**
   Unit photo—NASA photo
6. **Terrorism and the Media**
   Unit photo—DoD photo by Tech. Sgt. Sean M. Worrell, U.S. Air Force
7. **Terrorism and Religion**
   Unit photo—M. Freeman/PhotoLink/Getty Images
8. **Women and Terrorism**
   Unit photo—Tim Hall/Getty Images
9. **Government Response**
   Unit photo—Royalty-Free/CORBIS
10. **Future Threats**
    Unit photo—USCG photo by PA2 Kyle Niemi

Copyright

Cataloging in Publication Data
Main entry under title: Annual Editions: Violence and Terrorism. 2007/2008.
1. Violence and Terrorism—Periodicals. I. Badey, Thomas J., *comp.* II. Title: Violence and Terrorism.
ISBN-13: 978–0–07–351619–6 ISBN-10: 0–07–351619–8 658'.05 ISSN 1096–4274

Tenth Edition

Cover image © Alan Crosthwaite/Acclaim Images/fStop Images
Printed in the United States of America 1234567890QPDQPD9876 Printed on Recycled Paper

Editors/Advisory Board

Members of the Advisory Board are instrumental in the final selection of articles for each edition of ANNUAL EDITIONS. Their review of articles for content, level, currentness, and appropriateness provides critical direction to the editor and staff. We think that you will find their careful consideration well reflected in this volume.

Staff

EDITORIAL STAFF

Larry Loeppke, Managing Editor
Jay Oberbroeckling, Developmental Editor
Jade Benedict, Developmental Editor
Nancy Meissner, Editorial Assistant

PERMISSIONS STAFF

Lenny J. Behnke, Permissions Coordinator
Lori Church, Permissions Coordinator
Shirley Lanners, Permissions Coordinator

TECHNOLOGY STAFF

Luke David, eContent Coordinator

MARKETING STAFF

Julie Keck, Senior Marketing Manager
Mary Klein, Marketing Communications Specialist
Alice Link, Marketing Coordinator
Tracie Kammerude, Senior Marketing Assistant

PRODUCTION STAFF

Beth Kundert, Production Manager
Trish Mish, Production Assistant
Kari Voss, Lead Typesetter
Jean Smith, Typesetter
Karen Spring, Typesetter
Sandy Wille, Typesetter
Tara McDermott, Designer
Maggie Lytle, Cover Graphics

Preface

In publishing ANNUAL EDITIONS we recognize the enormous role played by the magazines, newspapers, and journals of the public press in providing current, first-rate educational information in a broad spectrum of interest areas. Many of these articles are appropriate for students, researchers, and professionals seeking accurate, current material to help bridge the gap between principles and theories and the real world. These articles, however, become more useful for study when those of lasting value are carefully collected, organized, indexed, and reproduced in a low-cost format, which provides easy and permanent access when the material is needed. That is the role played by ANNUAL EDITIONS.

Successful elections in Afghanistan and Iraq have failed to bring about peace or reduce the threat of terrorism. Attacks on coalition and government forces are once again on the rise. Continuing violence in Latin America and Europe and new violence in Asia and Africa remind us that the Global War on Terrorism is far from over. Violence and terrorism affect our lives and will continue to affect our lives well into the twenty-first century. Political, economic, social, ethnic and religious strife, fueled by the availability of weapons, advances in technology, and an ever-present international media, set the stage for the future of violence and terrorism. The only real defense against terrorism is to try to understand terrorism. Thus, *Annual Editions: Violence and Terrorism* continues to address some basic questions: Why does terrorism occur? How does it occur? Who are the terrorists? How can or should governments respond?

The selections for this edition of *Annual Editions: Violence and Terrorism* were chosen to reflect a diversity of issues, actors, and points of view. This revision incorporates many new articles that reflect the changes that have occurred since the previous edition was published. While, as always, influenced by recent events, this volume endeavors to maintain sufficient regional and topical coverage to provide students with a broad perspective and a basis for understanding contemporary political violence. Articles for this introductory reader were chosen from a variety of sources and reflect diverse writing styles. It is our hope that this broad selection will provide easy accessibility at various levels of understanding and will thus stimulate interest and discussion. In addition to the aforementioned considerations, elements such as timeliness and readability of the articles were important criteria used in their selection.

This anthology is organized into ten units. Unit 1 attempts to address the complex task of conceptualizing terrorism. It underlines the difficulty of finding a commonly accepted definition of the problem. Unit 2 examines the methods employed by terrorists. It focuses on current terrorist tactics used by terrorist organizations. Unit 3 examines the role of state-sponsors in international terrorism. Focusing primarily on so called “rogue states” this section sheds light on the complex and changing relationship between sponsor states and terrorist organizations. Unit 4 provides a quick overview of some of the major actors in contemporary international terrorism. Shifting the focus to the domestic front, Unit 5 examines terrorism in America. In addition to articles on domestic terrorism, it includes articles that examine the potential impact of policies to combat terrorism on personal freedom and civil liberties. Unit 6 highlights the role that the media play in terrorism and points to potential consequences of increased terrorist-media interaction. Unit 7 examines the complex relationship between terrorism and religion and how religion is used to justify contemporary political violence. Unit 8 looks at the role of women, who continue to play an important and increasingly active role in contemporary political violence. Unit 9 examines the methods and policies governments use to counter the threat of terrorism. Finally, Unit 10 explores trends, projections, and future threats in international terrorism.

This anthology provides a broad overview of the major issues associated with political violence and terrorism. It is our hope that *Annual Editions: Violence and Terrorism* will introduce students to the study of terrorism and serve as a stimulus for further exploration of this vital topic.

I would like to thank the many scholars who provided feedback and submitted suggestions for articles to be included in this volume. I am also grateful to a group of undergraduate students who helped in the development of this anthology. I am particularly grateful to Eric Loftin Smith who has worked as my research assistant on a number of projects over the past three years. I would also like to thank Erin Bumgarner, Bobby Graves, Caitlynn Husz, Yana Kanoukoeva, Megan Kittle, and Paul Patterson for their help in sorting through and reviewing the numerous articles that were submitted for consideration. These students provided valuable insights and above all a critical students' perspective, which made my job much easier. I hope that you, the reader, will take the time to fill out the article rating form in the back of this anthology so we can continue to improve future editions.

Thomas J. Badey
Editor

Contents

UNIT 1
The Concept of Terrorism

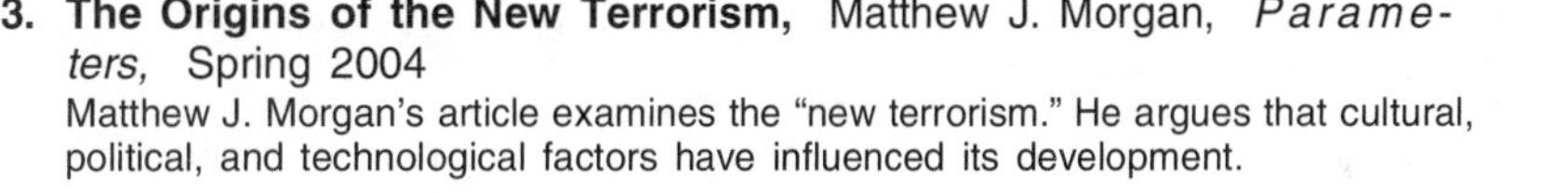

UNIT 2
Tactics of Terrorism

The concepts in bold italics are developed in the article. For further expansion, please refer to the Topic Guide and the Index.

UNIT 3
State-Sponsored Terrorism

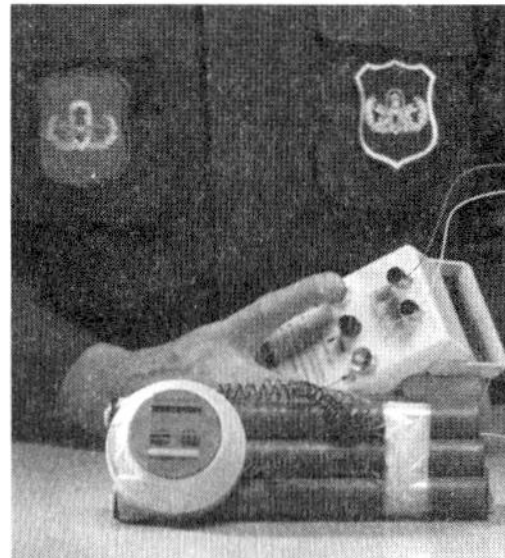

UNIT 4
International Terrorism

The concepts in bold italics are developed in the article. For further expansion, please refer to the Topic Guide and the Index.

UNIT 5
Terrorism in America

The concepts in bold italics are developed in the article. For further expansion, please refer to the Topic Guide and the Index.

UNIT 6
Terrorism and the Media

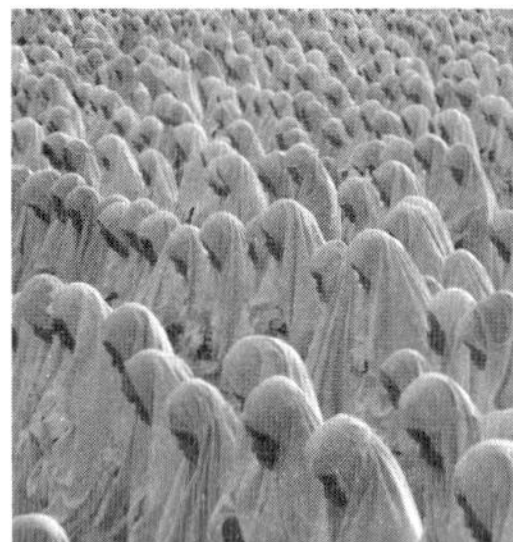

UNIT 7
Terrorism and Religion

The concepts in bold italics are developed in the article. For further expansion, please refer to the Topic Guide and the Index.

UNIT 8
Women and Terrorism

UNIT 9
Government Response

The concepts in bold italics are developed in the article. For further expansion, please refer to the Topic Guide and the Index.

UNIT 10
Future Threats

The concepts in bold italics are developed in the article. For further expansion, please refer to the Topic Guide and the Index.

Topic Guide

This topic guide suggests how the selections in this book relate to the subjects covered in your course. You may want to use the topics listed on these pages to search the Web more easily.

On the following pages a number of Web sites have been gathered specifically for this book. They are arranged to reflect the units of this *Annual Edition.* You can link to these sites by going to the student online support site at *http://www.mhcls.com/online/.*

ALL THE ARTICLES THAT RELATE TO EACH TOPIC ARE LISTED BELOW THE BOLD-FACED TERM.

Internet References

The following internet sites have been carefully researched and selected to support the articles found in this reader. The easiest way to access these selected sites is to go to our student online support site at *http://www.mhcls.com/online/*.

AE: Violence and Terrorism 07/08

The following sites were available at the time of publication. Visit our Web site—we update our student online support site regularly to reflect any changes.

General Sources

DefenseLINK (U.S. government)
http://www.defenselink.mil
The Department of Defense's public affairs online service provides DoD news releases and other public affairs documents. This is a gateway to other DoD agencies (i.e., Secretary of Defense, Army, Navy, Air Force, Marine Corps).

International Network Information Center at University of Texas
http://inic.utexas.edu
This gateway has many pointers to international sites, organized into African, Asian, Latin American, Middle East, Russian, and East European subsections.

U.S. Central Intelligence Agency Home Page
http://www.cia.gov
This site includes publications of the CIA, such as the *1996 World Fact Book; 1995 Fact Book on Intelligence; Handbook of International Economic Statistics, 1996;* and *CIA Maps*.

U.S. White House
http://www.whitehouse.gov
This official Web page for the White House includes information on the President and Vice President and What's New. See especially The Virtual Library and Briefing Room (today's releases) for Hot Topics and latest Federal Statistics.

UNIT 1: The Concept of Terrorism

MIPT Terrorism Knowledge Base
http://www.tkb.org/Home.jsp
Developed by the National Memorial Institute for the Prevention of Terrorism (MIPT), the Terrorism Knowledge Base offers in-depth information on terrorist incidents, groups, and trials.

Political Science Resources/International Relations
http://www.lib.umich.edu/govdocs/psintl.html
The Documents Center of the University of Michigan contains material relating to violence and terrorism under several headings, including Peace and Conflict and Human Rights. The site includes simulations.

Terrorism: Background and Threat Assessment Links
http://www.fas.org/irp/threat/terror.htm
This site provides documents covering a broad range of topics on Terrorism.

The Terrorism Research Center
http://www.terrorism.com
The Terrorism Research Center is dedicated to informing the public of the phenomena of terrorism and information warfare. This site features essays and thought pieces on current issues, as well as links to other terrorism documents, research, and resources. Navigate the site by clicking on the area of interest.

UNODC - Terrorism Definitions
http://www.unodc.org/unodc/terrorism_definitions.html
The lack of agreement on a definition of terrorism has been a major obstacle to meaningful international countermeasures. Cynics have often commented that one state's "terrorist" is another state's "freedom fighter." This site provides the consensus on the definition of "terrorism" by the United Nations Council.

UNIT 2: Tactics of Terrorism

FrontPage Magazine—Ecoterrorism and Us
http://www.frontpagemag.com/Articles/Printable.asp?ID=1277
Columnist Robert Locke examines the motivations behind modern ecoterrorism and signals a warning for possible future terrorist actions.

JCSS Military Resources
http://www.tau.ac.il/jcss/lmas.html
The Jaffe Center for Strategic Studies at Tel-Aviv University lists five different groups of Web site Directories on low-intensity warfare and terrorism.

Terrorist Groups Profiles
http://web.nps.navy.mil/~library/tgp/tgpndx.htm
Material from the U.S. Department of State's publication *Patterns of Global Terrorism* includes information on terrorist groups and a chronology of terrorist incidents.

The Irish Republican Army
http://users.westnet.gr/~cgian/irahist.htm
This essay offers a brief history of the armed struggle in Irish politics against English rule. It also provides further links to more sites on Irish history and copies of public statements by the IRA.

UNIT 3: State-Sponsored Terrorism

Council for Foreign Relations
http://www.cfr.org/issue/458/state_sponsors_of_terrorism.html
This site provides a regional update on terrorist activities around the world. It provides an overview to a country's/region's ties to terrorism and more.

International Institute for Terrorism and Counterterrorism
http://www.ict.org.il/inter_ter/st_terror/State_t.htm
ICT is unique in that it focuses solely on the subject of counter-terrorism. All of its efforts and resources are dedicated to approaching the issue of terrorism globally—that is, as a strategic problem that faces not only Israel but other countries as well.

Security Resource Net's Counter Terrorism
http://nsi.org/terrorism.html
This site, the National Security Institute, includes Terrorism Legislation and Executive Orders, Terrorism Facts, Commentary, Precautions, and other terrorism-related sites.

State Department's List of State Sponsors of Terrorism
http://www.state.gov/s/ct/rls/crt/
This site contains the U.S. Department of State's list of State Sponsors of Terrorism.

UNIT 4: International Terrorism

Arab.Net Contents

http://www.arab.net/sections/contents.html

Web links to 22 Arab countries ranging from Algeria through Yemen. It includes a search engine.

International Association for Counterterrorism and Security Professionals

http://www.iacsp.com/index.html

The International Association for Counterterrorism and Security Professionals was founded in 1992 to meet security challenges facing the world as it enters an era of globalization in the twenty-first century. The Web site includes a detailed overview of state-sponsored terrorism.

The International Policy Institute for Counter-Terrorism

http://www.ict.org.il

ICT is a research institute and think tank dedicated to developing innovative public policy solutions to international terrorism. The Policy Institute applies an integrated, solutions-oriented approach built on a foundation of real-world and practical experience.

International Rescue Committee

http://www.intrescom.org

Committed to human dignity, the IRC goes to work in the aftermath of state violence to help people all over the world. Click on Resettlement Programs, IRC Fact Sheet, Emergency Preparedness and Response, and links to other sites.

United Nations Website on Terrorism

http://www.un.org/terrorism/

This site gives information on what the United Nations is doing to counter terrorism.

United States Institute of Peace

http://www.usip.org/library/topics/terrorism.html

This site contains links by topical categories to resources primarily in English providing information on terrorism/counter-terrorism.

UNIT 5: Terrorism in America

America's War Against Terrorism

http://www.lib.umich.edu/govdocs/usterror.html

This Web site by the University of Michigan provides a news chronicle of the September 11, 2001 attacks and the war against terrorism.

Department of Homeland Security

http://www.dhs.gov/dhspublic/index.jsp

The home page for the Federal Bureau of Investigation includes up-to-date news and information.

FBI Homepage

http://www.fbi.gov

The home page for the Federal Bureau of Investigation includes up-to-date news and information and a section on terrorism.

ISN International Relations and Security Network

http://www.isn.ethz.ch

This is a one-stop information network for security and defense studies.

The Militia Watchdog

http://www.adl.org/mwd/m1.asp

This page is devoted to monitoring U.S. right-wing extremism, including abortion clinic bombings and neo-Nazi militias.

The Hate Directory

http://www.bcpl.lib.md.us/~rfrankli/hatedir.htm

This site has a list of hate groups on the Web, groups that advocate violence against, separation from, defamation of, deception about, or hostility toward others based on race, religion, ethnicity, gender, or sexual orientation.

UNIT 6: Terrorism and the Media

Institute for Media, Peace and Security

http://www.mediapeace.org

This Web page from the University for Peace is dedicated to examining interactions between media, conflict, peace, and security.

Terrorism Files

http://www.terrorismfiles.org

This is an up-to-date Web source for news and editorials covering terrorism and current events.

The Middle East Media Research Institute

http://www.terrorismfiles.org

The Middle East Media Research Institute (MEMRI) explores the Middle East through the region's media. MEMRI bridges the language gap, which exists between the West and the Middle East, providing timely translations of Arabic, Persian, and Turkish media, as well as original analysis of political, ideological, intellectual, social, cultural, and religious trends in the Middle East.

UNIT 7: Terrorism and Religion

FACSNET: "Understanding Faith and Terrorism"

http://www.facsnet.org/issues/faith/terrorism.php3#

This site is guided by the most current research from the nation's top institutions. FACS translates their knowledge and tailors it specifically for the educational needs of journalists.

Islam Denounces Terrorism

http://www.islamdenouncesterrorism.com

This Web site was launched to reveal that Islam does not endorse any kind of terror or barbarism and that Muslims share the sorrows of the victims of terrorism. It includes many references to the Koran that preaches tolerance and peace.

Religious Tolerance Organization

http://www.religioustolerance.org/curr_war.htm

This site provides some insight on civil unrest and warfare caused by religious belief.

SITE Institute

http://www.siteinstitute.org/

SITE provides interested parties with well-documented and comprehensive reports on terrorist entities and the individuals and organizations supporting them.

UNIT 8: Women and Terrorism

Free Muslims Against Terrorism Jihad

http://www.freemuslims.org/news/article.php?article=140

This site provides information and links such as Press Corner; Resources; and a Blog on the Muslim community.

Foreign Policy Association - Terrorism

http://www.fpa.org/newsletter_info2478/newsletter_info.htm

This page is a comprehensive source of information about terrorism and a gateway to the vast amount of information on the subject.

Israel Ministry of Foreign Affairs—The Exploitation of Palestinian Women for Terrorism

http://www.mfa.gov.il/mfa/go.asp?MFAH0ll10

This official Web site of the Israeli government chronicles the use of women by Arab terrorists as agents of terror.

Women, Militarism, and Violence

http://www.iwpr.org/pdf/terrorism.pdf

Dr. Amy Caiazza's paper, "Why Gender Matters in Understanding September 11: Women, Militarism, and Violence", analyzes women's roles as victims, supporters, and opponents of violence, terrorism, and militarism and proposes policy recommendations.

UNIT 9: Government Response

Coalition for International Justice

http://www.cij.org/index.cfm?fuseaction=homepage

This site provides all kinds of information about the investigation and prosecution of war crimes in the former Yugoslavia and in Rwanda, including some audio and video files.

Counter-Terrorism Page

http://counterterrorism.com

This site contains a summary of worldwide terrorism events, terrorist groups, and terrorism strategies and tactics, including articles from 1989 to the present of American and international origin, plus links to related Web sites, pictures, and histories of terrorist leaders.

ReliefWeb

http://www.reliefweb.int

This is the UN's Department of Humanitarian Affairs clearinghouse for international humanitarian emergencies. It has daily updates.

The South Asian Terrorism Portal

http://www.satp.org/

This site provides the current happenings of the intelligence community in South Asia.

UNIT 10: Future Threats

Centers for Disease Control and Prevention—Bioterrorism

http://www.bt.cdc.gov

The CDC Web site provides news, information, guidance, and facts regarding biochemical agents and threats.

Nuclear Terrorism

http://www.nci.org/nci/nci-nt.htm

The Nuclear Control Institute's Web site includes a Quick Index to articles on nuclear terrorism and a bibliography.

We highly recommend that you review our Web site for expanded information and our other product lines. We are continually updating and adding links to our Web site in order to offer you the most usable and useful information that will support and expand the value of your Annual Editions. You can reach us at: *http://www.mhcls.com/annualeditions/*.

UNIT 1
The Concept of Terrorism

Unit Selections

1. **Ghosts of Our Past**, Karen Armstrong
2. **An Essay on Terrorism**, Marc Nicholson
3. **The Origins of the New Terrorism**, Matthew J. Morgan

Key Points to Consider

- Why do Islamic fundamentalists feel threatened by secular society?
- How are “ethnic-based movements” different from “millennial groups?”
- What factors have contributed to the development of a “new terrorism?”

Student Website

www.mhcls.com/online

Internet References

Further information regarding these websites may be found in this book's preface or online.

MIPT Terrorism Knowledge Base
http://www.tkb.org/Home.jsp

Political Science Resources/International Relations
http://www.lib.umich.edu/govdocs/psintl.html

Terrorism: Background and Threat Assessment Links
http://www.fas.org/irp/threat/terror.htm

The Terrorism Research Center
http://www.terrorism.com

UNODC - Terrorism Definitions
http://www.unodc.org/unodc/terrorism_definitions.html

Defining and conceptualizing terrorism is an essential first step in understanding it. Despite volumes of literature on the subject there is still no commonly agreed upon definition of terrorism. The application of former Supreme Court Justice Potter Steward's famous maxim "I know it when I see it" has led to definitional anarchy. The U.S. government, in its efforts to fight a Global War on Terrorism, has further confounded the definitional problem by a myriad of confusing statements and policies.

Terrorists have also exacerbated this problem. They often portray themselves as victims of political, economic, social, religious, or psychological oppression. By virtue of their courage, their convictions, or their condition, terrorists see themselves as the chosen few, representing a larger population, in the struggle against the perceived oppressors. The actions of the oppressor—real or imagined—against the population they claim to represent, serve as motivation and moral justification for their use of violence. Existing institutional mechanisms for change are deemed either illegitimate or are in the hands of the oppressors. Hence, they portray themselves as freedom-fighters, as violence becomes the primary means of asserting their interests and the interests of the people they claim to represent.

While arguments among academics and policymakers about how terrorism should be defined continue, most would agree that terrorism involves three basic components: the perpetrator, the victim, and the target of the violence. The perpetrator commits violence against the victim. The victim is used to communicate with or send a message to the intended target. The target is expected to respond to the perpetrator. Fear is used as a catalyst to enhance the communication and elicit the desired response.

Defining the problem is an essential first step in the accumulation of statistical data. Definitions impact not only the collection and collation of data but also their analysis and interpretation. Ultimately, definitions have a profound effect on threat perceptions and policies developed to counterterrorism.

The articles in this section provide some insights into terrorist motivations and potential causes of violence. They introduce the reader to the definitional debate. The first article by Karen Armstrong explores the potential causes of violence. Article two provides an introduction to and a broad overview of terrorism and political violence. The final article in this section by Matthew Morgan examines what has been described by some as the "new terrorism." Morgan argues that cultural, political, and technological factors have influenced the development of terrorism.

GHOSTS OF OUR PAST

To win the war on terrorism, we first need to understand its roots

BY KAREN ARMSTRONG

ABOUT A HUNDRED YEARS AGO, almost every leading Muslim intellectual was in love with the West, which at that time meant Europe. America was still an unknown quantity. Politicians and journalists in India, Egypt, and Iran wanted their countries to be just like Britain or France; philosophers, poets, and even some of the *ulama* (religious scholars) tried to find ways of reforming Islam according to the democratic model of the West. They called for a nation state, for representational government, for the disestablishment of religion, and for constitutional rights. Some even claimed that the Europeans were better Muslims than their own fellow countrymen since the Koran teaches that the resources of a society must be shared as fairly as possible, and in the European nations there was beginning to be a more equitable sharing of wealth.

So what happened in the intervening years to transform all of that admiration and respect into the hatred that incited the acts of terror that we witnessed on September 11? It is not only terrorists who feel this anger and resentment, although they do so to an extreme degree. Throughout the Muslim world there is widespread bitterness against America, even among pragmatic and well-educated businessmen and professionals, who may sincerely deplore the recent atrocities, condemn them as evil, and feel sympathy with the victims, but who still resent the way the Western powers have behaved in their countries. This atmosphere is highly conducive to extremism, especially now that potential terrorists have seen the catastrophe that it is possible to inflict using only the simplest of weapons.

Even if President Bush and our allies succeed in eliminating Osama bin Laden and his network, hundreds more terrorists will rise up to take their place unless we in the West address the root cause of this hatred. This task must be an essential part of the war against terrorism.

We cannot understand the present crisis without taking into account the painful process of modernization. In the 16th century, the countries of Western Europe and, later, the American colonies embarked on what historians have called "the Great Western Transformation." Until then, all the great societies were based upon a surplus of agriculture and so were economically vulnerable; they soon found that they had grown beyond their limited resources. The new Western societies, though, were based upon technology and the constant reinvestment of capital. They found that they could reproduce their resources indefinitely, and so could afford to experiment with new ideas and products. In Western cultures today, when a new kind of computer is invented, all the old office equipment is thrown out. In the old agrarian societies, any project that required such frequent change of the basic infrastructure was likely to be shelved. Originality was not encouraged; instead people had to concentrate on preserving what had been achieved.

So while the Great Western Transformation was exciting and gave the people of the West more freedom, it demanded fundamental change at every level: social, political, intellectual, and religious. Not surprisingly, the period of transition was traumatic and violent. As the early modern states became more centralized and efficient, draconian measures were often required to weld hitherto disparate kingdoms together. Some minority groups, such as the Catholics in England and the Jews in Spain, were persecuted or deported. There were acts of genocide, terrible wars of religion, the exploitation of workers in factories, the despoliation of the countryside, and anomie and spiritual malaise in the newly industrialized mega-cities.

Successful modern societies found, by trial and error, that they had to be democratic. The reasons were many. In order to preserve the momentum of the continually expanding economy, more people had to be involved—even in a humble capacity as printers, clerks, or factory workers. To do these jobs, they needed to be educated, and once they became educated, they began to demand political rights. In order to draw upon all of a society's resources, modern countries also found they had to bring outgroups, such as the Jews and women, into the mainstream. Countries like those in Eastern Europe that did not become secular, tolerant, and democratic fell behind. But those that did fulfill these norms, including Britain and France, became so powerful that no agrarian, traditional society, such as those of the Islamic countries, could stand against them.

In the West, we have completed the modernizing process and have forgotten what we had to go through. We view the Islamic countries as inherently backward and do not realize we're seeing imperfectly modernized societies.

Today we are witnessing similar upheaval in developing countries, including those in the Islamic world, that are making their own painful journey to modernity. In the Middle East, we see constant political turmoil. There have been revolutions, such as the 1952 coup of the Free Officers in Egypt and the Islamic Revolution in Iran in 1979. Autocratic rulers predominate in this region because the modernizing process is not yet sufficiently advanced to provide the conditions for a fully developed democracy.

In the West, we have completed the modernizing process and have forgotten what we had to go through, so we do not always understand the difficulty of this transition. We tend to imagine that we have always been in the van of progress, and we see the Islamic countries as inherently backward. We have imagined that they are held back by their religion, and do not realize that what we are actually seeing is an imperfectly modernized society.

The Muslim world has had an especially problematic experience with modernity because its people have had to modernize so rapidly, in 50 years instead of the 300 years that it took the Western world. Nevertheless, this in itself would not have been an insuperable obstacle. Japan, for example, has created its own highly successful version of modernity. But Japan had one huge advantage over most of the Islamic countries: It had never been colonized. In the Muslim world, modernity did not bring freedom and independence; it came in a context of political subjection.

Modern society is of its very nature progressive, and by the 19th century the new economies of Western Europe needed a constantly expanding market for the goods that funded their cultural enterprises. Once the home countries were saturated, new markets were sought abroad. In 1798, Napoleon defeated the Mamelukes, Egypt's military rulers, in the Battle of the Pyramids near Cairo. Between 1830 and 1915, the European powers also occupied Algeria, Aden, Tunisia, the Sudan, Libya, and Morocco—all Muslim countries. These new colonies provided raw materials for export, which were fed into European industry. In return, they received cheap manufactured goods, which naturally destroyed local industry.

This new impotence was extremely disturbing for the Muslim countries. Until this point, Islam had been a religion of success. Within a hundred years of the death of the Prophet Muhammad in 632, the Muslims ruled an empire that stretched from the Himalayas to the Pyrenees. By the 15th century, Islam was the greatest world power—not dissimilar to the United States today. When Europeans began to explore the rest of the globe at the beginning of the Great Western Transformation, they found an Islamic presence almost everywhere they went: in the Middle East, India, Persia, Southeast Asia, China, and Japan. In the 16th century, when Europe was in the early stages of its rise to power, the Ottoman Empire [which ruled Turkey, the Middle East, and North Africa] was probably the most powerful state in the world. But once the great powers of Europe had reformed their military, economic, and political structures according to the modern norm, the Islamic countries could put up no effective resistance.

Muslims would not be human if they did not resent being subjugated this way. The colonial powers treated the natives with contempt, and it was not long before Muslims discovered that their new rulers despised their religious traditions. True, the Europeans brought many improvements to their colonies, such as modern medicine, education, and technology, but these were sometimes a mixed blessing.

Thus, the Suez Canal, initiated by the French consul Ferdinand de Lesseps, was a disaster for Egypt, which had to provide all the money, labor, and materials as well as donate 200 square miles of Egyptian territory gratis, and yet the shares of the Canal Company were all held by Europeans. The immense outlay helped to bankrupt Egypt, and this gave Britain a pretext to set up a military occupation there in 1882.

Railways were installed in the colonies, but they rarely benefited the local people. Instead they were designed to further the colonialists' own projects. And the missionary schools often taught the children to despise their own culture, with the result that many felt they belonged neither to the West nor to the Islamic world. One of the most scarring effects of colonialism is the rift that still exists between those who have had a Western education and those who have not and remain perforce stuck in the premodern ethos. To this day, the Westernized elites of these countries and the more traditional classes simply cannot understand one another.

After World War II, Britain and France became secondary powers and the United States became the leader of the Western world. Even though the Islamic countries were no longer colonies but were nominally independent, America still controlled their destinies. During the Cold War, the United States sought allies in the region by supporting unsavory governments and unpopular leaders, largely to protect its oil interests. For example, in 1953, after Shah Muhammad Reza Pahlavi had been deposed and forced to leave Iran, he was put back on the throne in a coup engineered by British Intelligence and the CIA. The United States continued to support the Shah, even though he denied Iranians human rights that most Americans take for granted.

Fundamentalists are convinced that modern, secular society is trying to wipe out the true faith and religious values. When people feel that they are fighting for their very survival, they often lash out violently.

Saddam Hussein, who became the president of Iraq in 1979, was also a protégé of the United States, which literally allowed him to get away with murder, most notably the chemical attack against the Kurdish population. It was only after the invasion in 1990 of Kuwait, a critical oil-producing state, that Hussein incurred the enmity of America and its allies. Many Muslims resent the way America has continued to support unpopular rulers, such as President Hosni Mubarak of Egypt and the Saudi royal family. Indeed, Osama bin Laden was himself a protégé of the West, which was happy to support and fund his fighters in the struggle for Afghanistan against Soviet Russia. Too often, the Western powers have not considered the long-term consequences of their actions. After the Soviets had pulled out of Afghanistan, for example, no help was forthcoming for the devastated country, whose ensuing chaos made it possible for the Taliban to come to power.

When the United States supports autocratic rulers, its proud assertion of democratic values has at best a hollow ring. What America seemed to be saying to Muslims was: "Yes, we have freedom and democracy, but you have to live under tyrannical governments." The creation of the state of Israel, the chief ally of the United States in the Middle East, has become a symbol of Muslim impotence before the Western powers, which seemed to feel no qualm about the hundreds of thousands of Palestinians who lost their homeland and either went into exile or lived under Israeli occupation. Rightly or wrongly, America's strong support for Israel is seen as proof that as far as the United States is concerned, Muslims are of no importance.

In their frustration, many have turned to Islam. The secularist and nationalist ideologies, which many Muslims had imported from the West, seemed to have failed them, and by the late 1960s Muslims throughout the Islamic world had begun to develop what we call fundamentalist movements.

Fundamentalism is a complex phenomenon and is by no means confined to the Islamic world. During the 20th century, most major religions developed this type of militant piety. Fundamentalism represents a rebellion against the secularist ethos of modernity. Wherever a Western-style society has established itself, a fundamentalist movement has developed alongside it. Fundamentalism is, therefore, a part of the modern scene. Although fundamentalists often claim that they are returning to a golden age of the past, these movements could have taken root in no time other than our own.

Fundamentalists believe that they are under threat. Every fundamentalist movement—in Judaism, Christianity, and Islam—is convinced that modern, secular society is trying to wipe out the true faith and religious values. Fundamentalists believe that they are fighting for survival, and when people feel their backs are to the wall, they often lash out violently. This is especially the case when there is conflict in the region.

The vast majority of fundamentalists do not take part in acts of violence, of course. But those who do utterly distort the faith that they purport to defend. In their fear and anxiety about the encroachments of the secular world, fundamentalists—be they Jewish, Christian, or Muslim—tend to downplay the compassionate teachings of their scripture and overemphasize the more belligerent passages. In so doing, they often fall into moral nihilism, as is the case of the suicide bomber or hijacker. To kill even one person in the name of God is blasphemy; to massacre thousands of innocent men, women, and children is an obscene perversion of religion itself.

Osama bin Laden subscribes roughly to the fundamentalist vision of the Egyptian ideologue Sayyid Qutb, who was executed by President Nasser in 1966. Qutb developed his militant ideology in the concentration camps in which he, and thousands of other members of the Muslim Brotherhood, were imprisoned by Nasser. After 15 years of torture in these prisons, Qutb became convinced that secularism was a great evil and that it was a Muslim's first duty to overthrow rulers such as Nasser, who paid only lip service to Islam.

Bin Laden's first target was the government of Saudi Arabia; he has also vowed to overthrow the secularist governments of Egypt and Jordan and the Shiite Republic of Iran. Fundamentalism, in every faith, always begins as an intra-religious movement; it is directed at first against one's own countrymen or co-religionists. Only at a later stage do fundamentalists take on a foreign enemy, whom they feel to lie behind the ills of their own people. Thus in 1998 bin Laden issued his fatwa against the United States. But bin Laden holds no official position in the Islamic world; he simply is not entitled to issue such a fatwa, and has, like other fundamentalists, completely distorted the essential teachings of his faith.

The Koran insists that the only just war is one of self-defense, but the terrorists would claim that it is America which is the aggressor. They would point out that during the past year, hundreds of Palestinians have died in the conflict with Israel, America's ally; that Britain and America are still bombing Iraq; and that thousands of Iraqi civilians, many of them children, have died as a result of the American-led sanctions.

None of this, of course, excuses the September atrocities. These were evil actions, and it is essential that all those implicated in any way be brought to justice. But what can we do to prevent a repetition of this tragedy? As the towers of the World Trade Center crumbled, our world changed forever, and that means that we can never see things in the same way again. These events were an "apocalypse," a "revelation"—words that literally mean an "unveiling." They laid bare a reality that we had not seen clearly before. Part of that reality was Muslim rage, but the catastrophe showed us something else as well.

In Britain, until September 11, the main news story was the problem of asylum seekers. Every night, more than 90 refugees from the developing world make desperate attempts to get into Britain. There is now a strong armed presence in England's ports. The United States and other Western countries also have a problem with illegal immigrants. It is almost as though we in the First World have been trying to keep the "other" world at bay. But as the September Apocalypse showed, if we try to ignore the plight of that other world, it will come to us in devastating ways.

So we in the First World must develop a "one world" mentality in the coming years. Americans have often assumed that they were protected by the great oceans surrounding the United States. As a result, they have not always been very well-informed about other parts of the globe. But the September Apocalypse and the events that followed have shown that this isolation has come to an end, and revealed America's terrifying vulnerability. This is deeply frightening, and it will have a profound effect upon the American psyche. But this tragedy could be turned to good, if we in the First World cultivate a new sympathy with other peoples who have experienced a similar helplessness: in Rwanda, in Lebanon, or in Srebrenica.

We cannot leave the fight against terrorism solely to our politicians or to our armies. In Europe and America, ordinary citizens must find out more about the rest of the world. We must make ourselves understand, at a deep level, that it is not only Muslims who resent America and the West; that many people in non-Muslim countries, while not condoning these atrocities, may be dry-eyed about the collapse of those giant towers, which represented a power, wealth, and security to which they could never hope to aspire.

We must find out about foreign ideologies and other religions like Islam. And we must also acquire a full knowledge of our own governments' foreign policies, using our democratic rights to oppose them, should we deem this to be necessary. We have been warned that the war against terror may take years, and so will the development of this "one world" mentality, which could do as much, if not more, than our fighter planes to create a safer and more just world.

Karen Armstrong is the author of The Battle for God: A History of Fundamentalism *and* Islam: A Brief History.

An Essay on Terrorism

Terrorist movements have rarely, if ever, succeeded militarily; when they succeeded, it was by bringing a superior power to the bargaining table....

By Marc E. Nicholson

We reflexively condemn terrorism after each new outrage—in Northern Ireland, Israel, Indonesia, and elsewhere—without a real attempt to understand and dissect it. Dissection is clinical, stripped of emotion, and does not imply approval: I emphasize the point lest any be tempted to view this essay as an apologia. It is not. It is an attempt to examine how some terrorists pursue a political goal beyond pure malice; why their tactics, if bloody, may be the most effective path open to them and have worked on occasion; how the familiar Western distinction between civilian and military combatants is ethically questionable in the modern age; and how, above all, we must distinguish in the future between movements we may be able to address by negotiation and those which we must annihilate.

Do terrorists' means justify their ends? That is a moral question with an answer that differs little in practical context from the decision by a national state to wage war. Such a state decision entails the unintended but wholly predictable consequence of the deaths by "collateral damage" of many civilians, as well as the equally predictable demise of enemy and friendly soldiers who are no less human than the civilian targets of terrorism.

It is a moral fiction to draw a sharp distinction between resort to force by states and employment of force by subnational, including terrorist, groups. Both cases bring death and entail the use of violence. The chief distinction is a surface legitimacy to the state premised on little more than its greater longevity and organized control of territory. Thus these varied actors—state and non-state—are better judged and distinguished ultimately by the morality of their ends, not by their a priori "status." If it were otherwise, would not the insurgents of the American Revolution have been damned in their time?

A separate but closely related issue: the stress on the distinction between human beings called soldiers (the first casualties of warring states) and civilians (the frequent first casualties of terrorist groups) is to deem the former as dispensable cannon fodder while asserting Marques of Queensbury rules protecting the latter. That violates modern morals. All lives are precious, and the fact that soldiers in theory "accept the risk" of the job is no dispensation for their lives. In a modern democratic state, soldiers can be categorized as civilians, not a separate caste. The civilian electorates who govern the state more than soldiers are responsible for the decisions of the government they elect, for its application of armed force, and thus for the negative consequences, and thus also for the fact that they are the logical targets of pressure for change.

This, of course, raises the question of means vs. ends and is at the core of the conventional moral critique of terrorism; indeed, terrorism's means define it. To the extent a consensus definition of terrorism exists, it may be described as the deliberate killing of non-military personnel in order to pursue a claimed political goal through exertion of pressure on a society. The literature is rife with other definitions, but their core comes down to this: murderous attacks on civilians for political purposes.

Terrorists who lash out from hatred but without concrete and achievable political goals, including those whose political goals are so sweeping as to be delusional—such as Al-Qaeda members "acting out" the multiple failures of Middle Eastern societies—are practically, if not philosophically, nihilists with nowhere to go. Their acts are pointless. They are a psychotic, not a political, phenomenon and the only reasonable answer is the use of force to kill or incarcerate them, while seeking in the longer term to address the social pathologies which produce new recruits.

But there are other "terrorist" movements now and in history with genuinely political aims, which resorted to violent tactics because the latter were the most effective available. The anti-colonial struggles of the 1950s and '60s that gave birth to numerous new nations in some instances relied in part on terrorism.

The "grand daddy" of such groups was none other than the Irgun faction of the Zionist movement in Israel, which engaged in bombings and assassinations (including of a senior UN official) to press the end of British occupation. The Irgun was far from the decisive factor in achieving Israeli independence, and was opposed by many in the Zionist movement, but it made a contribution and that contribution to independence eventually absolved its leader of his past and he went on to become prime minister of Israel, Menachem Begin.

The pattern is familiar. Terrorist movements have rarely, if ever, succeeded militarily; when they succeeded, it was by bringing a superior power to the bargaining table; and if the movement's leaders were ultimately successful and judged to be on the right side of history, they were cleansed of their past.

Who resorts to terrorism and why? Terrorism is the tool of the weak, used by disaffected groups or minorities to oppose the rule and (as they see it) the oppression of an established and militarily superior power. Because it is resistance on the cheap, terrorism often emerges out of civil society rather than state sponsorship, because oppressed civilian groups, lacking control over governmental machinery, can summon little or no regular military force able to confront their "oppressor" in conventional military terms.

Thus they resort to "hit and run" or even suicidal attacks, and may choose soft non-military targets to pressure the government they seek to influence. Whatever the morality of slaughtering innocents, this strategy can make sense in military/political terms: why fail in frontal armed assault against a far superior state-sponsored military apparatus? The goal instead is to so upset the civilian economic and social life of an adversary state as to force negotiations on more equal terms.

The specific methods of a given terrorist group depend on the nature of the regime it opposes. In democracies (e.g., the conflict in Northern Ireland), terrorists seek to wear down the voting majority until it is so sick of strife and uncertainty as to consent to a political solution by meeting the minority's demands in part or in whole. (Of course, there is always the possibility of backlash, as is evident in the case of Israel, where terrorism against the body politic successfully put the PLO on the map but more recently proved self-defeating by feeding Israeli doubt that Palestinians could ever be appeased short of the destruction of the Israeli state.)

In autocratic states (e.g., Egypt), which are less subject to public opinion and relatively indifferent to civilian casualties, terrorist groups seek more to disrupt national economies—in particular by scaring off foreign tourism and investment—to the point where governments are goaded to concessions because the damage to the nation's economic life threatens the (corrupt) elites' ability to sustain their rule.

While autocratic states may eventually crack under such strains, democratic governments are more immediately susceptible to terrorism, at least if the "cause" is plausible, because terrorism strikes common people who in democracies have influence to prod their governments towards negotiation if the pain becomes too great and the minority's grievances are perceived as not unreasonable, even if their methods are condemnable. Thus, though it seems perverse, one may argue that terrorism in some cases is more justified, or at least more effective, when directed against democratic governments. Terrorist movements in such states typically arise from confrontation between an oppressed minority and a dominant majority (e.g., Northern Ireland; Israel/Palestine).

Civilians guide such a state, the state commands the military, the military applies force, including death, to its opponents. Should the ultimate civilian authors of those consequences be exempt from pressure while their military servants (in fact their fellow citizens) take the brunt of the polity's decisions? The democratic nation in the modern age, certainly since World War I, is a nation in arms. Every citizen has a role in deciding its fate through the vote or by military effort expressed in mass mobilization or industrial support of the war machine. Thus, every citizen must accept the consequences of state policies.

We resist that notion of equal responsibility and we hate the idea of terrorism. Why? Because terrorism seeks to alter the status quo and shake complacent (dominant) populations or elites out of their complacency. It threatens our comfortable and insulated everyday lives...including the moral barrier we have sought to erect by the increasingly strained distinction between military combatants and the civilians who ultimately direct them in a democratic state. It puts electorates squarely up against the lethal consequences of their own voting decisions.

Or, if you prefer, it acknowledges the civilian electorate as politically influential agents who are targeted by terrorists seeking to influence or blackmail their political decisions. In the democratic West, terrorism is a handmaiden of democracy: everyman has the power, so everyman is now a target. And stoically accepting that fact, accepting our responsibility as citizens without whimpering or whining as potential combatants and agents of resistance is, in my view, required now as an act of patriotism on the part of participants in the modern democratic state. To plead overly the distinction between military combatants and civilian "victims" is an abdication of our responsibility as citizens. In that respect we are coming closer to the model of the ancient Greek city states which gave birth to democracy: our physical safety is more directly bound to the future of our polity than it has been in a long time...and it should be.

All governments condemn terrorism. But they sometimes give in to it and even later, if sometimes grudgingly, applaud its exponents, provided the latter's underlying cause was just and politically successful: Witness the ANC and Nelson Mandela in South Africa.

There is a life cycle to successful terrorist movements. They begin weak in their actions and condemned by "responsible" authorities. If they represent a serious and widely shared grievance, they may grow stronger, more effective (more lethal), and still more condemned. At some point, that very effectiveness can turn condemnation into reluctant acceptance of them by states as a negotiating partner. They have won a place at the table by the classic means any actor ever has in politics: by demonstrating the capacity to exert force or other influence.

That is a critical moment for such terrorist movements. Can their leadership shift from the role of hunted opponents to the role of accepted statesmen; can they shift from a narrow military/terrorist focus to a broader political vision, which inevitably implies compromise rather than maximalist rhetoric? That in the past has defined the difference between the success or failure of a number of such movements. An example of success: Nelson Mandela in South Africa. An example of failure: Yassir Arafat in Israel/Palestine. The roles of guerrilla leader and visionary statesman call for different qualities in an individual; not all terrorist/guerrilla leaders are personally capable of the transition.

The classic era of terrorists with a nationalist vision appears on the decline, since many of them have realized their goals in the post-World War II period. Increasingly we confront instead violently psychotic millenialist groups which must be extirpated rather than engaged. Nonetheless, some ethnic-based movements will continue to arise, perhaps with terrorist components, seeking in the traditional mode independence or autonomy for more or less narrowly defined populations. It behooves us to recognize the difference between those movements and irreconcilable millenialist groups and, where appropriate, to suspend our moral qualms and adopt our tactics and even negotiate with the former.

We will have enough on our hands as it is in dealing with the "wretched of the earth" in the coming century: given the ever-widening gap between rich and poor, we can expect many more terrorist movements based on pure frustration and psychosis. We will have to put them down insofar as they affect us. So, as a matter of pure economy, it behooves us to recognize where we are dealing instead with genuine political movements, albeit using terrorist means, which may be dealt with more cheaply (if holding our noses) by negotiation.

Born in California in 1950, Marc E. Nicholson graduated from Yale University, served in the U. S. Army in West Germany, and entered the Foreign Service in 1975. He had tours as a political and political/military officer in Brasilia, Lisbon, Bangkok, and Washington before retiring in 2000 to Washington, DC, where he now lives and works as a part-time consultant.

The Origins of the New Terrorism

MATTHEW J. MORGAN

The suicidal collision of hijacked commercial airliners into the World Trade Center and the Pentagon on 11 September 2001 was the most destructive terrorist attack in world history. Before the deaths of approximately 3,000 people in those attacks, the most devastating single terrorist attack had claimed the lives of about 380 people. The 2001 disaster took place at a time when experts had been defining a new form of terrorism focused on millennial visions of apocalypse and mass casualties. The catastrophic attacks confirmed their fears.

The State Department's *Patterns of Global Terrorism*, published in early 2002, revealed that terrorist attacks have scaled back in number in recent years, even though more casualties have occurred.[1] The late 1980s were a high point for the number of terrorist attacks, with the incidence of attacks exceeding 600 annually in the years 1985–88.With the exception of 1991, the number of terrorist attacks after 1988 decreased to fewer than 450 every year, reaching their recent low point in the years 1996–98, when the number of attacks was about 300. The number of attacks has increased slightly since 1998, when there were 274 attacks, but the level has not reached the number realized in any of the years of the 1980s. This report is not a linear progression from a large number to a small number of attacks, but the trend revealed is one of a decreasing incidence. Yet even if the frequency has decreased, the danger has not.

Osama bin Laden and the al Qaeda network of international terrorists are the prime examples of the new terrorism, but Islamic radicalism is not the only form of apocalyptic, catastrophic terrorism. Aum Shinrikyo, the Japanese religious cult, executed the first major terrorist attack using chemical weapons on a Tokyo subway in 1995. The bombing of the Murrah Federal Building in Oklahoma revealed similar extremism by American right-wing militants. Other plots by Christian Identity terrorists have shown similar mass-casualty proclivities.

Nadine Gurr and Benjamin Cole labeled nuclear-biological-chemical (NBC) terrorism as the "third wave of vulnerability" experienced by the United States beginning in 1995. (The first two waves were the Soviet test of the atomic bomb in 1949 and the escalating nuclear arms race that followed.[2]) David Rapoport made a similar assessment that religiously motivated modern terrorism is the "fourth wave" in the evolution of terrorism, having been preceded by terrorism focused on the breakup of empires, decolonialization, and anti-Westernism.[3]

The National Commission on Terrorism found that fanaticism rather than political interests is more often the motivation now, and that terrorists are more unrestrained than ever before in their methods.[4] Other scholarly sources have reached similar conclusions. Terrorism is increasingly based on religious fanaticism.[5] Warnings about the dangers of nontraditional terrorism were raised frequently in pre-2001 literature.[6] For instance, Ashton Carter, John Deutch, and Philip Zelikow declared in the pages of *Foreign Affairs* in 1998 that a new threat of catastrophic terrorism had emerged.[7] Earlier concerns about alienating people from supporting the cause are no longer important to many terrorist organizations. Rather than focusing on conventional goals of political or religious movements, today's terrorists seek destruction and chaos as ends in themselves. Yossef Bodansky's *Bin Laden* quotes from S. K. Malik's *The Quranic Concept of War:*

> Terror struck into the hearts of the enemies is not only a means, it is in the end in itself. Once a condition of terror into the opponent's heart is obtained, hardly anything is left to be achieved. It is the point where the means and the ends meet and merge. Terror is not a means of imposing decision upon the enemy; it is the decision we wish to impose upon him.[8]

Today's terrorists are ultimately more apocalyptic in their perspective and methods. For many violent and radical organizations, terror has evolved from being a means to an end, to becoming the end in itself. The National Commission on Terrorism quoted R. James Woolsey: "Today's terrorists don't want a seat at the table, they want to destroy the table and everyone sitting at it.[9]

Some analysts argue that the evolution of terrorism represents continuity rather than change, that mass-casualty bombings have long been characteristic of terrorist methods, and that radical extremism has always dominated terrorist motivations.[10] Walter Laqueur's most recent book warns against trying to categorize or define terrorism at all because there are "many terrorisms," and he emphasizes the particularities of various terrorist movements and approaches.[11] (Laqueur, however, recognizes some evolving strains of terrorism, especially the Islamist variant.) Bruce Hoffman discussed the definition of terrorism at length in his 1998 book, *Inside Terrorism*, and his final definition includes "political change" as the desired end-state of terrorist activity.[12] This would be more consistent with traditional means-end constructions of terrorism. Richard Falkenrath pointed out in a pre-9/11 article that mass-casualty terrorism is still an aberrant occurrence.[13] A recent survey of terrorism suggests historical and intellectual links between the fascism of fanatical Islamist terrorism today and the totalitarian movements of the 20th century, further emphasizing continuity rather than change.[14]

Most recent scholarship, however, has taken the perspective that contemporary terrorism represents a significant departure from the past. Various factors have led to the development of this new type of terrorism. Paul Wilkinson pondered the increase in indiscriminateness among terrorists, and he posited several possible reasons accounting for this upsurge.[15] First, the saturation of the media with images of terrorist atrocity has raised the bar on the level of destruction that will attract headline attention. Second, terrorists have realized that civilian soft targets involve lower risk to themselves. Finally, there has been a shift from the politically-minded terrorist to the vengeful and hard-line fanatic.

While Wilkinson's factors accurately describe developments in terrorist strategy and tactics, there are more fundamental forces at work. The world has undergone a variety of changes on several levels. While it is impossible to link all social changes to terrorism today, it is possible to track several distinct factors that have converged to evolve a form of terrorism that is unprecedented in the level of threat it poses around the world. This article will explore these factors from cultural, political, and technological perspectives.

Cultural Factors

Islamic radicalism is the most notorious form of the new culture of terrorism, but it is far from the only variety of cultural trends motivating terrorist activity. Numerous cults, whose emergence in many cases has been synchronized with the turn of the new millennium, have also posed an increasing threat. Finally, the American religious right has been active with escalating and destructive objectives, although law enforcement presence has restrained these groups.

"Islamic radicalism is not the only form of apocalyptic, catastrophic terrorism."

It is important to distinguish religious terrorists from those terrorists with religious components, but whose primary goals are political. Religiously motivated terrorist groups grew sixfold from 1980 to 1992 and continued to increase in the 1990s. Hoffman asserted that "the religious imperative for terrorism is the most important characteristic of terrorist activity today."[16] This may not be as much an entirely new phenomenon as a cyclic return to earlier motivations for terror. Until the emergence of political motives such as nationalism, anarchism, and Marxism, "religion provided the only acceptable justifications for terror."[17] However, terrorism in modern times has not, until recent years, been so dominated by religious overtones. At the time when modern international terrorism first appeared, of the 11 identifiable terrorist groups, none could be classified as religious.[18]

Today's terrorists increasingly look at their acts of death and destruction as sacramental or transcendental on a spiritual or eschatological level. The pragmatic reservations of secular terrorists[19] do not hold back religious terrorists. Secular terrorists may view indiscriminate violence as immoral. For religious terrorists, however, indiscriminate violence may not be only morally justified, but constitute a righteous and necessary advancement of their religious cause. In addition, the goals of secular terrorists are much more attuned to public opinion, so senseless violence would be counterproductive to their cause, and hence not palatable to them. As Hoffman observed, the constituency itself differs between religious and secular terrorists. Secular terrorists seek to defend or promote some disenfranchised population and to appeal to sympathizers or prospective sympathizers. Religious terrorists are often their own constituency, having no external audience for their acts of destruction.[20]

Aum Shinrikyo has been included in typologies of terrorism that include radical Islamists as part of a group of religiously motivated organizations that attack symbols of the modern state.[21] In many ways, the dynamics of cultist followings make groups such as Aum Shinrikyo (also known as Aleph) more dangerous than religious terrorists rooted in conventional and broadly based religious traditions or denominations. There is no constituency of more moderate adherents to share common beliefs with the radical group while at the same time posing a restraining influence. For the fundamentalist Islamic or Christian radical, authoritative figures from either of those religions can condemn violence and de-legitimize the terrorist, at least in the eyes of the average faithful.

Another feature of religious cults that makes them incredibly dangerous is the personality-driven nature of these groups. Cultist devotion to one leader leaves followers less able to make their own moral decisions or to consult other sources of reasoning. If that leader is emotionally or mentally unstable, the ramifications can be catastrophic. The more dangerous religious terrorist groups from traditional faiths may often share this feature of the cult: a charismatic leader who exerts a powerful influence over the members of the group.

According to many analysts, Aum Shinrikyo demonstrated its comparatively more threatening potential in its sarin attack in the Tokyo subway. As D.W. Brackett wrote, "A horrible bell had tolled in the Tokyo subway.... Terrorists do not follow rules of engagement in their operations but they do absorb the lessons to be learned from successful acts of violence."[22] If for no other reason than providing an example to others, Aum Shinrikyo has gained notoriety as one of the more dangerous terrorist elements. Despite setbacks such as the incarceration of key leadership figures, the group continues to pose future threats. The ability of Aum Shinrikyo to recruit individuals with a high level of education and technical knowledge also has been a significant aspect of the threat posed by the cult.[23]

In the past, cults were not viewed as national security threats; they were more dangerous to unwary individuals who might succumb to the cult's influence. Even the emergence of cultist mass suicides did not alter this perception. However, the recent appearance of cults willing and able to adopt destructive political goals has revised the more benign view of the cult phenomenon. Since cults are often fundamentally based on the violence of coercion, they can be accustomed to the mindset necessary to adopt terrorist methods. Although cults more often practice a mental violence with psychological control and extreme invasions of privacy, they do occasionally engage in physical abuse. The most dangerous cults are also fascinated by visions of the end of the world—which, like radicals from more mainstream religions, cultists often believe that they are instrumental in bringing about. The nature of the cult's mythical figure can also be indicative of the level of threat. A vengeful deity is more threatening than a suffering savior. This sign is somewhat unpredictable, however, because cults can switch their principal myths as circumstances change.[24] In summary, cults are a particularly dangerous form of religious terrorism because they can appear quickly without warning, have no rational goals, and can become agitated due to the apprehension and hostility with which they are viewed by the society at large.

Whether initiated by cultists or by extremists from more established religions, the violence of religious terrorists can be particularly threatening in comparison with that of the political terrorists of earlier years. As Hoffman notes, "For the religious terrorist, violence is a divine duty … executed in direct response to some theological demand … and justified by scripture."[25] Religion can be a legitimizing force that not only sanctions but compels large-scale violence on possibly open-ended categories of opponents.[26] Terrorist violence can be seen as a divinely inspired end in itself. One explanation that has been proffered to account for violent Islamic extremism views revenge as the principal goal of the terrorists.[27] This reasoning makes political change or conventional political objectives irrelevant, and it is consistent with observations that violence is itself the objective. Fundamentalist Islam "cannot conceive of either coexistence or political compromise. To the exponents of Holy Terror, Islam must either dominate or be dominated."[28] A recent study that traced the Islamic theological doctrine to the Middle Ages noted recent philosophical developments that explain the preponderance of religious mass-casualty terrorism coming from adherents of Islam.[29]

Remarkably, a recent analysis of bin Laden's fatwa, published in *Studies of Conflict and Terrorism*, found that the content of the fatwa was "neither revolutionary nor unique, as it encapsulates broad sentiments in the Muslim world, especially that of Islam's being on the defensive against foreign secular forces and modernization."[30] However, some of the content of the fatwa does fall directly within the paradigm of contemporary religious terrorism. Consider the following excerpts:

> Praise be to God, who revealed the book, controls the clouds, defeats factionalism, and says in his book: "But when the forbidden months are past, then fight and slay the pagans wherever ye find them.…"
>
> On that basis, and in compliance with God's order, we issue the following fatwa to all Muslims:
>
> The ruling to kill the Americans and their allies—civilians and military—is an individual duty for every Muslim who can do it in any country in which it is possible to do it.[31]

In an article published shortly after 9/11, Steven Simon and David Benjamin noted that many al Qaeda attacks, including the major planning phase of the 9/11 attacks, took place during favorable times for the Palestinians in the Middle East peace process, and that no foreign policy changes by the US government could possibly have appeased the bin Ladenist radical.[32]

While Islamic terrorists are the most notorious of today's violent radicals, others such as right-wing Christian extremists also exhibit many characteristics of the new terrorism. Mark Juergensmeyer, in his book *Terror in the Mind of God: The Global Rise of Religious Violence*, identified three elements that Islamists, radical Christians, and other religious terrorists share: They perceive their objective as a defense of basic identity and dignity; losing the struggle would be unthinkable; and the struggle is in deadlock and cannot be won in real time or in real terms.[33]

In the past, right-wing Christian terrorists conducted racially motivated or religiously motivated acts of violence discriminately against chosen victims, and confrontation with the state was limited to instances when the state interfered with the political or religious agenda of the terrorist groups.[34] Today, some such groups are directly hostile to the government, which adherents believe is engaged in a widespread conspiracy threatening the existence of the "white Christian way of life." A recent FBI strategic assessment of the potential for domestic terrorism in the United States focused on such groups as Christian Identity and other ultraconservative movements associated with Christian fundamentalism.[35] The most extreme of these fanatics attribute a subhuman status to people of color, which in their eyes mitigates any moral compunction to avoid harming such individuals. In addition, they view themselves in a perpetual battle with the forces of evil (as manifested through non-white races and a powerful, sinister government) that must culminate in the apocalyptic crisis predicted by the Book of Revelations. The Christian terrorists view it as their duty to hasten the realization of this divine plan, which permits and even exhorts them to greater levels of violence. That violence is directed against existing social structures and governments, which are viewed to be hopelessly entangled with such "dark forces" as Jewry, enormous financial conglomerates, and international institutions trying to form an ominous "new world order."

While Christian violence in the United States has been discriminately focused for decades against racial minorities and "immoral" targets, it recently has expanded into attempted bombings and poisoning municipal water supplies.[36] These indiscriminate attacks demonstrate a willingness to tolerate greater levels of collateral damage in efforts to generate mass levels of casualties. The bombing of the Murrah Federal Building in Oklahoma City was the pinnacle of this trend, and although Timothy McVeigh accepted responsibility for that attack, some speculate that there was additional involvement by other conservative militia or Christian terrorists.[37] Effective domestic law enforcement in the United States has largely prevented these groups from achieving widespread violence on the level of Oklahoma City, making that incident a tragic exception among a larger number of foiled plots.

"At the same time that globalization has provided a motivation for terrorism, it has also facilitated methods for it."

While there is certainly no cooperation between foreign Islamist and US-domestic Christian radicals, there is a disquieting similarity in their views. August Kreis of the paramilitary group Posse Comitatus responded to the collapse of the World Trade Center towers with this disconcerting rant: "Hallelu-Yahweh! May the WAR be started! DEATH to His enemies, may the World Trade Center BURN TO THE GROUND!"[38] Jessica Stern's recent book, *Terror in the Name of God: Why Religious Militants Kill*, which compiles interviews with international terrorists conducted over five years, does not begin with an example from the Guantanamo Bay detention facility or the streets of the Middle East.[39] Her introductory example is a former Christian terrorist in a Texas trailer park. While Islamic terrorism is the most salient threat to the United States, it is not the only danger posed by the new trend of a culture of religious violence and extremism.

A cluster of several cultural features among new international terrorist groups indicates the high level of threat. These aspects include a conception of righteous killing-as-healing, the necessity of total social destruction as part of a process of ultimate purification, a preoccupation with weapons of mass destruction, and a cult of personality where one leader dominates his followers who seek to become perfect clones.[40] These aspects taken together represent a significant departure from the culture of earlier terrorist groups, and the organizations that these characteristics describe represent a serious threat to the civilized world.

Political and Organizational Factors

A number of developments on the international scene have created conditions ripe for mass-casualty terrorism. Gross inequalities in economic resources and standards of living between different parts of the world are a popular reason given for the ardency and viciousness of contemporary terrorists,[41] although governmental collapse in "failed states" as a breeding ground for terrorists presents a more convincing variation on this logic.[42] However, there is no "comprehensive explanation in print for how poverty causes terror," nor is there a "demonstrated correlation between the two."[43] The intrusion of Western values and institutions into the Islamic world through the process of free-market globalization is an alternative explanation for the growth of terrorism, which is the weaker party's method of choice to strike back.[44] The process of globalization, which involves the technological, political, economic, and cultural diminution of boundaries between countries across the world, has insinuated a self-interested, inexorable, corrupting market culture into traditional communities. Many see these forces as threatening their way of life. At the same time that globalization has provided a motivation for terrorism, it has also facilitated methods for it.

One of the major consequences of globalization has been a deterioration of the power of the state.[45] The exponential expansion of non-governmental organizations (NGOs), regional alliances, and international organizations has solidified this trend. Although certainly not a conventional humanitarian-based NGO like the Red Cross or Doctors without Borders, al Qaeda has distinguished itself as among the most "successful" of non-governmental organizations in pursuing its privately-funded global agenda. The trend among terrorists to eschew direct connections with state sponsors has had several advantages for the enterprising extremist. Terrorist groups are more likely to maintain support from "amorphous constituencies," so extreme methods are more acceptable because such methods can be used without fear of alienating political support.[46] Harvey Kushner described this development as a growth of "amateur" groups as direct state sponsorship has declined.[47] Lawrence Freedman pointed out that the Taliban-ruled Afghanistan was not so much a state sponsor of terrorism as it was a "terrorist-sponsored state."[48]

Terrorists do, however, continue to enjoy the benefits of indirect state sponsorship. Although the opportunity for state sponsorship has arguably diminished as a result of the Bush Administration's war on terror that has been prosecuted in the aftermath of the 9/11 attacks, state sponsorship remains widespread. In fact, developments in counterterrorist measures may propagate some dangerous trends of modern terrorism. When terrorists cannot rely on direct state sponsorship, they may become less accountable and harder to track. States must conceal their involvement by exercising less control and thus maintain less-comprehensive intelligence of radical terrorist organizations. Many states have been on the US government list of state sponsors for more than a decade, including Cuba, Iran, Iraq, North Korea, Libya, and Syria. More recently, Sudan and Afghanistan became government sponsors of terrorism. Many state sponsors cooperate with one another to promote terrorist violence, making terrorist activity further disconnected from the foreign policy of any single state. Iran has funded training camps in Sudan, and the Palestinian Islamic Jihad has received support from both Iran and Syria.[49]

Further exacerbating the problem is the method of funding, which often has no measures for accountability. Iran's support for terrorist organizations can include no particular target selection, and it occasionally results, with the funds disappearing, in no terrorist attacks.[50] This unpredictability is tolerated by state sponsors because of the occasional destructive payoff and the obfuscation of evidence connecting the state to the terrorist. Iran has consciously created a decentralized command structure because of these advantages.[51] A further advantage of maintaining arm's length from extremist operatives is for self-protection. The government intelligence organization of Sudan evidently monitored Osama bin Laden while he lived in that country, apparently to prevent his activists from eventually doing harm to even that extremist government.[52]

While the American operations in Afghanistan and Iraq have diluted the threat from those states, other sponsors have possibly been left off official lists for political reasons. (It has been frequently argued that inclusion of a state on the list of state sponsors of terrorism reflects its relationship with the United States.[53]) Pakistani intelligence reportedly has been involved in sponsoring violent terrorists, both in Afghanistan and in the contentious Kashmir. Additionally, the Kingdom of Saudi Arabia has been at the center of controversy over sponsorship and proliferation of radicalism and violence. Laurent Murawiec, an analyst at the RAND Corporation, attracted public attention by pointing out the dangers of Saudi support for radical Islamists and specifically Osama bin Laden in a briefing to the Defense Policy Board in 2002. While no official publication of the RAND Corporation documents this analysis, Murawiec highlighted evidence of Saudi support for the Islamist agenda through Islamic educational venues and financial backing.

So while globalization has helped remove many of the restraints that state sponsorship once imposed, terrorists can still enjoy the funding and protection that sponsorship provides. Another factor of globalization that benefits terrorism is targeting: "In today's globalizing world, terrorists can reach their targets more easily, their targets are exposed in more places, and news and ideas that inflame people to resort to terrorism spread more widely and rapidly than in the past."[54] Among the factors that contribute to this are the easing of border controls and the development of globe-circling infrastructures, which support recruitment, fund-raising, movement of materiel, and other logistical functions.

In addition to international political changes, developments in organizational practice have enhanced the lethality of terrorists. As corporations have evolved organizationally, so have terrorist organizations. Terrorist groups have evolved from hierarchical, vertical organizational structures, to more horizontal, less command-driven groups. John Arquilla, David Ronfeldt, and Michele Zanini note that terrorist leadership is derived from a "set of principles [that] can set boundaries and provide guidelines for decisions and actions so that members do not have to resort to a hierarchy—'they know what they have to do.'" The authors describe organizational designs that may "sometimes appear acephalous (headless), and at other times polycephalous (Hydra-headed)."[55] Paul Smith observed that the multi-cellular structure of al Qaeda gave the organization agility and cover and has been one of its key strengths.[56] This flexibility has allowed al Qaeda to establish bases using indigenous personnel all over the world. It has infiltrated Islamic nongovernmental organizations in order to conceal operations.[57] Jessica Stern recently commented on al Qaeda's ability to maintain operations in the face of an unprecedented onslaught:

> The answer lies in the organization's remarkably protean nature. Over its life span, al Qaeda has constantly evolved and shown a surprising willingness to adapt its mission. This capacity for change has consistently made the group more appealing to recruits, attracted surprising new allies, and—most worrisome from a Western perspective—made it harder to detect and destroy.[58]

Technological Factors

In addition to the cultural and religious motivations of terrorists and the political and organizational enabling factors, technology has evolved in ways that provide unprecedented opportunities for terrorists. The collapse of the Soviet Union and the possibility of proliferation of nuclear weapons to nonstate users is the primary factor that has significantly increased the danger of nuclear terrorism.[59] However, nonnuclear weapons of mass destruction and information technology also have created opportunities for terrorists that are in many ways more threatening than radiological terrorism because these alternatives are more probable.

Some theorists have argued that weapons of mass destruction do not represent a weapon of choice for most terrorists, even in these changing times. Stern writes that "most terrorists will continue to avoid weapons of mass destruction (WMD) for a variety of reasons," preferring the "gun and the bomb."[60] Brian Jenkins agreed that most terrorist organizations are technologically conservative, but he also noted that the self-imposed moral restraints which once governed terrorist actions are fading away.[61] As the trends in the preceding sections reach fullness, increasing the proclivity toward mass-casualty terrorism, terrorists may turn more to these weapons that will better fit their objectives and moralities.

Walter Laqueur's *New Terrorism* emphasizes the availability of very powerful weapons of mass destruction as the major current danger facing the industrialized world.[62] Aside from the nuclear variety of WMD, biological and chemical weapons pose serious dangers. Biological weapons are limited because human contact is required to spread the effects, but as the Asian brush with Severe Acute Respiratory Syndrome (SARS) demonstrated, the associated panic and uncertainty can take a large economic and political toll—not to mention the cost in human suffering for those exposed to the pathogen, perhaps without knowing how or even whether they have been infected. Biological weapons can come in a variety of forms, including viruses, bacteria, and rickettsia (bacteria that can live inside host cells like viruses).

Chemical toxins differ from biological weapons in that they are nonliving pathogens and require direct infection and contact with the victim. This negates the continual spread of the weapon, but it entails more direct and possibly more damaging effects. Chemical agents appear in several types: choking agents that damage lung tissue, blood agents that cause vital organs to shut down, blister agents (also known as vesicants) that damage the skin, and—most lethal—nerve agents. Various methods allow the agent to infect its victim, including inhalation, skin absorption, and ingestion into the digestive tract. Exacerbating the danger is the fact that many deadly chemicals, or their components, are commercially available.

The State Department's annual report on terrorism asserted that the events of 11 September 2001 confirmed the intent and capability of terrorist organizations to plan and execute mass-casualty attacks. The report also stated that these unprecedented attacks may lead to an escalation of the scope of terrorism in terms of chemical, biological, radiological, or nuclear methods.[63] The report further cited evidence discovered in military raids of Afghan terrorist facilities, the use of poison by Hamas to coat shrapnel in improvised explosives, and an unnamed group arrested in Italy with maps of the US embassy and possessing a compound capable of producing hydrogen cyanide. Activities of cults such as Aum Shinrikyo and American terrorist plans to poison municipal water facilities provide further evidence of the WMD threat.

Another key development is recent advances in communications and information technology. This technology provides both assistance to the terrorists and an opportunity for targeting as industrialized societies place greater reliance on information infrastructures. Terrorists will likely avoid dismantling the internet because they need the technology for their own communication and propaganda activities. Accordingly, terrorists may be more interested in "systemic disruption" rather than the total destruction of information networks.[64] While the consequences of a major disruption of American or global information infrastructures could be catastrophic financially or socially, terrorists have not shown the inclination or capability to undertake massive strikes in this area. There have been limited attacks along these lines, but the major use of information technology has been as an aid for terrorists rather than as a target of their activity. The reported use of the internet and e-mail by al Qaeda to coordinate the strikes on the World Trade Center and the Pentagon provides a dramatic example of this sort of coordination. As Paul Pillar noted, "Information technology's biggest impact on terrorists has involved the everyday tasks of organizing and communicating, rather than their methods of attack."[65]

Technology also has increased the ability of terrorists to conduct mass-casualty attacks. As noted earlier, the worst single terrorist attack before 9/11 claimed the lives of about 380 people. The yield of contemporary radiological, chemical, and biological weapons could dwarf that number, given the goals of today's terrorists as exemplified by the World Trade Center and Pentagon attacks, the Oklahoma City bombing, the sarin gas attack on the Tokyo subway, and other, less-successful attacks of the past decade. Technological developments and their availability as spread by the globalized market economy have unavoidably expanded the dangers of terrorism in the new century.

Conclusions

The practice of terrorism has undergone dramatic changes in recent years. The categorical fanaticism that is apparent in terrorist organizations across a spectrum of belief systems is a major part of this change. In the past, terrorists were more likely to be dominated by pragmatic considerations of political and social change, public opinion, and other such factors. Today, a phenomenon that was a minute rarity in the past—terrorists bent on death and destruction for its own sake—is more commonplace than ever. In addition, the statelessness of today's terrorists removes crucial restraints that once held the most extreme terrorists in check or prevented them from reaching the highest levels in their organizations. Terrorists can still enjoy the funding and shelter that only a national economy can mobilize, but they are on their own to a greater degree in greater numbers than in the past. Organizationally, terrorists are using the nonhierarchical structures and systems that have emerged in recent years. Finally, the potential availability of nuclear, chemical, and biological WMD technology provides the prospect that these trends could result in unprecedented human disasters.

Terrorism has quantitatively and qualitatively changed from previous years. Whether it is Gurr and Coleman's "third wave of vulnerability" or Rapoport's "fourth wave of terrorism," contemporary terrorism is a significant departure from the phenomenon even as recently as during the Cold War. The US *National Security Strategy* has recognized terrorism, in the memorable phrase "the crossroads of radicalism and technology," as the predominant security threat in the post-Cold War world. The cataclysmic impact of 9/11 on both the American strategic consciousness and the international security environment can scarcely be overstated. Those attacks resulted from a combination of cultural, political, and technological factors and were a revelation to the world of the emergence of the new terrorism.

NOTES

1. US Department of State, *Patterns of Global Terrorism 2001* (Washington: GPO, May 2002), p. 171. The statistical review in the State Department's report does not cover total casualties; it tracks only Americans, and the casualty reporting is not as longitudinal as the number of attacks. The casualties of terrorist incidents are tracked for the previous five years versus the previous 20 years.
2. Nadine Gurr and Benjamin Cole, *The New Face of Terrorism: Threats from Weapons of Mass Destruction* (New York: I. B. Tauris, 2002).
3. David C. Rapoport, "The Fourth Wave: September 11 and the History of Terrorism," *Current History*, December 2001, pp. 419–24.
4. National Commission on Terrorism, *Countering the Changing Threat of International Terrorism: Report of the National Commission on Terrorism* (Washington: GPO, 2000).
5. Walter Laqueur, "Terror's New Face," *Harvard International Review*, 20 (Fall 1998), 48–51.
6. Richard A. Falkenrath, Robert D. Newman, and Bradley A. Thayer, *America's Achilles' Heel: Nuclear, Biological, and Chemical Terrorism and Covert Attack* (Cambridge, Mass.: MIT Press, 1998); Philip B. Heymann, *Terrorism and America: A Commonsense Strategy for a Democratic Society* (Cambridge, Mass.: MIT Press, 1998); Bruce Hoffman, *Inside Terrorism* (New York: Columbia Univ. Press, 1998); Brad Roberts, ed., *Terrorism with Chemical and Biological Weapons: Calibrating Risks and Responses* (Alexandria, Va.: Chemical and Biological Arms Control Institute, 1997); and Jessica Stern, *The Ultimate Terrorists* (Cambridge, Mass.: Harvard Univ. Press, 1999).
7. Ashton Carter, John Deutch, and Philip Zelikow, "Catastrophic Terrorism," *Foreign Affairs,* 77 (November/ December 1998), 80–94.
8. S. K. Malik, *The Quranic Concept of War* (Lahore, India: Wajidalis, 1979), quoted in Yossef Bodansky, *Bin Laden* (Roosevelt, Calif.: Prima Publishing, 1999), p. xv.
9. National Commission on Terrorism, *Countering the Changing Threat*, p. 2.
10. Chris Quillen, "A Historical Analysis of Mass Casualty Bombers," *Studies in Conflict and Terrorism,* 25 (September/October 2002), 279–92.
11. Walter Laqueur, *No End to War: Terrorism in the Twenty-First Century* (New York: Continuum, 2003).
12. Bruce Hoffman, *Inside Terrorism* (New York: Columbia Univ. Press, 1998).
13. Richard Falkenrath, "Confronting Nuclear, Biological and Chemical Terrorism," *Survival,* 40 (Autumn 1998), 52.
14. Paul Beuman, *Terror and Liberalism* (New York: W. W. Norton, 2003).
15. Paul Wilkinson, *Terrorist Targets and Tactics: New Risks to World Order*, Conflict Study 236 (Washington: Research Institute for the Study of Conflict and Terrorism, December 1990), p. 7.
16. Hoffman, *Inside Terrorism*.
17. David C. Rapoport, "Fear and Trembling: Terrorism in Three Religious Traditions," *American Political Science Review*, 78 (September 1984), 668–72.
18. Bruce Hoffman, "'Holy Terror': The Implications of Terrorism Motivated by a Religious Imperative," *Studies in Conflict and Terrorism*, 18 (October-December 1995), 271–84.
19. Brian M. Jenkins, *The Likelihood of Nuclear Terrorism,* P-7119 (Santa Monica, Calif.: RAND, July 1985).
20. Hoffman, "'Holy Terror,'" p. 273.

21. Mark Juergensmeyer, "Terror Mandated by God," *Terrorism and Political Violence*, 9 (Summer 1997), 16–23.
22. D. W. Brackett, *Holy Terror: Armageddon in Tokyo* (New York: Weatherhill, 1996), pp. 5–7.
23. David Kaplan and Andrew Marshall, *The Cult at the End of the World* (New York: Crown Publishers, 1996), p. 74.
24. Stern, *The Ultimate Terrorists*, p. 72.
25. Hoffman, *Inside Terrorism*, p. 20.
26. Hoffman, "'Holy Terror,'" p. 280.
27. Gavin Cameron, *Nuclear Terrorism* (Basingstoke, Eng.: Macmillan, 1999), p. 139.
28. Amir Taheri, *Holy Terror: The Inside Story of Islamic Terrorism* (London: Hutchinson, 1987), p. 192.
29. Daniel Benjamin and Steven Simon, *The Age of Sacred Terror* (New York: Random House, 2002).
30. Magnus Ranstorp, *Studies in Conflict & Terrorism*, 21 (October–December 1998), 321–32.
31. Shaikh Osama Bin Muhammad Bin Laden, Ayman al Zawahiri, Abu-Yasir Rifa'I Abroad Taha, Shaikh Mir Hamzah, and Fazlul Rahman, "The World Islamic Front's Statement Urging Jihad Against Jews and Crusaders," *London al-Quds al-Arabi*, 23 February 1998.
32. Steven Simon and Daniel Benjamin, "The Terror," *Survival*, 43 (Winter 2001), 12.
33. Mark Juergensmeyer, *Terror in the Mind of God: The Global Rise of Religious Violence* (Berkeley: Univ. of California Press, 2000).
34. Gurr and Cole, *The New Face of Terrorism*, p. 144.
35. Federal Bureau of Investigation, *Project Megiddo* (Washington: GPO, 20 October 1999), `http://permanent.access.gpo.gov/lps3578/www.fbi.gov/library/megiddo/megiddo.pdf`.
36. Gurr and Cole, *The New Face of Terrorism.*
37. See Gore Vidal, *Perpetual War for Perpetual Peace: How We Got to Be So Hated* (New York: Verso, 2002) for an exposition of the point of view that the Murrah Federal Building bombing could not have possibly occurred without a larger support structure.
38. Daniel Levitas, *The Terrorist Next Door: The Militia Movement and the Radical Right* (New York: Thomas Dunne Books, 2002).
39. Jessica Stern, *Terror in the Name of God: Why Religious Militants Kill* (New York: HarperCollins, 2003).
40. Robert J. Lifton, *Destroying the World to Save It: Aum Shinrikyo, Apocalyptic Violence, and the New Global Terrorism* (New York: Metropolitan Books, 1999).
41. James D.Wolfensohn, "Making the world a Better and Safer Place: The Time for Action is Now," *Politics, 22* (May 2002), 118–23; Andrew S. Furber, "Don't Drink the water ..." *British Medical Journal*, 326 (22 March 2003), 667; Jan Nederveen Pieterse, "Global Inequality: Bringing Politics Back In, *Third World Quarterly*, 23 (December 2002), 1023–46.
42. Karin von Hippel, "The Roots of Terrorism: Probing the Myths," *Political Quarterly*, 73 (August 2002), 25–39.
43. Michael Mousseau, "Market Civilization and Its Clash With Terror," *International Security*, 27 (Winter 2003), 6.
44. Mousseau, "Market Civilization"; Audrey Kurth Cronin, "Behind the Curve: Globalization and International Terrorism," *International Security*, 27 (Winter 2003), 30–58.
45. Charles W. Kegley, Jr., and Gregory A. Raymond, *Exorcising the Ghost of Westphalia: Building World Order in the New Millennium* (Upper Saddle River, N.J.: Prentice Hall, 2002).
46. Stern, *The Ultimate Terrorists.*
47. Harvey W. Kushner, ed., *The Future of Terrorism: Violence in the New Millennium* (Thousand Oaks, Calif.: Sage Publications, 1998).
48. Lawrence Freedman, "The Third World War?" *Survival*, 43 (Winter 2001), 61–88.
49. James Adams, *The New Spies* (London: Hutchinson, 1994), pp.180, 184.
50. Ibid., p. 180.
51. Taheri, *Holy Terror*, pp. 100–01.
52. Frank Smyth, "Culture Clash, bin Laden, Khartoum and the War Against the West," *Jane's Intelligence Review*, October 1998, p. 22.
53. Adrian Guelke, *The Age of Terrorism* (London: I. B. Tauris, 1998), p. 148.
54. Paul R. Pillar, "Terrorism Goes Global: Extremist Groups Extend their Reach Worldwide," *The Brookings Review*, 19 (Fall 2001), 34–37.
55. John Arquilla, David Ronfeldt, and Michele Zanini, "Networks, Net war, and Information-Age Terrorism," in *Countering the New Terrorism*, ed. Ian O. Lesser et al., MR-989-AF (Santa Monica, Calif.: RAND, 1999), p. 51.
56. Paul J. Smith, "Transnational Terrorism and the al Qaeda Model: Confronting New Realities," *Parameters*, 32 (Summer 2002), 37.
57. Ibid., p. 37.
58. Jessica Stern, "The Protean Enemy," *Foreign Affairs*, 82 (July/August 2003).
59. Brian M. Jenkins, "Will Terrorists Go Nuclear? A Reappraisal," in Kushner, *The Future of Terrorism*, pp. 225–49.
60. Stern, *The Ultimate Terrorist*, p. 70.
61. Jenkins, "Will Terrorists Go Nuclear?"
62. Walter Laqueur, *The New Terrorism: Fanaticism and the Arms of Mass Destruction* (New York: Oxford Press, 2000).
63. US Department of State, *Patterns of Global Terrorism*, p. 66.
64. Arquilla, Ronfeldt, and Zanini, "Networks, Net war, and Information-Age Terrorism."
65. Pillar, "Terrorism Goes Global."

Captain Matthew J. Morgan is the Commander of the Headquarters and Headquarters Operations Company (HHOC), 125th Military Intelligence Battalion, at Schofield Barracks, Hawaii. Following command, Captain Morgan will deploy to Operation Enduring Freedom in Afghanistan on the Joint Task Force intelligence staff.

UNIT 2
Tactics of Terrorism

Unit Selections

Key Points to Consider

- Who are likely targets of assassination attempts? How have they responded?
- How have terrorist organizations responded to the decline in state sponsorship?
- What are some common misconceptions about suicide bombers?

Student Website

www.mhcls.com/online

Internet References

Further information regarding these websites may be found in this book's preface or online.

FrontPage Magazine—Ecoterrorism and Us
http://www.frontpagemag.com/Articles/Printable.asp?ID=1277

JCSS Military Resources
http://www.tau.ac.il/jcss/lmas.html

Terrorist Groups Profiles
http://web.nps.navy.mil/~library/tgp/tgpndx.htm

The Irish Republican Army
http://users.westnet.gr/~cgian/irahist.htm

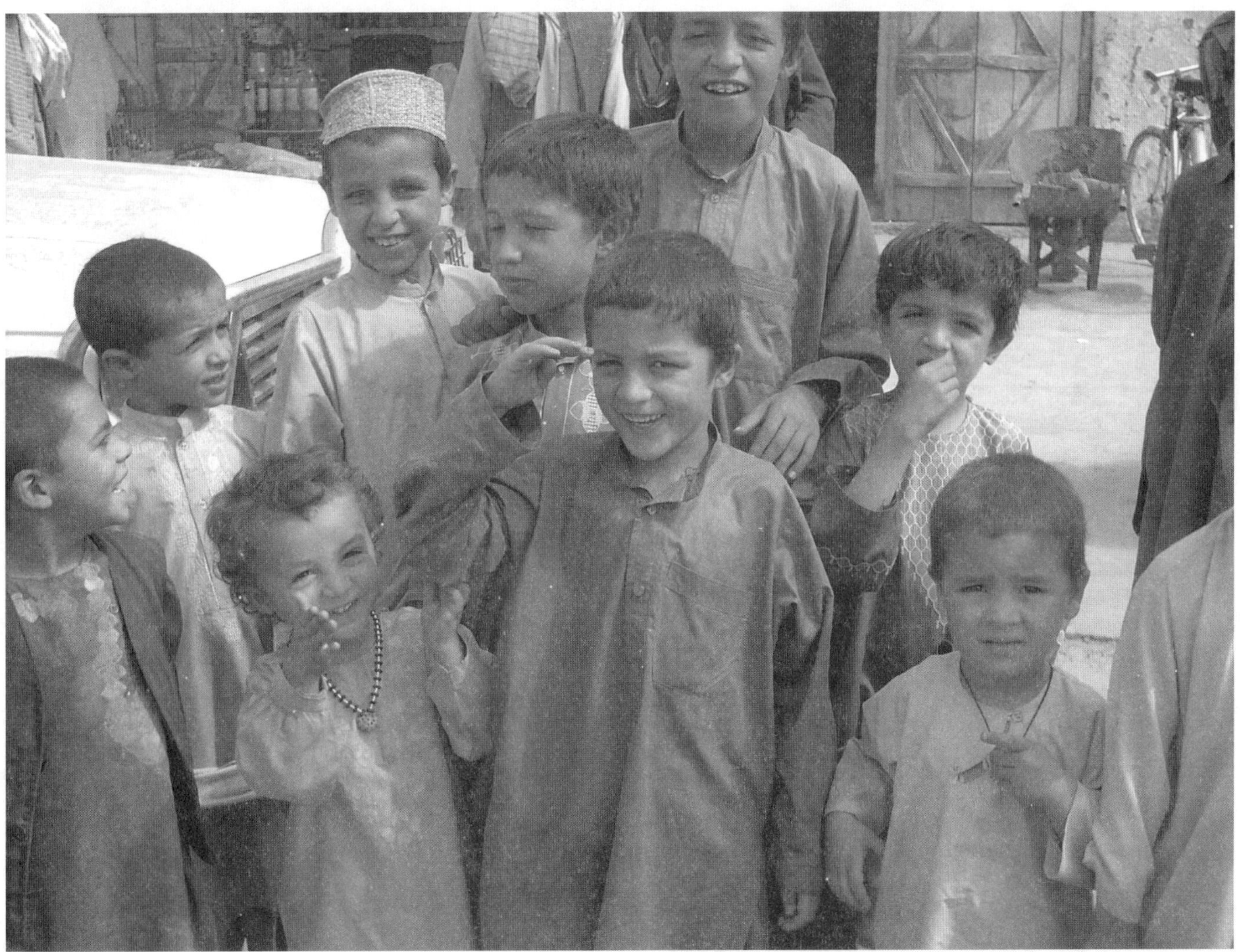

The tactics of terrorism appear to be universal. While ideologies and motivations vary, terrorist organizations in different parts of the world often use similar methods. It remains unclear whether this is the consequence of increased communications among terrorist organizations or the result of greater access to information in an age of global media. Some argue that terrorists simply tend to be conservative in their selection of tactics, relying on tactics that have proven successful rather than risking failure. Regardless of the underlying reasons, the tactics used by terrorists have remained remarkably consistent. While they have increased in size and sophistication, bombs are still the primary tool employed by terrorist organizations. On the average, bombs are used in over two-thirds of all terrorist attacks around the world. In addition to bombings, kidnapping, hostage taking, hijacking, armed attacks, and arson are tactics commonly employed by terrorist organizations. To finance these activities terrorists are increasingly resorting to organized crime and drug-trafficking.

The articles in this unit highlight some contemporary terrorist tactics. Greenless and McBeth look at a potential shift in tactics from mass killings to targeted assassinations. They note that Jemaah Islamiah is targeting Western ambassadors, businessmen, and Indonesian public figures in a new wave of violence. The next article highlights the potential link between organized crime and terrorist organizations. It argues that the smuggling and sale of illegal drugs has become an important source of funding for terrorist organizations. Finally, Scott Atran discusses the difficulty of categorizing the motivations of suicide bombers and offers some concrete suggestions for defusing the threat.

Terrorists' New Tactic: Assassination

Bombs are indiscriminate killers of Muslims and Westerners alike, so Southeast Asian terrorists are now targeting high-profile individuals. The REVIEW *reveals a disturbing shift in the terrorist threat*

Donald Greenlees and John McBeth

AFTER A 10-MONTH lull in terrorist attacks on Western targets in Indonesia, Jemaah Islamiah, the extremist group linked to the Al Qaeda terrorist network, is readying a new tactic: assassinations of public figures, diplomats and business people, according to intelligence gathered recently by Western security agencies.

In a disturbing shift in the nature of the terrorist threat in Indonesia, there are mounting signs that JI is turning away from large-scale car bombs and towards simpler methods that are potentially just as effective, say a number of Western police, security officials and analysts.

They tell the REVIEW that British and Australian intelligence organizations have collected specific and credible information—much of it from communications intercepts—that a group of JI operatives, trained to carry out assassinations, has been infiltrated into Indonesia in recent weeks. The group arrived through the Indonesian province of East Kalimantan from Mindanao, in the southern Philippines. No further information about the group was disclosed.

Top of the target list, the Western security sources say, are the American, British and Australian ambassadors, as well as other senior officials from these embassies. There are also serious concerns about potential attacks on senior foreign business executives, particularly in the mining and energy industries, and Indonesian public figures.

"We have seen they have the ability to build bombs and we have evidence of their capability in using weapons and we know they are well-trained and have Afghan backgrounds, so it is clear they have the capacity of carrying out such an attack," says Ansyaad M'bai, head of the counter-terrorism desk at the Ministry of Political and Security Affairs. "If the target is an important figure, then it would be just as effective as a bomb."

The most likely form of attack would be the shooting of targeted individuals while they are in their cars heading to or from work. In response, the United States, British and Australian embassies are insisting that staff vary routes to and from work and their times of departure. "I don't think it is a big leap to see [terrorists] going away from large types of events for more selected targeting," says one U.S. official.

Publicly, U.S. Ambassador to Indonesia Ralph Boyce has remained silent on the latest threat, but a May 21 embassy message reminded Americans to observe "vigilant personal security precautions." And pointing to the May 21 bombing in Bangladesh that wounded the British high commissioner there, British Ambassador to Jakarta Charles Humfrey says: "British ambassadors in countries where there have been terrorist attacks are considered to be at risk."

Western security officials say there are no indications that JI has given up on the idea of large-scale bombings, but they believe the terrorists are looking at other ways of hitting foreign interests, particularly economic interests, without causing heavy Muslim casualties.

According to leading terrorism expert Sidney Jones, leaders of JI met for a strategy rethink after last August's bombing of the JW Marriott Hotel in Jakarta. The meeting, in Solo, central Java, used a conference of the Indonesian Mudjahidin Council, a hardline Muslim group, as cover. The JI leaders were mostly unhappy about the outcome of the Marriott blast, which killed 11 Indonesians and only one foreigner. "The assessment was that it was a failure in implementation," says Jones, the Indonesia director of the International Crisis Group, a not-for-profit organization that works to resolve conflicts.

Fears over a switch in tactics by Jemaah Islamiah have prompted the three embassies to toughen consular warnings on travel to Indonesia. Embassy officials have also held a series of meetings with the expatriate community to remind them that the danger of terrorist violence remains very real. The new threat comes amid concerns that Indonesia's flurry of parliamentary and presidential elections, stretching between April and September, is sapping the political will to deal with terrorism. Few candidates for the presidency have mentioned the issue.

But in an incident that alarmed Western diplomats, four gunmen killed prosecutor Ferry Silalahi on May 26 in central Sulawesi as he drove home from church. Silalahi had prosecuted three men charged with a role in the October 2000 Bali bombings that killed 202 people and were blamed on Jemaah Islamiah. Central Sulawesi, riven by sectarian strife in recent years, is a known centre of JI operations and once was the site of an Al Qaeda training camp.

For the small expatriate community, now used to living with searches at shopping malls, office buildings and hotels, the assassination of individuals would have as terrifying an impact as a hotel bombing. Says one Western police official: "You get just as much news out of taking out an ambassador as a bombing, and it's easier to do."

Targeting individuals would be a particular concern for foreign businesses. In the past two years, security professionals have simply advised corporate clients to beef up security in their offices and homes and avoid high-risk locations. But if the target is an individual executive, providing protection is tougher.

LIVE TARGETS

New intelligence indicates that Jemaah Islamiah is targeting:

- The U.S., British and Australian ambassadors, plus other senior embassy officials
- Foreign business executives, especially of mining and energy companies
- Indonesian public figures

Some big oil-and-gas companies have hired bodyguards for their senior executives. These companies, used to working in precarious environments, say the security risks of doing business in Indonesia are unlikely to affect investment or staff decisions. Still, there is no disguising their concern. "If you look at the [embassy] warnings, the only thing you can do that you are not doing now is pull out," says a security adviser to one big Western energy company.

Unease over the terrorist threat has grown in the past two weeks following attacks by Islamic militants on an oil company's office and an expatriate-housing complex in Saudi Arabia, and the killing of two Westerners there. "We have to remind people that terrorism can comes in different forms. We've seen that from Saudi Arabia," says Ambassador Humfrey.

The attacks in Saudi Arabia heighten fears of copycat attacks in Indonesia. In March a memo appeared on an Al Qaeda Web site attributed to Abdulaziz al-Mokrin, allegedly Al Qaeda's leader in Saudi Arabia, urging attacks on "unprotected soft targets and ... individuals" in Indonesia, citing businessmen, diplomats, scientists, soldiers and tourists. Police raids on Jemaah Islamiah safe houses in the past two years have uncovered at least two significant caches of automatic weapons.

All the major U.S. oil companies—including ExxonMobil, Unocal and ConocoPhillips—were on a target list found in a JI safe house in Semarang, central Java, last year. In the past month, counter-terrorism chief M'bai says, there have been "indications" of a terrorist attack planned against the Canadian-owned International Nickel Co. (Inco) in southeast Sulawesi. Warned by Western security agencies, Inco's foreign mine executives were evacuated to Jakarta.

Last August, Indonesian troops seized a 30-year-old Yemeni national, believed to be a recruiter for Al Qaeda, after he illegally boarded a bus taking workers to the giant Freeport copper-and-gold mine in Papua. He was taken to Jakarta, then disappeared from police custody. Videotape of Freeport was used in an Al Qaeda propaganda film made in 2002, accusing U.S. multinationals of exploiting Muslim workers.

Western security officials say a range of other factors increases anxiety about a resumption of terrorist attacks on Westerners in Indonesia. They include anger among Muslims over the abuse of Iraqi prisoners by U.S. guards, and the rearrest and impending second trial of alleged former JI spiritual leader Abu Bakar Bashir.

Indonesian investigators say a lack of political will and Indonesia's weak legal system mean they face an uphill struggle in their efforts to track JI and five or six other homegrown militant groups that feed off their own ideology. And in newly democratic Indonesia, "you can't just round up people because they espouse a particular philosophy," notes Jones, the terrorism expert.

Much of the investigators' criticism has been heaped on President Megawati Sukarnoputri, whose ineffective leadership is the main reason for her declining popularity ahead of the July 5 first round of the country's first direct presidential elections. "What we need is a national leader to make a clear decision on terrorism," one senior Indonesian police officer told the REVIEW. "If Megawati is re-elected, she will change nothing. When it comes to any issue regarding Islam, she won't make a decision. This is a crucial issue."

Among the presidential candidates, only Susilo Bambang Yudhoyono, the former coordinating minister for political and security affairs, appears to have a coherent policy for dealing with the terrorist threat. "He is the only one who fully understands how it is viewed by the international community and what the constraints are in dealing with counter-terrorism in Indonesia," says the senior police officer. "The other candidates use only rhetoric."

Yudhoyono's approach centres not on more legislation, but on strengthening the legal system and improving co-ordination among law-enforcement agencies—measures that go beyond Islamic militancy and address the other serious problems of social injustice and flagging foreign investment.

Counter-terrorism officials also speak of an urgent need to change the thinking of Indonesian judges, to get them to understand that terrorism is "an extraordinary crime that demands an extraordinary solution." Efforts to win a long jail term for radical cleric Bashir at his first trial foundered largely on the judges' refusal to accept teleconference testimony from detained terrorist suspects in Malaysia and Singapore, due to a legal technicality.

One counter-terrorism official also criticizes the judges for not being more proactive in questioning witnesses during the trial. "We don't want to intervene in the judicial process," he explains, "but we do believe these are issues that should be discussed outside the courtroom."

Outlawing various militant groups doesn't appear to be an option. In the world's largest Muslim country, that would only drive Islamic radicals underground. The government has yet to prosecute any of the militants on the often-provable charge of preaching hatred, an offence under Article 160 of the Criminal Code. It is, however, studying the legal systems in France, Germany and Spain to look for new ways to tackle terrorism.

MORE MILITANT GROUPS

Terrorism experts believe it is a mistake simply to focus on JI. "What has to be changed is the militancy and the ideology [that attract recruits to the organization]," warns the senior police officer. "The militants will only gather strength if the next leaders can't improve the situation. This is very disappointing to the people in the field."

Jones also thinks it is a mistake to see Indonesian militancy as monolithic. Although the militants may be interwoven in one way or another, she identifies several different home-grown groups outside JI with a capacity for independent action. They include:

- Radical members of the Ngruki network, named after the central Java boarding school run by Bashir. Most of the militants broke away after a split in the Ngruki alumni in 1995 when Bashir was living in exile in Malaysia.
- Followers of Darul Islam, the Muslim organization which fought for the creation of an Islamic state in the 1950s. JI founder Abdullah Sungkar, who died in late 1999, helped to revive remnants of Darul Islam in the early 1980s. Many of the current militants are either relatives or descendants of former Darul Islam leaders.
- Groups of veterans from Afghanistan and Mindanao training camps who operate independently from JI.

Although police have the names of more than 350 Indonesians who have received instruction in everything from infantry weapons to map-reading, booby traps and explosives, many others have still not been identified "What do we do with them?" says M'bai. "We can't find all of them. We know these are dangerous people [but] we can't touch them unless there is clear evidence they are directly involved in terrorist acts."

Police say an accidental blast at a house in the Jakarta suburb of Cimanggis in March shows how radical groups have proliferated. Twelve people were arrested in connection with the explosion, but investigators have yet to determine where they fit in the terrorism picture.

The blast also showed that training and recruitment continues unabated, with evidence pointing to the emergence of a second-generation Laskar Kos, the JI operations arm formerly headed by Riduan Isamuddin, or Hambali, who was captured in Thailand last year and handed over to the U.S. Indonesian police believe he has been moved recently from Diego Garcia to Guantanamo Bay in Cuba. Warns Jones: "JI is very much an alive and ongoing organization, even if a lot of the leadership is incapacitated."

PAYING FOR TERROR

How jihadist groups are using organized-crime tactics—and profits—to finance attacks on targets around the globe

David E. Kaplan

The first blast struck at 1:25 p.m., shattering the walls of the Bombay Stock Exchange, leaving a grisly scene of broken bodies, shattered glass, and smoke. Next to be hit was the main office of the national airline, Air India, followed by the Central Bazaar and major hotels. At the international airport, hand grenades were thrown at jets parked on the tarmac. For nearly two hours the carnage went on, as unknown assailants wreaked havoc upon one of the world's largest cities. In all, 10 bombs packed with plastic explosives rocked Bombay, killing 257 and injuring over 700.

What happened in Bombay on that day, March 12, 1993, was a chilling precursor to the 9/11 terrorist attacks, a careful choreography of death and destruction, aimed at the heart of a nation's financial center and intended to maximize civilian casualties. Engineered by Muslim extremists, the attacks were meant to exact revenge for deadly riots by Hindu fundamentalists that had claimed over a thousand lives, most of them Muslim. But more than vengeance was at work in Bombay. Indian police later recovered an arsenal big enough to spark a civil war: nearly 4 tons of explosives, 1,100 detonators, nearly 500 grenades, 63 assault rifles, and thousands of rounds of ammunition. Within days of the attacks, police had gotten their first break by tracing an abandoned van filled with a load of weapons. The trail soon led to a surprising suspect: not a terrorist but a gangster. And not just any gangster but an extraordinary crime boss, a man known as South Asia's Al Capone.

Virtually unknown in the West, Dawood Ibrahim is a household name across the region, his exploits known by millions. He is, by all accounts, a world-class mobster, a soft-spoken, murderous businessman from Bombay who now lives in exile, sheltered by India's archenemy, Pakistan. He is India's godfather of godfathers, a larger-than-life figure alleged to run criminal gangs from Bangkok to Dubai. Strong-arm protection, drug trafficking, extortion, murder-for-hire—all are stock-in-trade rackets, police say, of Dawood Ibrahim's syndicate, the innocuously named D Company.

"The world is seeing the birth of a new hybrid of organized-crime-terrorist organizations."

Dawood, as he is known in the Indian press, is very much on Washington's radar screen today. Two years ago, the Treasury Department quietly designated Ibrahim a "global terrorist" for lending his smuggling routes to al Qaeda, supporting jihadists in Pakistan, and helping engineer the 1993 attack on Bombay. He is far and away India's most wanted man, his name invoked time and again by Indian officials in their discussions of terrorism with U.S. diplomats and intelligence officers. As a result of those discussions, the FBI and the Drug Enforcement Administration each have active investigations into Dawood's far-flung criminal network, *U.S. News* has learned.

Understanding Dawood's operations is important, experts say, because they show how growing numbers of terrorist groups have come to rely on the tactics—and profits—of organized criminal activity to finance their operations across the globe. An inquiry by *U.S. News*, based on interviews with counterterrorism and law enforcement officials from six countries, has found that terrorists worldwide are transforming their operating cells into criminal gangs. "Transnational crime is converging with the terrorist world," says Robert Charles, the State Department's former point man on narcotics. Antonio Maria Costa, the head of the United Nations Office on Drugs and Crime, agrees: "The world is seeing the birth of a new hybrid of organized-crime-terrorist organizations. We are breaking new ground."

Blood money. Some scholars argue that terrorists and traditional crime groups both now exist on a single, violent plane, populated at one end by politically minded jihadists and at the other by profit-driven mobsters, with most groups falling somewhere in between. Mafia groups and drug rings in Colombia and the Balkans, for example, commit political assassinations and bomb police and prosecutors, while terrorist gangs in Europe and North Africa traffic in drugs and illegal aliens. Both crime syndicates and terrorist groups thrive in the same subterranean world of black markets and laundered money, relying on shifting networks and secret cells to accomplish their objectives. Both groups have similar needs: weapons, false documentation, and safe houses.

But some U.S. intelligence analysts see little evidence of this melding of forces. Marriages of convenience may exist, they say, but the key difference is one of motive: Terrorist groups are driven by politics and religion, while purely criminal groups have just one thing in mind—profit. Indeed, associating with terrorists, particularly since 9/11, can be very bad for business—and while crime syndicates may be parasitical, most do not want to kill their host.

What many intelligence analysts do see today, however, is terrorist organizations stealing whole chapters out of the criminal playbook—trafficking in narcotics, counterfeit goods, illegal aliens—and in the process converting their terrorist cells into criminal gangs.

The terrorist gang behind the train bombings in Madrid last year, for example, financed itself almost entirely with money earned from trafficking in hashish and ecstasy. Al Qaeda's affiliate in Southeast Asia, Jemaah Islamiyah, engages in bank robbery and credit card fraud; its 2002 Bali bombings were financed, in part, through jewelry store robberies that netted over 5 pounds of gold. In years past, many terrorist groups would have steered away from criminal activity, worried that such tactics might tarnish their image. But for hard-pressed jihadists, committing crimes against nonbelievers is increasingly seen as acceptable. As Abu Bakar Bashir, Jemaah Islamiyah's reputed spiritual head, reportedly said: "You can take their blood; then why not take their property?"

The implications are troubling because organized crime offers a means for terrorist groups to increase their survivability. A Stanford University study conducted after the 9/11 attacks looked at why some conflicts last so much longer than others. One key factor: crime. Out of 128 conflicts, the 17 in which insurgents relied heavily on "contraband finances" lasted on average 48 years—over five times as long as the rest. "If the criminal underworld can keep terrorist coffers flush," says Charles, the former State Department official, "we will continue to face an enemy that would otherwise run out of oxygen."

The terrorists behind the Madrid train bombings paid their way by selling drugs.

The growing reliance on crime stems from the end of the Cold War, when state sponsorship of terrorism largely faded along with communism, forcing groups to become much more self-sufficient. Accelerating the trend, analysts say, is the crackdown since 9/11 on fundraising by Islamic radicals from mosques and charities, which has pushed their operations further toward racketeering. "The bottom line is if you want to survive today as a terrorist, you probably have to support yourself," says Raphael Perl, a counterterrorism specialist at the Congressional Research Service. The drug trade, in particular, has proved irresistible for many. Nearly half of the 41 groups on the government's list of terrorist organizations are tied to narcotics trafficking, according to DEA statistics.

Scams. The new face of terrorism can best be seen in western Europe. "Crime is now the main source of cash for Islamic radicals in Europe," says attorney Lorenzo Vidino, author of the new book *Al Qaeda in Europe*. "They do not need to get money wired from abroad like 10 years ago. They're generating their own as criminal gangs." European police and intelligence officials agree: The Continent's most worrisome cells, composed largely of immigrants from Morocco and Algeria, have in effect become racketeering syndicates. Their scams are as varied as the criminal world: Drugs, smuggling, and fraud are mainstays, but others include car theft, selling pirated CDs, and counterfeiting money. One enterprising pair of jihadists in Germany hoped to fund a suicide mission to Iraq by taking out nearly $1 million in life insurance and staging the death of one in a faked traffic accident. Some cells are loosely bound and based on petty crime; others, like the group behind the Madrid bombings, suggest a whole new level of sophistication.

The terrorists behind the Madrid attacks were major drug dealers, with a network stretching from Morocco through Spain to Belgium and the Netherlands. Their ringleader, Jamal "El Chino" Ahmidan, was the brother of one of Morocco's top hashish traffickers. Ahmidan and his followers paid for their explosives by trading hashish and cash with a former miner. When police raided the home of one plotter, they seized 125,800 ecstasy tablets—one of the largest hauls in Spanish history. In all, authorities recovered nearly $2 million in drugs and cash from the group. In contrast, the Madrid bombings, which killed 191 people, cost only about $50,000.

Similar reports of drug-dealing jihadists are coming out of France and Italy. In Milan, Islamists peddle heroin on the streets at $20 a hit and then hand off 80 percent of the take to their cell leader, according to Italy's *L'Espresso* magazine. The relationship, surprisingly, is not new. As early as 1993, says Vidino, French authorities warned that dope sales in suburban Muslim slums had fallen under the control of gangs led by Afghan war veterans with ties to Algerian terrorists. What is new is the scale of this toxic mix of jihad and dope. Moroccan terrorists used drug sales to fund not only the 2004 Madrid attack but the 2003 attacks in Casablanca, killing 45, and attempted bombings of U.S. and British ships in Gibraltar in 2002. So large looms the North African connection that investigators believe jihadists have penetrated as much as a third of the $12.5 billion Moroccan hashish trade—the world's largest—a development worrisome not only for its big money but for its extensive smuggling routes through Europe.

Along with drug trafficking, fraud of every sort is a growth industry for European jihadists. Popular scams include fake credit cards, cellphone cloning, and identity theft—low-level frauds that are lucrative but seldom attract the concerted attention of authorities. Some operatives are more ambitious, however. Officials point to the case of Hassan Baouchi, a 23-year-old ATM technician in France. Baouchi told police last year that he'd been held hostage by robbers who forced him to empty six ATMs of their cash—about $1.3 million. Investigators didn't quite buy Baouchi's story and soon put him under arrest; the money, they believe, has ended up with the Moroccan Islamic Combatant Group—the al Qaeda affiliate tied to the bombings in Casablanca and Madrid.

Another big racket for European jihadists is human smuggling. "North Africa and western Europe are somewhat like Mexico and the United States," says a U.S. counterterrorism agent. "But now imagine if Mexico were Muslim and jihadist cells were the ones moving aliens across the border." Jihadists do not dominate Europe's lucrative human smuggling trade, but they are surely profiting by it. Authorities in Italy suspect that one gang of suspected militants made over 30 landings on an island off Sicily, and that it moved thousands of people across the Mediterranean at some $4,000 a head. Particularly active is the Salafist Group for Call and Combat, an Algerian al Qaeda affiliate known by its French acronym, GSPC. Two years ago, German authorities dismantled another group moving Kurds into Europe, tied to al Qaeda ally Ansar al-Islam, a fixture of the Iraq insurgency. A recent Italian intelligence report notes that jihadists' work in human smuggling has brought them into contact with domestic and foreign criminal organizations. One partner in the trade, sources say, is the Neapolitan Camorra, the notorious Naples-based version of the Mafia, which operates safe houses for illegal aliens. Italian court records show contact between Mafia arms dealers and radical Islamists as early as 1998.

Prison recruits. The prime training ground for Europe's jihadist criminals may well be prison. There are no hard numbers, but as much as half of France's prison population is now believed to be Muslim. In Spanish jails, where Islamic radicals have recruited for a decade, the number has reached some 10 percent. Ahmidan, leader of the Madrid bombing cell, is thought to have been radicalized while serving time in Spain and Morocco. Prison was also the recruiting center for many of the 40-plus suspects nabbed by Spanish authorities last year for plotting a sequel to the Madrid bombings—an attack with a half ton of explosives on Spain's national criminal court. Nearly half the group had rap sheets with charges ranging from drug trafficking to forgery and fraud.

Al Qaeda's leadership, however, has proved more wary about jumping into the drug trade. Holed up in the forbidding mountain refuges of the Pakistan-Afghanistan border, Osama bin Laden and his remaining lieutenants have steered clear of the largest horde of criminal wealth in years: the exploding Afghan heroin trade. Press reports of bin Laden's involvement in the drug trade are flat wrong, say counterterrorism officials. Long ago, al Qaeda strategists reasoned that drug trafficking would expose them to possible detection, captives have told U.S. interrogators. They also don't trust many of the big drug barons, intelligence officials say, and have encouraged their members not to get involved with them.

Bin Laden continues to come up with funds raised from sympathetic mosques and other supporters, but the money no longer flows so easily. Pakistan and Saudi Arabia have banned unregulated fundraising at mosques, and western spy agencies now watch closely how the money flows from big Islamic charities. One result: Cash-strapped jihadists in the badlands of the border region are staging kidnappings-for-ransom and highway robberies, Pakistani officials tell *U.S. News*. "Those people are now feeling the pinch," says Javed Cheema, head of the Interior Ministry's National Crisis Management Cell. "We see a fertile symbiosis of terrorist organizations and crime groups." Some jihadists have joined in wholesale pillaging of Afghanistan's heritage by smuggling antiquities out of the country—a trade nearly as lucrative as narcotics. Among the items being sold clandestinely on the world market: centuries-old Buddhist art and other works from the pre-Islamic world. Apparently, al Qaeda's interest is not new; before 9/11, hijack ringleader Mohamed Atta approached a German art professor about peddling Afghan antiquities, Germany's Federal Criminal Police Office revealed this year. Atta's reason, reports *Der Spiegel* magazine: "to finance the purchase of an airplane."

"You can take their blood; then why not take their property?"

Al Qaeda may be avoiding the heroin trade, but nearly everyone else in the region—from warlords to provincial governors to the Taliban—is not. The reasons are apparent: Afghanistan's opium trade is exploding. The cultivation of opium poppies, from which heroin is made, doubled from 2002 to 2003, according to CIA estimates. Then, last year, that amount *tripled.* Afghanistan now provides 87 percent of the world's heroin. "We have never seen anything like this before," says Charles, the former State Department narcotics chief. "No drug state ever made this much dope and so quickly." The narcotics industry now makes up as much as half of Afghanistan's gross domestic product, analysts estimate, and employs upward of 1 million laborers, from farmers to warehouse workers to truck drivers. And now Afghans are adding industrial-level amounts of marijuana to the mix. U.N. officials estimate that some 74,000 to 86,000 acres of pot are being grown in Afghanistan—over five times what is grown in Mexico.

Corruption. And if al Qaeda itself is staying out of drugs, its allies certainly are not. The booming drug trade has given a strong second wind to the stubborn insurgency being waged by the Taliban and Islamist warlords like Gulbuddin Hekmatyar. Both the Taliban and Hekmatyar's Hezb-i-Islami army control key smuggling routes out of the country, giving them the ability to levy taxes and protection fees on drug caravans. Crime and terrorism experts are also alarmed over the corrosive, long-term effects of all the drug money, not just within Afghanistan but across the region. The ballooning dope trade is rapidly creating narco-states in central Asia, destroying what little border control exists and making it easier for terrorist groups to operate. Ancient smuggling routes from the Silk Road to the Arabian Sea are being supercharged with tons of heroin and billions of narcodollars. Within Afghanistan, drug-fueled corruption is pervasive; governors, mayors, police, and military are all on the take. A raid this year in strategically located Helmand province came up with a whopping 9 1/2 tons of heroin—stashed inside the governor's own office.

The smuggling routes lead from landlocked Afghanistan to the south and east through Pakistan, to the west through Iraq, and to the north through central Asia. Throughout the region the amounts of drugs seized are jumping, along with rates of crime, drug addiction, and HIV infection. Particularly hard hit are Afghanistan's impoverished northern neighbors, the former Soviet republics of Kirgizstan and Tajikistan. Widely praised demonstrations in Kirgizstan this year, which overthrew the regime of strongman Askar Akayev, have brought to power an array of questionable figures. "Entire branches of government are being directed by individuals tied to organized crime," warns Svante Cornell of the Central Asia-Caucasus Institute at Johns Hopkins University. "The whole revolution smells of opium."

"If you want to survive today as a terrorist, you probably have to support yourself."

Neighboring republics are little better off. Central Asia's major terrorist threat, the Islamic Movement of Uzbekistan, has largely degenerated into a drug mafia, officials say. In Kazakhstan the interior minister tried to investigate corruption by going undercover in a truck packed with 9 tons of watermelons, motoring 1,200 miles from the Kirgiz Republic to the Kazakh capital. His team had to pay bribes to 36 different police and customs officials en route—some as little as $1.50. (Others merely accepted their bribe in melons.) The cargo was never inspected. What is happening in Iran, meanwhile, is "a national tragedy," according to the U.N.'s Costa. So much Afghan dope is being shipped into the country that it now has the world's highest per capita rate of addiction. The ruling mullahs in Tehran have taken it seriously; Iranian security forces have fought deadly battles with drug traffickers along their border, losing some 3,600 lives in the past 16 years. But even as their troops fight, the corruption has reached high officials of the Iranian government, who are using drug profits as political patronage, sources tell *U.S. News*. "There are indications," says Cornell, "that hard-line conservatives are up to their ears in the Afghan opium trade."

Nor has Russia escaped the heroin boom's impact. From central Asia, growing amounts of Afghan heroin are entering the south of the country; drug-control officials report that large numbers of Russian military are on the take, even trucking the stuff in army vehicles. The level of corruption has, in turn, raised concern over the ultimate black market: in radioactive materials. Russia's "nuclear belt"—a chain of nuclear research and weapons sites—runs directly along those heroin smuggling routes. How bad are conditions in the area? Bemoans one intelligence expert: "We know so little."

"Even if a little bit ends up in insurgent hands, it doesn't take a lot to build a truck bomb."

Iraq, too, is starting to see its share of narcotics, but drugs are but a bit player in an insurgency that has also blurred the lines between terrorism and organized crime. Within Iraq's lawless borders exists an unsavory criminal stew composed of home-grown gangsters, ex-Baathists, and jihadists. "Terrorists and insurgents are conducting a lot of criminal activities, extortion and kidnapping in particular, as a way to acquire revenues," Caleb Temple of the Defense Intelligence Agency testified to Congress this July. Among the biggest cash cows: The insurgents take part in the wholesale theft of much of Iraq's gasoline supply, earning millions of dollars in a thriving black market. Extortion and protection are also rife, and kidnapping for ransom has ballooned into a major industry, with up to 10 abductions a day. Among those targeted: politicians, professors, foreigners, and housewives. Those with political value may find they've been sold to militants.

The insurgents are also key players in the graft and corruption that have enveloped Iraq. So much foreign aid money has disappeared that two U.S. intelligence task forces are now investigating its diversion to the insurgency, *U.S. News* has learned. Western aid agencies, Islamic charities, and U.S. military supply programs all have been targeted, analysts believe. Occupation authorities cannot account for nearly $9 billion of oil revenues it had transferred to Iraqi government agencies between 2003 and 2004, according to an audit by a special U.S. inspector general set up by Congress. "Even if a little bit ends up in insurgent hands," says one official, "it doesn't take a lot to build a truck bomb." The implications are troubling: The insurgents may be using America's own foreign aid to fund the killing of U.S. troops.

Back at home, U.S. officials are looking warily at the growing rackets of terrorist groups overseas and voice concern that the trend will grow here. "We see a lot of individual pockets of it in the United States," says Joseph Billy, deputy chief of the FBI's counterterrorism division. "Left unchecked, it's very worrisome—this is one we have to be aggressive on." Federal investigators have uncovered repeated scams here largely involving supporters of Hamas and Hezbollah, and they have traced tens of thousands of dollars back to those groups in the Middle East. "There's a direct tie," says Billy. The list of crimes includes credit card fraud, identity theft, the sale of unlicensed T-shirts—even the theft and resale of infant formula. Most of these U.S. rackets have been low level, but some, involving cigarette smuggling and counterfeit products, have earned their organizers millions of dollars.

Foreign fish. The big fish—the Indian mobsters, Moroccan hash dealers, and Afghan drug barons—are swimming overseas, however, and U.S. law enforcement is starting to train its sights on the worst of them. After being shut out of Afghanistan for two decades, the Drug Enforcement Administration is making progress in going after top Afghan traffickers tied to the Taliban. DEA agents are applying the same sort of "kingpin strategy" that helped to break the Medellin and Cali cartels in Colombia by targeting whole trafficking organizations. The agency has identified some 10 of these "high-value targets," led by Afghans who have amassed fortunes of as much as $100 million, officials say. Two already have fallen this year: In April, agents nabbed a man they've dubbed "the Pablo Escobar of Afghanistan," Taliban ally Bashir Noorzai, by enticing him to a New York meeting. Noorzai is said to have helped establish the modern Afghan drug trade, and so lucrative were his operations that the indictment against him calls for the seizure of $50 million in drug proceeds. Then in October, the Justice Department announced the extradition of Baz Mohammad, another alleged "Taliban-linked narcoterrorist," charged with conspiring to import over $25 million worth of heroin into the United States and other countries. Mohammad, according to an indictment, boasted that "selling heroin in the United States was a 'jihad' because they were taking the Americans' money at the same time the heroin was killing them." He is now awaiting trial, in a New York jail.

The DEA's experience, however, illustrates some of the problems in grappling with the nexus between organized crime and terrorism. Despite post-9/11 calls for cooperation, the DEA's ties to other U.S. agencies are often strained; the drug agency is not even considered part of the U.S. intelligence community. Pentagon officials, worried over "mission creep," routinely refuse to give DEA agents air support in Afghanistan. Other turf issues still plague the FBI, CIA, Homeland Security, and other agencies, making collaborative work on the crime-terrorism issue problematic. The intelligence community, for example, remains leery of seeing its people or information end up in court. "It's the same wall we saw between law enforcement and the intelligence world," says one insider. "Only now it's between terrorism and other crimes."

"We have stovepiped battling terrorism and organized crime," agrees Charles, the State Department's former top cop. "You cannot meet a complex threat like this without a similar response. And we don't have one." Indeed, interviews with counterterrorism and law enforcement officials in a half-dozen federal agencies suggest that cooperation across the government remains episodic at best and depends most often on personal relationships. "The incentives are all still against sharing," complains one analyst. "The leadership says yes, the policies say yes, but the culture says no. The bureaucracy has won."

Washington looks like a model of cooperation compared with Europe, however, where the walls between agencies and across borders stand even higher. "If we don't get on top of the criminal aspect and the drug connections, we will lose ground in halting the spread of these [terrorist] organizations," warns Gen. James Jones, head of the U.S. European Command, who has watched the rise in terrorist rackets with mounting concern. "You have to have much greater cohesion and synergy."

For a brief moment in the early '90s, American cops, spies, and soldiers did come together on a common target of crime and terrorism—Pablo Escobar and his Medellin cocaine cartel. Escobar's killers were blamed for the murder of hundreds of Colombian officials and the bombing of an Avianca airliner that killed 110. To get Escobar, firewalls between agencies came down, information was shared, and money and people were focused on destroying one of the world's most powerful crime syndicates. During the 1990s, transnational crime continued to be seen as a national security priority, but it fell off the map after 9/11. Until January of this year, the federal government's chief interagency committee on organized crime hadn't even met for three years. The intelligence community's reporting on the area, although boosted in the past year, remains a near-bottom priority. One knowledgeable source called the quality of work overseas on crime "sorely lacking" and said the best material comes from other governments. Domestically, meanwhile, years of neglect by the FBI of analysis and information technology have left the agency without much useful information in its files. "Everyone thinks we've got huge databases with all our materials on organized crime," says one veteran. "We've got nothing close."

Still, the growing criminal inroads by terrorist groups have raised alarms among a handful of tough-minded policymakers in Washington, and they are pushing for change. The National Security Council has begun work on a new policy on transnational crime that promises to make the crime-terrorism connection a top priority. The CIA's Crime and Narcotics Center is spearheading the work of a dozen agencies in revamping the government's overall assessment of international crime; their report, with special attention to the nexus with terrorism, is scheduled for release early next year. One program that has caught U.S. attention is underway in Great Britain, where London's Metropolitan Police now routinely monitor low-level criminal activity for ties to terrorism, checking over reports of fraud involving banks, credit cards, and travel documents. Similar work is being done by the feds' Joint Terrorism Task Forces in some U.S. cities. And Immigration and Customs Enforcement has prioritized cracking rings smuggling people from high-risk countries, with good success.

"Draining the swamp." In some ways, the deeper involvement by terrorists in traditional criminal activity may make it easier to track them. Criminal informants, who can be tempted with shortened prison time and money, are much easier to develop than the true believers who fill the ranks of terrorist groups. Acts of crime also attract attention and widen the chances that terrorists will make a mistake. Take, for example, a case uncovered this July, in which four men allegedly plotted to wage a jihad against some 20 targets in Southern California, including National Guard facilities, the Israeli Consulate, and several synagogues. Prosecutors said the planned attacks, led by the founder of a radical Islamic prison group, were being funded by a string of gas station robberies. The break in the case came not by an elite counterterrorist squad but by local cops who found a cellphone one of the robbers had lost during a gas station stickup.

Getting a handle on terrorism's growing criminal rackets will not prove easy, however. "Draining the swamp," as counterterrorism officials vow to do, may require more than even a seamless approach by intelligence and law enforcement can offer. Many of the worst groups owe their success to a pervasive criminality overseas, to failed states and no man's lands from Central Asia to North Africa to South America, where the rule of law remains an abstract concept. In other places, it is the governments themselves that are the criminal enterprises, so mired in corruption that entire countries could be indicted under U.S. antiracketeering laws. Together, they help make up a criminal economy that, like a parallel universe, runs beneath the legitimate world of commerce. This global shadow economy—of dirty money, criminal enterprises, and black markets—has annual revenues of up to $2 trillion, according to U.N. estimates, larger than the gross domestic product of all but a handful of countries. Without its underground bankers, smuggling routes, and fraudulent documents, al Qaeda and its violent brethren simply could not exist. But taking on a worldwide plague of crime and corruption might be more than the public bargained for. "Ultimately, cracking down means trying to impose order where there's instability, good governance where there's corruption and crime, and economic growth where there's poverty," says the University of Pittsburgh's Phil Williams, a consultant to the United Nations on crime and terrorism finance. "As long as the only routes of escape are violence and the black market, then organized crime and terrorism will endure as global problems."

Jihadists don't dominate Europe's human-smuggling trade, but they profit from it.

The Moral Logic and Growth of Suicide Terrorism

Scott Atran

Suicide attack is the most virulent and horrifying form of terrorism in the world today. The mere rumor of an impending suicide attack can throw thousands of people into panic. This occurred during a Shi'a procession in Iraq in late August 2005, causing hundreds of deaths. Although suicide attacks account for a minority of all terrorist acts, they are responsible for a majority of all terrorism-related casualties, and the rate of attacks is rising rapidly across the globe (see figures 1 and 2). During 2000–2004, there were 472 suicide attacks in 22 countries, killing more than 7,000 and wounding tens of thousands. Most have been carried out by Islamist groups claiming religious motivation, also known as jihadis. Rand Corp. vice president and terrorism analyst Bruce Hoffman has found that 80 percent of suicide attacks since 1968 occurred after the September 11 attacks, with jihadis representing 31 of the 35 responsible groups.[1] More suicide attacks occurred in 2004 than in any previous year,[2] and 2005 has proven even more deadly, with attacks in Iraq alone averaging more than one per day, according to data gathered by the U.S. military.[3] The July 2005 London and Sinai bombings, a second round of bombings at tourist destinations in Bali in October, coordinated hotel bombings in Jordan in November, the arrival of suicide bombings in Bangladesh in December, a record year of attacks in Afghanistan, and daily bombings in Iraq have spurred renewed interest in suicide terrorism, with recent analyses stressing the strategic logic, organizational structure, and rational calculation involved.[4]

Figure 1: Suicide Attacks Worldwide, Annualized by Decade

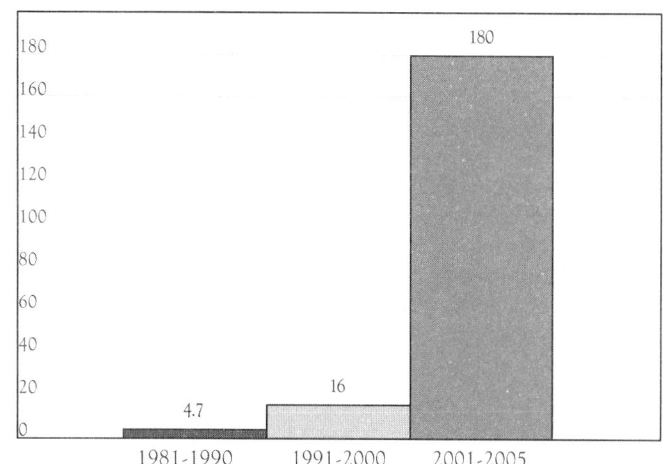

Figure 2: Suicide Attacks Worldwide, 2001–2005

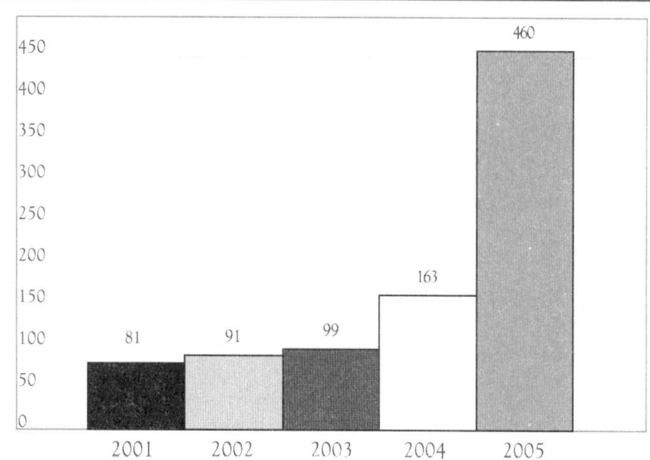

Whereas they once primarily consisted of organized campaigns by militarily weak forces aiming to end the perceived occupation of their homeland, as argued by University of Chicago political scientist Robert Pape in *Dying to Win: The Strategic Logic of Suicide Terrorism*,[5] suicide attacks today serve as banner actions for a thoroughly modern, global diaspora inspired by religion and claiming the role of vanguard for a massive, media-driven transnational political awakening. Living mostly in the diaspora and undeterred by the threat of retaliation against original home populations, jihadis, who are frequently middle-class, secularly well educated, but often "born-again" radical Islamists, including converts from Christianity, embrace apocalyptic visions for humanity's violent salvation. In Muslim countries and across western Europe, bright and idealistic Muslim youth, even more than the marginalized and dispossessed, internalize the jihadi story, illustrated on satellite television and the Internet with the ubiquitous images of social injustice and political repression with which much of the Muslim world's bulging immigrant and youth populations intimately identifies. From the suburbs of Paris to the jungles of Indonesia, I have interviewed culturally uprooted and politically restless youth who echo a stunningly simplified and decontextualized message of martyrdom for the sake of global jihad as life's noblest cause. They are increasingly as willing and even eager to die as they are to kill.

The policy implications of this change in the motivation, organization, and calculation of suicide terrorism may be as novel as hitherto neglected. Many analysts continue to claim that jihadism caters to the destitute and depraved, the craven and criminal, or those who "hate freedom." Politicians and pundits have asserted that jihadism is nihilistic and immoral, with no real program or humanity. Yet, jihadism is none of these things. Do we really understand the causes of today's suicide terrorism? Do suicide attacks stem mainly from a political cause, such as military occupation? Do they need a strong organization, such as Al Qaeda? What else could be done to turn the rising tide of martyrdom?

It's Not Just for Politics Anymore

In *Dying to Win*, Pape claims that foreign occupation is the root cause of suicide terrorism. The rise in attacks correlates with U.S. military occupation of countries whose governments tend to be authoritarian and unresponsive to popular demands. Analyzing data on 315 suicide attacks from 1980 to 2003, he asserts that the common thread linking 95 percent of all suicide attacks around the world is not religion or ideology, but rather a clear, strategic, political objective. They are organized campaigns to compel a modern democracy, principally the United States, to withdraw military forces from a group's perceived homeland. Al Qaeda supposedly fits this model, being driven primarily by the efforts of Osama bin Laden and those sympathetic to his cause to expel the United States from the Arab heartland—Saudi Arabia, Palestine, Egypt, and Iraq—and ultimately from all Muslim countries. On September 11, 2001, for example, 15 of the 19 suicide attackers came from Saudi Arabia, where nearly 5,000 U.S. combat troops were billeted at the time, with 7,000 more stationed elsewhere on the Arabian peninsula.

According to Pape's findings, suicide bombers come disproportionately from among the largely secular and educated middle classes that aspire to freedom and greater opportunities, yet see their hopes stymied by corrupt dictators and one-party elites acting in collusion with U.S. oil and other interests. On the surface, recent trends seem to bolster Pape's thesis. In October 2003, bin Laden warned in a televised video that European nations fighting in Iraq or Afghanistan would be fair game for attack. The next month, suicide bombings targeted the British consulate and HSBC Bank in Istanbul. In December 2003, jihadi Web sites were broadcasting "Iraqi Jihad, Hopes and Risks,"[6] a 42-page blueprint for strategically timing bombings to European political events, the first target of which became the March 2004 Spanish elections. (The bombers in these attacks blew themselves up with suicide belts later when cornered by police, and two Madrid plotters who escaped the police dragnet died as suicide bombers in Iraq in April 2005.)

In the last two years, suicide attackers have struck in 18 countries, mostly among U.S. allies linked to undemocratic regimes, such as Pakistan, Saudi Arabia, Uzbekistan, and Egypt, or in places with perceived occupations, such as the Palestinian territories, Chechnya, Kashmir, Afghanistan, and Iraq. In Iraq in 2004, there were more suicide attacks than in the entire world in any previous year of contemporary history, involving "martyrs" from 14 other Arab countries,[7] as well as volunteers from all over Europe. Nevertheless, Pape's basic data, correlations, and conclusions about the causes of terrorism are problematic, outdated in the wake of the September 11 attacks and sometimes deeply misleading.

In broad terms, statistical regularity or predictability alone can only indicate correlations but never demonstrates cause. This study relies exclusively on the computation of statistical trends rather than complementing them with judgments gleaned from nonrandom interviews with the human subjects of study themselves. This dichotomy is unnecessary and even pernicious.[8] Although one should not make the reciprocal mistake of taking personal accounts at face value, structured psychological interviews and systematic observations by anthropologists and other social scientists who participate in the lives of their informants can provide new and surprising alternatives to frame the collection of statistical data.[9] More specifically, at least four critical flaws are embedded in the conclusions drawn from the data.

First, there is a concern with sampling. Pape continues to argue that "the leading instigator of suicide attacks is the Tamil Tigers in Sri Lanka, a Marxist-Leninist group whose members are from Hindu families but who are adamantly opposed to religion."[10] Yet, in the last three years, the Tigers and the other main secular and nationalist groups in Pape's sample, such as Turkey's Kurdistan Worker's Party (PKK), have carried out very few attacks or none at all. The Tigers have carried out only two confirmed suicide attacks since the beginning of 2002. Although they perhaps remain the leading single organizer of suicide attacks (77 in total), there have been more suicide attacks by various Iraqi groups in 2005 (where more than 400 attacks killed more than 2,000 people) than in the entire history of the Tigers.[11] The Iraqi attacks have not been carried out through a particularly well-organized strategic operation, but rather via a loose, ad hoc constellation of many small bands that act on their own or come together for a single attack.[12] There also appear to be clear and profound differences between secular nationalist groups such as the Tamil Tigers, who fight to expel occupiers from their homeland, and global jihadis, who fight against perceived global domination. For example, Tamil suicide operatives are actively selected by recruiters and cannot withdraw from planned operations without fear of retaliation against their families,[13] whereas the martyrs of the Al Qaeda network are mostly self-recruiting and deeply committed to global ideology through strong network ties of friendship and kinship so that events anywhere in jihad's planetary theater may directly impact actions anywhere else.

There were more suicide attacks in the last two years than between 1980 and 2003.

Second, Pape's conclusions about suicide terrorism are both too narrow and too broad. On the narrow side, none of Pape's data indicate that conventional tactics are less useful than suicide attacks in cases where terrorists appear to have scored successes in liberating their homeland.[14] For example, Israel exited southern Lebanon 14 years after Hizballah ceased to use suicide tactics, and the United States withdrew from Somalia after casualties suffered under conventional attack. The category of suicide terrorism today also proves to be too broad and motley a category to draw reliable conclusions about motivations and goals. For example, although the Tigers deserve their due in any inquiry on contemporary suicide terrorism, their relevance to the global diffusion of martyrdom operations is questionable.

Marc Sageman, a forensic psychiatrist and former intelligence case officer, has traced the links among more than 400 jihadis with ties to Al Qaeda.[15] No significant demographic differences emerge between Sageman's global jihadi sample and Pape's more restricted sample of 71 terrorists who killed themselves between 1995 and 2004 in attacks sponsored or inspired by Al Qaeda. In other words, there seem to be no reliable differences between jihadi martyrs in particular and jihadis who fight in the name of global jihad more generally.

Third, rather than judging as Pape does the success of suicide tactics primarily by whether the sponsoring organization has helped to expel foreigners from its homeland, the broader strategic goal that suicide attacks seek may be to increase the sponsoring organization's political "market share" among its own potential supporters, that is, to broaden its political base among the population and narrow popular support for rival organizations.[16] For example, the Hamas campaign of suicide bombings actually caused Israel to reoccupy Palestinian lands during the second intifada and not to withdraw, but levels of popular support for Hamas increased to rival and at times even surpass support for the Fatah-dominated Palestinian Liberation Organization.[17] Inciting the withdrawal of foreign forces is only one means of accomplishing that goal. Thus, the net effect of the suicide attacks was not to expel foreign forces but to increase the appeal of Hamas. Similarly, the net result of the September 11 attacks was substantially greater U.S. and foreign intervention in Muslim territories, although Al Qaeda's profile rose to the top of the global jihadi ranks.

There is a lingering misconception that martyrs are being directed by 'Al Qaeda.'

Fourth, Pape's argument that suicide terrorism is unrelated or only marginally related to Salafi ideology employs an unfounded inference. Salafis believe that the *hadith* (oral traditions) and literal readings of the Qur'an are sufficient guides for social law and personal life. The most militant among them, the jihadis, believe that all contemporary majority-Muslim countries with the exception of Afghanistan under the Taliban have strayed from the true path of Islam and that the only way back is through violent jihad. Pape relies on statistics to show that "the presence of [U.S.] military forces for combat operations on the homeland of suicide terrorists is stronger than Islamic fundamentalism in predicting" whether someone will become an Al Qaeda suicide terrorist.[18] According to Pape's analysis, nearly one-third of Pakistan's 150 million people are estimated to be Salafi, but the country has produced only two suicide terrorists, whereas Morocco has no Salafis, according to a single secondary source from 1993, but has produced 12 suicide terrorists.

Some of these statistics can be contested, whereas others need to be updated. For example, although Salafism may still lack wide popular support in Morocco, its appeal has grown steadily among the discontented in that country since the 1970s and especially after the return of the Afghan alumni in the 1990s.[19] By the end of 2004, Pakistan had at least 10 suicide bombers in the country, apart from the dozens of suicide terrorists from Pakistan's Kashmiri groups that profess allegiance to global jihad. Similarly, the number of suicide terrorists in heavily Salafi Egypt has quadrupled over the figure presented in Pape's 2003 tables and has more than doubled in Indonesia. Uzbekistan, which had six suicide bombers in 2004 and which Pape listed as having no Salafi population, in fact has large numbers of mostly urban youth who now sympathize with Hizb ut-Tahrir, a radical Islamist movement largely unknown in Central Asia before the mid-1990s, whose proclaimed goal is jihad against the United States and the overthrow of existing political regimes and their replacement with a caliphate run on Salafi principles.

Most jihadis follow kin and colleagues more than they do orders from afar.

As a causal variable, however, "Salafi population in country of origin" is largely irrelevant to what is happening in the world today. Even if the sample's statistical reliability were to hold up for the years that Pape surveyed, suicide terrorism is a rapidly moving phenomenon that still involves relatively small numbers, the significance of which can shift dramatically in a relatively short period of time. Indeed, there were more suicide attacks in the last two years, roughly 600, than in Pape's entire sample between 1980 and 2003. The British-born bombers who attacked London in 2005 or the Malaysians who likely planned the latest Bali bombing belong to the fringe of a large and growing Muslim diaspora.

This changing jihadi landscape is revealed in the formation of the cell responsible for the 2004 Madrid train bombings. As early as October 2002, the substitute imam of the Takoua Mosque in Madrid was informing Spanish police under the codename "Cartagena" that a band of friends, unhappy with the mosque's seemingly moderate preachings, had begun calling themselves Al Haraka Salafiya (The Salafi Movement). According to Cartagena, they met "clandestinely, with no regularity or fixed place, by oral agreement and without any schedule, though usually on Fridays." Soon, the informal group of mostly homesick Moroccan descendants and émigrés "reached the conclusion that they had to undertake jihad." By November 2002, opinion within the group began to shift against "going to other countries to undertake jihad, when operations were possible in Morocco and Spain."[20] A detailed action plan only began to coalesce later the following year, however, around the time the Internet tract "Iraqi Jihad, Hopes and Risks" began to circulate a call for "two or three attacks … to exploit the coming general elections in Spain in March 2004" on the Global Islamic Media Front Web site,[21] to which the Madrid plotters had been systematically logging on since the spring of 2003. The police reports show that targeting trains to force Spain out of the coalition in Iraq was only a late goal emanating from an informal network dedicated to the simple but diffuse project of undertaking jihad to defend and advance a Salafist vision of Islam.

When Egyptian Bedouin are dying to kill European tourists and the Egyptians who cater to them; when British citizens blow themselves up along with other British because of the country's involvement in Iraq and Afghanistan; when jihadis exclusively target co-religionists linked to the secular government in Bangladesh, which is not a particularly close friend of the United States or its allies; when Malaysian bombers kill Australians and Balinese Hindus in Indonesia as "self-defense" in a "clash of civilizations" between Islam and the United States;[22] and when Arabs from more than a dozen countries rush to embrace death in Iraq in order to kill Shi'as, who are probably more supportive of Iran than they are of the United States, it is quite a stretch to identify the common thread as a secular struggle over foreign occupation of a homeland, unless "secular" covers transcendent ideologies, "foreign occupation" includes tourism, and "homeland" expands to at least three continents.

Al Qaeda as Bogeyman

Attributing the motivations to narrowly expelling foreign forces is not the only common misconception about today's suicide terrorists. There is a preferred interpretation and lingering misconception among leading pundits and politicians that contemporary martyrs around the world are being directed and organized by a specific "Al Qaeda" group. Al Qaeda leaders and operatives, they claim, are behind the coordinated bombings in London, Egypt, Iraq, and elsewhere. Yet, remnants of the mostly Egyptian hardcore around bin Laden have not managed a successful attack in three years, since the bombing in Djerba, Tunisia, in October 2002. In fact, they do not even know who many of the new terrorists are, much less be able to communicate with and direct them reliably.

Whither Al Qaeda? Only after the embassy bombings in Africa in 1998 did U.S. officials or anyone else begin referring to Al Qaeda as a worldwide terrorist organization led by bin Laden. It came from the somewhat contradictory testimony in *United States vs. Usama Bin Laden, et al.* in U.S. federal court in New York that same year. One detainee, Khalfan Khamis Mohamed, claimed never to have heard of an "Al Qaeda" group but recognized the term as a way of carrying out militant actions. Another, Jamal Ahmed Al-Fadl, said "the Al Qaeda group" had existed since "around '89."[23]

Especially after the September 11 attacks, nationalist Islamist movements from Morocco to Indonesia, which had jealously guarded their independence, spiritually united under the Al Qaeda logo. Intense public attention on Al Qaeda encouraged homegrown groups only tenuously connected with bin Laden, if at all, to claim responsibility for attacks in Al Qaeda's name, to be taken more seriously both by friends and foes. Abu Mus'ab Al-Zarqawi's Iraq-based jihadi group, whose violent, anti-Shi'a policy was never bin Laden's, is a good example.

"Al Qaeda" terrorist actions are now chiefly executed by self-forming cells of friends that swarm for attack, then disappear or disperse to form new swarms. Independent studies by Nixon Center scholar Robert Leiken[24] and Sageman show that more than 80 percent of known jihadis currently live in diaspora communities, which are often marginalized from the host society and physically disconnected from each other. Similar to the decentralized anarchist movement that terrorized the world a century ago, most jihadis follow kin and colleagues more than they do orders from afar. Their difficult-to-penetrate social networks consist of about 70 percent friends and 20 percent family.[25]

It is nonsense to claim that terrorists simply want to annihilate Western civilization.

Yearning for a sense of community and a deeper meaning in life, small groups of friends and family from the same area "back home" bond as they surf Islamist Web sites and seek direction from Al Qaeda's inspiration.[26] In the last five years, Web sites carrying Islamist messages have increased from less than 20 to more than 3,000,[27] with about 70 avowedly militant sites collectively forming a virtual jihadi university. In fact, an October 2005 posting by Ahmad Al-Wathiq bi-Llah, "deputy general emir" of the Al-Zarqawi–affiliated Global Islamic Media Front, reissued a 2003 announcement for the "Al Qaeda University of Jihad Studies ... a tangible reality for the enemies of the Nation and the Faith; a decentralized university without geographical borders, present in every place."[28] "Graduates," he explains, pass through "faculties" that advance the cause of a global caliphate through morale boosting and bombings and specialize in "electronic jihad, media jihad, spiritual and financial jihad."

Although veteran jihadis may sometimes help trigger the newer groups into action, even information for the do-it-yourself explosives used in the Madrid and London bombings is available on the Internet.[29] As in the case of the Madrid train bombers, even if all of the plotters are caught or kill themselves, it need not affect the ability of other groups to self-organize and stay motivated so that the movement does not die with them. The December 2003 Internet tract that foreshadowed the Madrid bombings inspired attacks that would force Spain's withdrawal from Iraq as a way to generate "huge pressure on the British presence, which [Prime Minister] Tony Blair could not overcome."[30] When Blair did not follow suit, however, it appears that a mix of homegrown talent—three cricket friends of Pakistani origin: one married, one in college, one "born again," later joined by a convert—fused by foreign-born incitement bombed London to press matters. Web sites such as that of the Global Islamic Media Front[31] that host these tracts have become the new organizational agents in jihadi networks, replacing physical agents such as bin Laden. Even functioning central actors such as Zarqawi are effectively only co-leaders with the media sites that increasingly control the distribution of knowledge and resources.

The edited snippets and sound bites favored by today's mass media have been used with consummate skill by jihadi leaders and ideologues, beginning with bin Laden himself. As a result, deeply local and historically nuanced interpretations of religious canon have been flattened and homogenized across the Muslim world and beyond, in ways that have nothing in particular to do with actual Islamic tradition but everything to do with a polar reaction to perceived injustice in the prevailing unipolar world. At the same time, the historical narrative, however stilted or fictitious, translates personal and local ties within and across small groups into a profound connection with the wider Muslim community (*ummah*).

Jihadis Are Not Nihilistic (Even If Apocalyptic)

A third misconception is that those who carry out attacks in the name of Al Qaeda or through its inspiration do so mostly because the terrorist is desperate or a nihilist who, in the words of President George W. Bush, "hates freedom, rejects tolerance, and despises dissent" and wants only to replace the current mildly corrupt and undemocratic regimes with the terrorist's own far more authoritarian and arbitrary form of "evil."[32] This is the thesis of the U.S. leadership.[33] It is hopelessly tendentious and willfully blind.[34]

It is nonsense to claim that Al Qaeda and its sympathizers have no morality and simply want to annihilate Western civilization. In general, charges of nihilism against an adversary usually reflect ignorance of the adversary's moral framework or an attempt publicly to cast it as simply evil to mobilize domestic support for war. Even bin Laden has never preached destruction of Western culture or else, as he has taunted, "Why didn't we attack Sweden?"[35] At every turn, bin Laden has earnestly sought moral justification for Al Qaeda's actions and demands.[36] This includes his invocation of a fatwah published in May 2003 by radical Saudi cleric Hamid bin Al-Fahd permitting the use of nuclear weapons to inflict millions of casualties on the United States, unless Washington changes its foreign policy toward the Middle East and elsewhere in the Muslim world.[37]

One important post–September 11 development in global network jihad is that Al Qaeda splinters no longer consider themselves to be territorially rooted in supporting populations. Unconstrained by concrete concerns for what will happen to any population that supports them, deracinated jihadis can conceive of any manner of attack, including one leading to fulfillment of Al-Fahd's apocalyptic vision. Al Qaeda deputy Ayman Al-Zawahiri continues to urge jihadis everywhere to inflict the greatest possible damage and maximum casualties on the West, regardless of the time and effort required or of the immediate consequences.

The key is *not* to change deeply held religious beliefs, but to channel them.

To some extent, organizations that sponsor suicide terrorism are motivated to fight policies they abhor by hard-nosed calculations of the material costs and benefits associated with martyrdom actions. Al-Zawahiri argues in his "testament," *Knights Under the Prophet's Banner*, that "[t]he method of martyrdom operations [i]s the most successful way of inflicting damage against the opponent and the least costly to the mujahideen in casualties."[38] Increasing their organization's political market share, jihadi leaders point to the sacrifice of their best and brightest as signals of their costly commitment to their community. In September 2004, Sheikh Hamed Al-Betawi, a spiritual leader of Hamas, told me that "[o]ur people do not own airplanes and tanks, only human bombs. Those who undertake martyrdom actions are not hopeless or poor, but are the best of our people, educated, successful. They are intelligent, advanced combat techniques for fighting enemy occupation."[39]

Yet, according to Harvard political scientist Jessica Stern, who has interviewed terrorists and those who sponsor them, holy wars are dependent first and foremost on redressing a deep pool of perceived humiliation, not just on military occupation per se and certainly not on simply nihilistic grounds.[40] Al-Zawahiri decries globalization, including tourism, as cultural domination that degrades Muslims. Groups claiming responsibility for the July 2004 suicide bombings of hotel and market areas in Sharm el-Sheikh, Egypt, for example, did not simply wish to purge Egypt of foreign influence. Rather, the plotters or those sympathetic to them considered "that this operation was in response to the crimes committed by the forces of international evil, which are spilling the blood of Muslims in Iraq, Afghanistan and Chechnya."[41]

Inspiration for targeting tourists can be found in the recent online musings of Abu Mus'ab Al-Suri, the new global jihadi Web "star" and principal theoretician of "leaderless resistance,"[42] who sets a top priority of targeting Western tourists.[43] This tactic, initiated by Zawahiri, has become a mainstay of suicide attacks linked to Al Qaeda's primary ally in Southeast Asia, Jemaah Islamiyah (JI), most recently with the September 2005 Bali bombings. JI lieutenant Riduan Isamuddin, also known as Hambali, first decided to intervene in a local dispute between Muslims and Christians in the remote Indonesian Maluku islands to extend his brand of jihad to all of Indonesia and then, after the September 11 attacks, to globalize the jihad by enlisting suicide bombers to hit Western targets and interests, including the 2002 Bali bombings that killed more than 200, mostly tourists, and the 2003 attack on Jakarta's J.W. Marriott hotel.

We can no longer ignore the moral values and group dynamics that drive suicide terrorism.

In Iraq, the theme of humiliation, such as well-publicized U.S. military actions at Abu Ghraib prison, is important to understanding the Islamists' rage. As an observation from interviewing terrorists and those who inspire and care for them, individuals who opt for suicide attacks often seem motivated by values and small-group dynamics that trump rational self-interest. Violation of such values leads to moral outrage and seemingly irrational vengeance ("get the offender, even if it kills us"). Adherence to sacred values, which provides the moral foundations and faith of every society or sect that has endured for generations,[44] ultimately leads to perceived moral obligations that appear to be irrational, such as martyrdom.[45] One is obliged to act "independently of the likelihood of success,"[46] as in acts of heroism or terrorism, because believers could not live with themselves if they did not.

When our research team interviews would-be suicide bombers and their sponsors and when we survey their supporters, for example, Hamas students or students in Indonesian madrassas that have produced suicide bombers, we pose questions such as, "What if your family were to be killed in retaliation for your action?" or "What if your father were dying and your mother discovered your plans and asked you to delay until the family could recover?" Almost all answer along the lines that, although duty to family exists, duty to God cannot be postponed. The typical response to the question "What if your action resulted in no one's death but your own?" is, "God will love you just the same." When I posed these questions to the alleged emir of JI, Abu Bakr Ba'asyir, in Jakarta's Cipinang prison in August 2005, he responded that martyrdom for the sake of jihad is the ultimate *fardh 'ain*, an inescapable individual obligation that trumps all others, including four of the five pillars of Islam (pilgrimage, almsgiving, fasting, prayer). This is a radically new interpretation of Islam, where only the profession of faith in Allah and his prophet counts as equal to jihad. What matters for Ba'asyir, as for most would-be martyrs and their sponsors whom I have interviewed, is the martyr's intention and commitment to God.[47] It is inspired by love of one's group and by rage at those who would humiliate it, but certainly not blind rage.

The power of faith is something many understand at home but few deem worthy of consideration for enemies abroad. Yet, responses from jihadis, as well as their actions, suggest that sacred values are not entirely sensitive to standard political or economic calculations regarding costs or payoffs that come with undertaking martyrdom actions, nor are they readily translatable from one culture to another. Especially in Arab societies, where the culture of honor applies even to the humblest family as it once applied to the noblest families of the southern United States,[48] witnessing the abuse of elders in front of their children, whether verbal insults at roadblocks or stripdowns during house searches, indelibly stains the memory and increases popular support for martyrdom actions.[49] This can be true even if the person is witness only to an injured party living thousands of miles away, for example, an Indonesian observing events in the Palestinian territories on television or though the Internet. What may be considered standard police practice in the United States may warrant undying calls for revenge in another society. Moreover, when sacred values are at stake, traditional calculations of how to defeat or deter an enemy, for example, by invoking a democratic vote, providing material incentives to defect, or threatening massive retaliation against supporting populations, might not succeed.

Policy Implications: Responding Transnationally

Despite common popular misconceptions, suicide terrorists today are not motivated exclusively or primarily by foreign occupation, they are not directed by a central organization, and they are not nihilistic. Most suicide terrorists today are inspired by a global jihadism which, despite atavistic cultural elements, is a thoroughly modern movement filling the popular political void in Islamic communities left in the wake of discredited Western ideologies co-opted by corrupt local governments. Appeals to Muslim history and calls for a revival of the caliphate are widespread and heartfelt. To some extent, jihadism is also a countermovement to the ideological and corresponding military thrust ensconced, for example, in the *National Security Strategy of the United States*, which enshrines liberal democracy as the "single sustainable model of national development right and true for every person, in every society—and the duty of protecting these values against their enemies."[50]

An alternative both to global jihadis and U.S. democracy promoters are those who support a more measured version of realism. This strategy would advocate returning to "offshore balancing," which would shift the burden of and recognition for managing the security of turbulent regions to regional powers.[51] Pape favors this approach as the only coherent strategy to minimize suicide terrorism and what it represents: unwanted domination. U.S. military superiority would permit a nod to others' primary interests in a region and help to maintain the latter's spheres of influence, while also allowing surgical interventions in cases where regional affairs got out of hand, for example, if oil supplies were threatened.

The policy implications of this change may be as novel as hitherto neglected.

Would ending military adventures and support for corroded, undemocratic regimes, as Pape's analysis suggests, end the menace from suicide attack? As a long-term strategy, ending belligerent intervention in other people's affairs is likely a sound suggestion. In the short term, however, the link between ceasing military intervention and terminating suicide terrorism is less compelling. Simply exiting Iraq or Afghanistan today would not likely solve matters. Because the United States was not involved in Iraq or Afghanistan at the time, they could not have figured into the motivations for the September 11 attacks or earlier Al Qaeda suicide operations. After Spain's exit from Iraq, the threats to that country from jihadi terrorists did not abate: a subsequent bomb plot against a high-speed train to Toledo, as well as a plot by an independent Algerian cell to blow up the Madrid office of an antiterrorism judge, were both foiled by authorities, but they existed. Even the fact that the United States has largely withdrawn its military forces from Saudi Arabia seems to have done absolutely nothing to appease bin Laden. Moreover, his sympathizers see this as a concession, perhaps even inciting further attacks.

Both the realist and idealist approaches focus on the increasingly transected nation-state system, that is, on building nations or maintaining alliances among states. The problem, however, is transnational, and the solutions are not likely subject to any sovereign control, even that of the United States. What may be needed are new and varied forms of transnational associations that reach across cities and cyberspace, where decisions by and interactions among states represent only one of several possible dimensions. How, then, to deal with this deeply faith-inspired, decentralized, and self-adjusting global jihadi "market," where any small group of friends can freely shop for ideas or even for personnel and materials and any can inflict such widespread damage? A successful counterstrategy would have to act on at least three different levels: changing the motivations of potential recruits, disrupting sponsors' organizations, and undermining popular support.

Recruits: Forget the Profiles, Understand the Cells

In targeting potential recruits for suicide terrorism, it must be understood that terrorist attacks will not be prevented by trying to profile terrorists. They are not sufficiently different from everyone else. Insights into homegrown jiahdi attacks will have to come from understanding group dynamics, not individual psychology. Small-group dynamics can trump individual personality to produce horrific behavior in otherwise ordinary people.

Most jihadi cells are small in number, with eight members being most common.[52] Although the members of each cell usually show remarkable homogeneity within the group (age, place of origin, residence, educational background, socioeconomic status, dietary preferences, etc.), there is little homogeneity across the jihadi diaspora. This renders attempts at profiling practically worthless. Cells are often spontaneously formed and self-mobilizing, with few direct physical contacts with other cells. Radicalization usually requires outside input from and interaction with the larger jihadi community. The Internet is taking over from the hands-on gurus of global jihad in radicalizing friends into pseudo-families for whom they will give their lives. Without the Internet, the extreme fragmentation and decentralization of the jihadi movement into a still functioning global network might not be possible.

This requires the careful monitoring rather than simple removal of existing jihadi Web sites. What is needed is a subtle infiltration of opportunities to create chat rooms, as well as physical support groups, which advance causes that can play to jihadi sentiments but that are not destructive, such as providing faith-based social services. The key is not to try to undermine the sacred values that inspire people to radical action or attempt to substitute one's own preferred values by forceful imposition or through propaganda. Our studies show that such tactics, as well as offers of instrumental incentives such as monetary compensation to forego martyrdom or make peace with an enemy, only incite further moral outrage and extreme behavior.[53] Rather, the aim should be to show how deeply held sacred values can be channeled into less-belligerent paths. What has struck me in my interviews with mujahideen who have rejected suicide bombing is that they remain very committed to Salafi principles, with firm and deep religious beliefs. Those who seem to succeed best in convincing their brethren to forsake wanton killing of civilians do so by promoting alternative interpretations of Islamic principles that need not directly oppose Salafi teachings.

In his recent book, *Unveiling Jemaah Islamiyah*, one of JI's former top leaders, Nasir Abas, refutes what he believes to be a tendentious use of the Qur'an and hadith to justify suicide bombing and violence against fellow Muslims and civilians. (JI, similar to many of the militant Salafi groups that are sympathetic to Al Qaeda, is riddled with internal divisions over the wisdom of killing fellow Muslims and civilians.) "Not one verse in the [Qur'an] contains an order for Muslims to make war on people of another religion," Abas writes, "or that killing women, children and civilians can ever be proper, just or balanced. [The contrary belief] has only created discord in the Muslim community and has led non-Muslims to regard Islam as sadistic and cruel."[54] He reasons that the best way to turn altruistic suicide bombers who believe that what they are doing is sacred away from violence may be by religiously promoting competing sacred values, such as spreading the faith and promoting equal economic opportunity, as well as social and political advancement through educational achievement and personal piety. Sincere alternative appeals to sacred values could undermine consensus for violent jihad. The United States and its allies should quietly encourage this process without strangling it in open embrace.

Organizations: Disrupting Decentralized Networks

In targeting organizations that sponsor suicide terrorism, police action and intelligence are crucial, but the present mode of operations is not encouraging. Although traditionally hierarchical forms of military and intelligence command and control were suitable for large-scale operations against the Taliban and Al Qaeda's global organization, they are patently less effective now, even though classical means are still needed to prevent sanctuaries from reemerging. Impressions, such as the one that Central Intelligence Agency director Porter Goss reaffirmed in his inaugural address before Congress, that some specific group called Al Qaeda is out there planning bigger and better attacks are misleading.[55] Ever more hierarchical and centralized management under a new national intelligence director to deal with increasingly decentralized terrorist networks may also be precisely the wrong way to go. According to former attorney general Edwin Meese, "It [will not] help matters to have a national intelligence director whose job is to prepare briefs to bring to the president every day or simply to coordinate intelligence products. What we could use is a facilitator to bring people and ideas together, not another operative."[56]

It may take a broad and elastic web of the diverse talent, tolerance, and spare conformity of our democracies, unbound from any nation's hegemony, calls to ruthlessness, or rigid rules of hierarchy, to snare the virtual hand of born-again jihad that guides suicide missions. Informal bonds may need to grow among diverse experts with idiosyncratic personal skills and the operational branches fighting terrorism, so that a phone call from an expert or operator in one country to another country can trigger specific responses without plodding through official channels—much the way globally networked jihadis now operate. This would help to convert fairly static responses into a dynamic system that would throw open the flow of information that would allow the intelligence and military communities' technological advantages to keep ahead of jihadi innovations without being mired in existing or reformed bureaucracies.

Society: Diverting Sympathy for Martyrs

In targeting popular support for suicide terrorism, it must be understood that terrorists depend, often quite consciously, on prodding those attacked into committing atrocious retaliatory acts, either deliberately or by a willingness to tolerate indefinitely horrific degrees and amounts of collateral damage to innocent bystanders. Historically and today, it is desecration of sacred places and perceived humiliation, even more than death and destruction, that has moved people to embrace violence.[57]

One strategic alternative is for the United States to do more of what it did in Aceh in early 2005 for tsunami victims, providing constructive investments of soft power that can generate longer-term relief from the need to use destructive and usually snowballing forms of hard power.[58] U.S. tsunami relief arguably has been the one significant victory since the September 11 attacks in the struggle to prevent the enlistment of future terrorists for jihad (unfortunately, Kashmiri jihadi groups are the primary providers of assistance to the victims of Pakistan's recent earthquake). According to recent surveys conducted by the Pew Global Attitudes Project, Indonesian views of the United States, which were largely favorable before the U.S. invasion of Iraq, plummeted to 15 percent immediately afterward.[59] Yet, the study shows that, since the rescue and relief role that the United States played in concert with other nations, international associations, and nongovernmental organizations after the tsunami, favorable attitudes toward the United States among Indonesians have risen to 38 percent.[60] Concurrently, popular support for combating terrorism has doubled from 23 percent in 2003 to 50 percent in 2005. Indonesia is now one of the few nations where a majority (59 percent, up from 25 percent in 2003) believes that the United States can act in other nations' interests. Although these polls indicate that popular sentiment remains volatile, the bright side is that such instability suggests that the anti-Americanism that helps sustain the jihadi cult of martyrdom could yet be reversed.

Embracing Complexity

Those who believe suicide terrorism can be explained by a single political root cause, such as the presence of foreign military forces or the absence of democracy, ignore psychological motivations, including religious inspirations, which can trump rational self-interest to produce horrific or heroic behavior in ordinary people. Those who believe that some central organization such as the old Al Qaeda directs such suicide terrorists ignore the small-group dynamics involving friends and family that form the diaspora cell of brotherhood and camaraderie on which the rising tide of martyrdom actions is based. Finally, those who simply dismiss jihadis as nihilists risk developing policies based on faulty assumptions that seek to challenge deeply held religious beliefs, rather than more effectively channel them to less-violent expressions.

Simple explanations and solutions, based mostly on familiar research and policy paradigms but no first-hand knowledge or field experience capable of challenging them, may be more appealing and easier to grasp. They are liable to fail, however, because they ignore the underlying moral values and group dynamics that drive jihadis to suicide terrorism. "Know thine enemy" is not a call for therapy but for a better understanding of who out there is dying to kill and why. Understanding that can help decisionmakers devise organizational and ideological solutions to defuse the threat of martyrdom.

Notes

1. Bruce Hoffman, "Security for a New Century," Washington, D.C., September 23, 2005 (briefing for Senate Foreign Affairs Committee staff).
2. Scott Atran, "Suicide Terrorism Database, 2004," `http://www.sitemaker.umich.edu/satran/files/suicide_terrorism_database_2004.xls`.
3. Neil Macdonald, "Suicide Attack Every Day in the New Iraq," *Financial Times*, July 14, 2005, `http://news.ft.com/cms/s/7c74abf8-f495-11d9-9dd1-0000e2511c8.html`; Chris Tomlinson, "Fewer Suicide Bombings in Iraq in November," Associated Press, December 1, 2005.
4. Jim Giles and Michael Hopkin, "Psychologists Warn of More Suicide Attacks in the Wake of London Bombs," *Nature*, July 21, 2005, pp. 308–309.
5. Robert Pape, *Dying to Win: The Strategic Logic of Suicide Terrorism* (New York: Random House, 2005).
6. Reuven Paz, "Qa'idat al-Jihad, Iraq and Madrid," *PRISM Special Dispatches on Global Jihad*, March 13, 2004, `http://www.e-prism.org/images/PRISM_Special_dispatch_no_12.pdf`.
7. Reuven Paz, "Arab Volunteers Killed in Iraq," *PRISM Special Dispatches on Global Jihad*, March 3, 2005, `http://www.e-prism.org/images/PRISM_no_1_vol_3_Arabs_killed_in_Iraq.pdf`.
8. For representations of these two poles in the field of terrorism studies, see Marc Sageman, *Understanding Terror Networks* (Philadelphia: University of Pennsylvania Press, 2004); Jessica Stern, *Terror in the Name of God* (New York: HarperCollins, 2003). Both approaches yield important insights that may be enhanced by a coordination of efforts, which our common project with the National Science Foundation aims to accomplish.
9. For detailed examples of how cultural reframing of hypotheses can thoroughly undermine the presumed validity of statistical trends, no matter how significant and reliable those trends appear to be, see Scott Atran, Douglas Medin, and Norbert Ross, "The Cultural Mind: Ecological Decision Making and Cultural Modeling Within and Across Populations," *Psychological Review* 112 (2005): 744–776.
10. Robert Pape, "Blowing Up an Assumption," *New York Times*, May 18, 2005, p. A29.
11. Pape does not consider coordinated attacks launched around the same time as separate attacks. Even so, according to S.W.A.B. Daulagala, Sri Lanka's chief inspector of police, in total, 262 Tamil Tigers launched suicide attacks. Data presented to the Technical Experts Workshop on Suicide Terrorism, Organization for Security and Cooperation in Europe, May 20, 2005.
12. Dexter Filkins, "Profusion of Rebel Groups Helps Them Survive in Iraq," *New York Times*, December 2, 2005, p. A1.
13. Selliah Ignasius Yoganathan, "Rise and Decline of Suicide Terrorism in Sri Lanka," paper presented to the NATO Advanced Workshop on "The Strategic Threat from Suicide Terrorism," Lisbon, June 11, 2004.
14. Joshua Sinai, "The Unsettling Lure of Suicide Terrorism," *Washington Times*, June 19, 2005, `http://www.washingtontimes.com/books/20050618-115922-1289r.htm`.
15. Sageman, *Understanding Terror Networks* (updated with data presented to World Federation of Scientists Permanent Monitoring Panel on Terrorism, Erice, Italy, May 8, 2005).
16. Mia Bloom, *Dying to Kill: The Allure of Suicide Terrorism* (New York: Columbia University Press, 2005).
17. See Palestinian Center for Policy and Survey Research, `http://www.pcpsr.org/survey/survey.html` (political tracking polls of Palestinian popular opinion for the last seven years).
18. Pape, *Dying to Win*, p. 103.
19. Anouar Boukhars, "Origins of Militancy and Salafism in Morocco," *Jamestown Terrorism Monitor*, June 17, 2005, `http://jamestown.org/terrorism/news/article.php?articleid=2369722`.
20. "Las notas del Confidente 'Cartagena' Preuban Que la Policia Controlaba la Cúpula del 11-M" [Notes from the police informant "Cartagena" prove that the police were controlling the ringleader of March 11], `http://www.belt.es/noticias/2005/junio/01/11-M.asp`.
21. See Paz, "Qa'idat al-Jihad, Iraq and Madrid."
22. Scott Atran, "The Emir: An Interview With Abu Bakr Ba'asyir, Alleged Leader of the Southeast Asian Jemaah Islamiyah Organization," *Jamestown Foundation Spotlight on Terrorism*, September 15, 2005, `http://jamestown.org/terrorism/news/article.php?articleid=2369782`.
23. *United States of America v. Usama bin Laden, et al.*, February 6, 2001, `http://cryptome.org/usa-v-ubl-02.htm` (defendant Jamal Ahmed Al-Fadl); *United States of America v. Usama bin Laden, et al.*, March 19, 2001, `http://cryptome.org/usa-v-ubl19.htm` (discussion of Khalfan Khamis Muhamed). Members of Jemaah Islamiyah who studied in Afghanistan with Abdullah Azzam (Bin Laden's mentor and originator of the term al-Qaeda al-sulbah [the strong base]) and who over the years met with bin Laden and hosted September 11 mastermind Khalid Sheikh Muhammad, also tell me they never heard the term "Al Qaeda" applied to an organization until after September 11.
24. Robert Leiken, "Bearers of Global Jihad? Immigration and National Security After 9/11," March 25, 2004, `http://www.nixoncenter.org/publications/monographs/Leiken_Bearers_of_Global_Jihad.pdf`.
25. Sageman, *Understanding Terror Networks*. Our research team has been updating Sageman's database with several hundred entries, and we find the demographic and networking trends fairly constant.
26. For more information on youth and the Internet, see "European Youth Ditching TV and Radio for Web," *European Tech*

Wire, June 24, 2005, http://www.europeantechwire.com/etw/2005/06/24/. For more information on interpersonal communications through the Internet, see Paul Resnick and Richard Zeckhauser, "Trust Among Strangers in Internet Interactions," in *Advances in Applied Microeconomics* 11, ed. Michael Baye (Amsterdam: Elsevier Science, 2002).

27. Luis Miguel Ariza, "Virtual Jihad: The Internet as the Ideal Terrorism Recruiting Tool," *Scientific American*, January 2006, http://www.sciam.com/article.cfm?articleID=000B5155-2077-13A8-9E4D83414B7F0101.
28. Al-Farouq jihadi forum, October 7, 2005, http://www.Al-farouq.com/vb/.
29. Raymond Bonner, Don Van Natta, and Stephen Grey, "Investigators So Far Find Little Foreign Involvement," *New York Times*, July 31, 2005, p. A1.
30. Scott Atran, "The Jihadist Mutation," *Jamestown Terrorism Monitor*, March 25, 2004, http://www.jamestown.org/publications_details.php?volume_id=400&issue_id=2929&article_id=23646.
31. See http://online2005.100free.com/.
32. Office of the Press Secretary, The White House, "President Addresses Nation, Discusses Iraq, War on Terror," June 28, 2005, http://www.whitehouse.gov/news/releases/2005/06/20050628-7.html.
33. Stephen Hadley and Frances Fragos Townsend, "What We Saw in London," *New York Times*, July 23, 2005, p. A13.
34. For a discussion of the "hates freedom" thesis, see Scott Atran, "Mishandling Suicide Terrorism," *The Washington Quarterly* 27, no. 3 (Summer 2004): 67–90.
35. Al Jazeera, October 29, 2003. See Al Jazeera, November 28, 2003 (Ayman Al-Zawahiri).
36. Michael Scheuer, *Imperial Hubris* (Dulles, Va.: Potomac Books, 2004).
37. Reuven Paz, "Yes to WMD: The First Islamist *Fatwah* on the Use of Weapons of Mass Destruction," *PRISM Special Dispatches on Global Jihad* 1, no. 1 (May 2003), http://www.e-prism.org/images/PRISM%20Special%20dispatch%20no%201.doc.
38. Ayman Al-Zawahiri, *Knights Under the Prophet's Banner*, serialized in Al-Sharq Al-Awsat (London), December 2–10, 2001, trans. Foreign Broadcast Information Service, FBIS-NES-2001-1202.
39. Sheikh Hamed Al-Betawi, interview with author, Nablus, West Bank, September 2004.
40. Jessica Stern, "Beneath Bombast and Bombs, a Caldron of Humiliation," *Los Angeles Times*, June 6, 2004, p. M1.
41. "[Al Qaeda]–Linked Militants Claim Responsibility for Resort Bomb," *Ireland On-Line*, July 23, 2005 (quoting Abdullah Azzam Brigades, "Al Qaeda in Syria and Egypt").
42. Mustafa Setmariam Nasar (also known as Abu Mus'ab Al-Suri), *Da'wah lil-Muqawamah Al-Islamiyyah Al-'Alamiyyah* [A call for the Islamic global resistance], http://www.fsboa.com/vw/index.php?subject=7&rec=27&tit=tit&pa=0.
43. Reuven Paz, "Al-Qaeda's Search for New Fronts: Instructions for Jihadi Activity in Egypt and Sinai," *PRISM Special Dispatches on Global Jihad* 3, no. 7, October 2005, http://www.e-prism.org/images/PRISM_no_7_vol_3_-_The_new_front_in_Egypt_and_Sinai.pdf. See Abu Mus'ab Al-Suri, lecture, October 5, 2005, http://z15.zupload.com:download.php?file=getfile&filepath=1233 (posted by the Global Islamic Media Front).
44. Scott Atran, *In Gods We Trust* (New York: Oxford University Press, 2002).
45. Alan Fiske and Phillip Tetlock, "Taboo Tradeoffs: Reactions to Transactions That Transgress the Spheres of Justice," *Political Psychology* 18 (1997): 255–297.
46. Max Weber, *Economy and Society* (Berkeley, Calif.: University of California Press, 1978).
47. Atran, "Emir."
48. R. Nisbett and D. Cohen, *The Culture of Honor: Psychology of Violence in the South* (Boulder, Colo.: Westview Press, 1996); J. Peristiani, ed., *Honor and Shame: The Values of Mediterranean Society* (Chicago: University of Chicago Press, 1966).
49. See Atran, "Mishandling Suicide Terrorism" (discussion of support for suicide attack rising as a function of roadblocks).
50. *National Security Strategy of the United States*, September 2003, introduction, http:// www.whitehouse.gov/nsc/nss.html.
51. Christopher Layne, "From Preponderance to Offshore Balancing: U.S. Grand Strategy in the Twenty-first Century," *International Security* 22, no. 1 (Summer 1997): 86–124; Christopher Layne, "Offshore Balancing Revisited," *The Washington Quarterly* 25, no. 2 (Spring 2002): 233–248.
52. Sageman, data presented to World Federation of Scientists Permanent Monitoring Panel on Terrorism, Erice, Italy, May 8, 2005.
53. Scott Atran, "Risk in the Wild: Reassessing Terrorist Threats From the Field" (presentation, American Association for the Advancement of Science, St. Louis, Mo., February 19, 2006).
54. Nasir Abas, *Membongkar Jamaah Islamiyah* [Unveiling Jemmah Islamiyah] (Jakarta: Grafindo Khazanah Ilmu, 2005), pp. 221, 316.
55. Porter J. Goss, "Global Intelligence Challenges 2005: Meeting Long-Term Challenges With a Long-Term Strategy," testimony before the Senate Select Committee on Intelligence, February 16, 2005, http://www.cia.gov/cia/public_affairs/speeches/2004/Goss_testimony_ 02162005.html.
56. Edwin Meese, remarks to the Critical Incident Analysis Group, Charlottesville, Va., April 3, 2005; Edwin Meese, e-mail communication with author, April 6, 2005.
57. See Ron Hassner, "Fighting Insurgencies on Sacred Ground," *The Washington Quarterly* 29, no. 2 (Spring 2006): 149–166.
58. Scott Atran, "In Indonesia, Democracy Isn't Enough," *New York Times*, October 5, 2005, p. A29.
59. Pew Research Center, "U.S. Image Up Slightly, but Still Negative," *Pew Global Attitudes Project*, June 2, 2003, http://pewglobal.org/reports/display. php?ReportID=185.
60. Pew Research Center, "Views of a Changing World," *Pew Global Attitudes Project*, June 23, 2005, http://pewglobal.org/reports/display.php?ReportID=247.

***Scott Atran** is director of research in anthropology at France's Centre National de la Recherche Scientifique (CNRS) and a visiting professor of psychology and public policy at the University of Michigan.* *Background research was supported by the U.S. Air Force Office of Scientific Research, the National Science Foundation, and CNRS. The author would like to thank Khalil Shikaki, Rohan Gunaratna, and Noor Huda Ismail for help in the field and Marc Sageman, Reuven Paz, Jessica Stern, and Bruce Hoffman for relevant data and ideas.*

UNIT 3
State-Sponsored Terrorism

Unit Selections

Key Points to Consider

- How should the United States respond to Iran's sponsorship of international terrorism?
- What impact will Syrian missile capability have on U.S. interests in the region?
- How can governments respond to "society-sponsored" terrorism?
- What is the link between Hugo Chavez and the FARC?

Student Website

www.mhcls.com/online

Internet References

Further information regarding these websites may be found in this book's preface or online.

Council for Foreign Relations
http://www.cfr.org/issue/458/state_sponsors_of_terrorism.html

International Institute for Terrorism and Counterterrorism
http://www.ict.org.il/inter_ter/st_terror/State_t.htm

Security Resource Net's Counter Terrorism
http://nsi.org/terrorism.html

State Department's List of State Sponsors of Terrorism
http://www.state.gov/s/ct/rls/crt/

The role of states in international terrorism has long been the subject of debate. It is clear that states often support foreign groups with similar interests. This support can take a number of forms. States may provide political support, financial assistance, safe-havens, logistic support, training, or in some cases even weapons and equipment to groups that advocate the use of political violence. State support for terrorist organizations, however, does not necessarily translate into state control over terrorism.

Nevertheless, since the passage of the Export Administration Act of 1979, the United States government has sought to make some states responsible for the actions of groups they support. Section 6 (j) of the Export Administration Act requires the publication of an annual list of state sponsors of terrorism and thus provides the basis for the contemporary U.S. anti-terrorism and sanctions policy. This list currently includes Cuba, Iran, Libya, North Korea, Sudan, and Syria. Not surprisingly, this list includes only states perceived to be, for a wide variety of reasons, a threat to U.S. interests. States in which the United States has significant political or economic interests are, regardless of their record on terrorism, deliberately excluded.

In the first article in this unit, Gary Sick examines Iran's sponsorship of terrorist organizations. He contends that terrorism is being used as a means to spread Khomeini-style revolutions throughout the world. Sick offers a number of policy recommendations to address this problem. Next, Lee Kass focuses on Syria's ambitions to develop its weapons of mass destruction (WMD) capability. He points to Syria's links with terrorist organizations and argues that it will become more difficult to confront Syrian sponsorship of international terrorism. Fareed Zakaria challenges our understanding of state-sponsorship of terrorism with his assertion that terrorist groups do not need to rely on state-sponsors for their survival. He argues that rather than being sponsored by states, some terrorist groups sponsor the states in which they reside. Finally, Thor Halvorssen discusses evidence of a potential connection between Venezuelan President Hugo Chavez and the Revolutionary Armed Forces of Colombia (FARC). Halvorssen reasons that, in spite of Chavez's consistent denials, he continues to support the FARC.

Iran: Confronting Terrorism

Gary Sick

Charges of terrorist activities have plagued Iran from the earliest days of the Islamic revolution to the present. More than any other factor, they have interfered with Iran's ability to establish a responsible foreign policy image. Yet, terrorism is murky and highly ambiguous. As penalties for terrorism escalate, terrorists try to mask their identities; determining who planned and executed an act of terror is extremely difficult, and it is often virtually impossible to establish with any certainty the policy motives behind such acts. Iran is a particularly complex case.

Iran has a split personality. Some parts of its government—the presidency, the Majlis (parliament), and the functional ministries—though far from a fully functioning democracy, are held accountable for their policies and actions through public review and frequent elections. A second set of government institutions, including the Supreme Leader (*velayat-e faqih*), oversight committees such as the Guardian Council and the Expediency Council, and the security services, are dominated by a conservative clergy who are officially above reproach, essentially accountable only to themselves. These institutions have veto power over government policies and command a shadowy but potent network of influence and protection that grew out of the revolution, permeating Iran's national security structure and economy. The tension between these two unevenly balanced power centers affects Iranian policy at all levels so that, at times, Iran appears to be pursuing different or even contradictory objectives.

Since at least the mid-1990s, the main objectives of the elected government have been to attract foreign political and economic support. Especially since President Muhammad Khatami's election in 1997, Iran has played a significant and constructive role at the United Nations, normalized its relations with its neighbors in the Persian Gulf region, and moved much closer toward mutually respectful relations with the European Union. At the same time, some unaccountable elements of Iran's power structure have seemed unwilling to accept this normalization process and have clung to a very different agenda of destabilization, revolutionary vengeance, and violent intimidation, including terrorist acts. The two sets of policies, often directly contradictory, reflect the struggle that lies at the very heart of the Iranian revolutionary experience.

At times, Iran appears to be pursuing different, or even contradictory, objectives.

The triumph of the Iranian revolution in February 1979 kindled a burst of radical actions by Iran that deserve to be called terrorism.[1] These include kidnappings sanctioned and sponsored by the government itself, such as the taking of American hostages in the first years of the revolution, and reputed Iranian support for and suspected direct involvement in Hizballah operations in Lebanon, including the bombings of U.S. installations and hostage-taking throughout the 1980s. During the Iran-Iraq War, Iran pursued a strategy of maritime terror, using unmarked gunboats and floating mines to attack noncombatant shipping. Numerous assassinations of enemies abroad in the late 1980s and 1990s were widely and persuasively attributed to Iranian official sponsorship, and Iran was accused of sponsoring operations by other militant organizations, such as the Argentinean bombings of 1992 and 1994 and the 1996 Khobar Towers bombing, attributed to Hizballah organizations in Lebanon and Saudi Arabia. Iran is currently suspected of supporting terrorist acts against Israel through its support of radical Palestinian factions.

Given the ambiguities of the public record, if not the intelligence data on which it is based, Iran's actual behavior may be better, worse, or substantially different from the brief survey presented here. We may never have all the facts about many of the terrorist incidents of which Iran is accused. Assuming, however, that the following discussion of Iran's record on terrorism and the main driving forces of that record are at least roughly accurate, certain conclusions can be drawn about Iranian policy on terrorism, the direction in which it is headed today, and possible U.S. responses. Iran undoubtedly behaves differently today than it did nearly a quarter century ago. Iran's postrevolutionary policies of hostage-taking and rebellion promotion among its neighbors have been abandoned, as have its wartime shipping attacks and targeted assassinations of enemies. Today, Iran's promotion of violence seems to be increasingly focused on support for radical anti-Israeli groups in Palestine. This shift calls for a different and more creative set of responses on the part of the United States.

Iran's Historical Motivations for Terrorism

Exporting the Revolution

The capture of the U.S. embassy in Tehran in 1979 by a band of students and the imprisonment of a large group of U.S. diplomats and private citizens for 444 days with the explicit acquiescence of the Iranian government set the tone for Iran's relations with the United States and many other countries. The United States and much of the world regarded this act as the quintessential example of state-supported terrorism. It traumatized the U.S. public and darkened the lens through which the United States would view the Islamic Republic of Iran and all of its policies and actions during the decades that followed.

In the years immediately after the revolution, Iranian militants—with or without the official support of the government—attempted to export the revolution by stirring up radical Islamist discontent in Bahrain, Saudi Arabia, and other Gulf states. A botched attempt by Iranian supporters to assassinate senior Iraqi officials, including Tariq Aziz, in April 1980 was one of the catalysts that persuaded Saddam Hussein to invade Iran in September of that year.

Iran's ambassador to Syria in the early 1980s, Ali Akbar Mohtashemi, provided financing and support for the creation of Hizballah ("Party of God"), the Lebanese political party and resistance movement.[2] Hizballah is widely believed to have been associated with the bombings of the U.S. Marines barracks and the U.S. embassy in Lebanon in 1983, although its leadership denies the charge, as well as the killing and hostage-taking of Americans and others throughout the 1980s. Its success in conducting a guerrilla war in southern Lebanon against Israel, ultimately leading to Israel's departure in 2000, won widespread admiration in the Islamic world and made Hizballah a source of inspiration and training for militant organizations throughout the Middle East, many of which adopted the same name. Iran takes pride in its continued support for Hizballah as a national resistance organization but denies having operational control over decisionmaking. In recent years, Iran has openly called on Hizballah to display "prudence and self-restraint" to prevent Israel from finding a pretext to attack Lebanon again.[3]

Enemies of the State

Just before he died in 1989, Ayatollah Ruhollah Khomeini, the father of the Iranian revolution, issued his famous *fatwa* against Salman Rushdie. Khomeini regarded Rushdie's depiction of the prophet Muhammad and other Islamic subjects in *The Satanic Verses* as blasphemous, and the *fatwa* in effect incited the general Muslim community to murder Rushdie. It also seemed to signal the beginning of an assassination campaign against individuals associated with Rushdie's book as well as other "enemies of the revolution." The rash of killings that followed included Kurdish leader Abdol Rahman Qasemlu in Vienna in 1989, former Iranian prime minister and opposition leader Shapour Bakhtiar in Paris in 1991, four Iranian Kurds in Berlin in 1992, and several leaders of the opposition Mujahideen-e Khalq movement. In addition, two bombings in Argentina—the Israeli embassy in March 1992 and a Jewish community center in July 1994—were attributed to the Lebanese Hizballah organization, allegedly with Iranian assistance.

To be sure, Iran may often be falsely accused. Many of these crimes were never solved, and the degree of Iranian official responsibility may be overstated. For its part, Iran flatly and unequivocally denied any role in these incidents. A German court that formally investigated the 1992 Berlin murders, however, implicated the highest levels of the Iranian government and indicted the minister of intelligence, Ali Fallahian, for his role. An Argentinean court officially concluded in 2003 that officials in the Iranian embassy provided unspecified support to Hizballah for the 1994 bombing of the Jewish Community Center.

Iran's past reputation for supporting terrorism, the incendiary rhetoric of its ultraconservative clerical leaders, and its almost total lack of transparency concerning issues of national security have created an environment in which it is easy to believe the worst. In fact, Iran's behavior since the revolution has allowed its opponents to accuse it of almost anything and to find a receptive audience for their claims. Iran's vigorous denial in all of the aforementioned cases ultimately undermined its credibility because the formula never varied, even when the evidence was quite incriminating, and there was never any visible effort by Iran to investigate the circumstances or to punish any of the individuals who might have been involved.

Maritime Terrorism

During the Iran-Iraq War (1980-1988), Iranian gunboats—usually small speedboats with hand-held grenade launchers and other weapons—attacked commercial shipping in the Gulf. Iran also seeded the waters of the shipping lanes with floating mines. These tactics were usually regarded as acts of war, and they have not figured into the terrorism charges against Iran. The case can be made, however, that they represented a form of maritime terrorism.

That Iran used these strikes to retaliate against Iraqi air attacks against its own shipping is obvious. Iran could not retaliate in kind because all Iraqi ports were closed and there were no Iraqi ships in the Gulf. Instead, Iran sent unmarked speedboats to fire at commercial ships en route to Arab ports on the unspoken but entirely valid assumption that countries such as Kuwait and Saudi Arabia were serving as a supply channel for Iraq.

> Iran undoubtedly behaves differently today than it did nearly a quarter century ago.

Although Iran never formally acknowledged that its military forces were behind these attacks, Iran undoubtedly organized and sponsored them. They were not truly acts of war because they were conducted by nonuniformed personnel against unarmed civilians of noncombatant states; they more closely resembled drive-by shootings or the mining of a busy thoroughfare. These attacks, which threatened the region's shipping lanes, eventually led to direct military clashes between the United States and Iran in the Gulf.[4] They are significant here because they indicated Iran's willingness to use unconventional, even terrorist, methods to pursue a political and military strategy, even if that meant confronting the United States.

Rafsanjani and the Al-Khobar Bombings

Khomeini's death was perhaps an even greater challenge for Iran than war with Iraq. This event brought a new generation of revolutionaries to the top leadership positions and produced substantial changes in the constitution, even though it did not seriously threaten the regime or cause any dramatic shift in policy. Iran's competing foreign policies, however, were dramatically visible during the presidency of Ali Akbar Hashemi Rafsanjani (1989–1997). Rafsanjani's systematic efforts to build constructive political as well as commercial ties with the West were sabotaged repeatedly by a policy that appeared to be driven by revolutionary vengeance and executed by shadowy forces. Tehran never publicly identified the perpetrators or publicly held them accountable, presumably because they enjoyed the protection of individuals at or near the top of the conservative power structure.

A major terrorist event during the last few years of the Rafsanjani presidency was the June 1996 bombing of the U.S. military barracks at Al-Khobar in the eastern province of Saudi Arabia that killed 19 U.S. servicemen and wounded 372. Five years later, the Bush administration issued an indictment that identified Saudi Hizballah as responsible for carrying out the attack and asserted that Iran had "inspired, supported, and directed" Hizballah organizations in Saudi Arabia, Lebanon, Kuwait, and Bahrain since the early 1980s.[5] The indictment specifically identified Iranian contact and exchange of information with various Saudi Hizballah groups during 1993 and 1994, but it contained no evidence of Iranian contact with any of the Saudi perpetrators during the year prior to the Al-Khobar operation and no evidence of Iranian involvement in the operation itself.[6] When the June 2001 indictment was issued, Attorney General John Ashcroft indicated quite clearly that it contained only those charges that the administration believed would stand up in court.[7]

> The border between Iran, Afghanistan, and Pakistan is far from secure.

The Al-Khobar case is crucially important to understanding Iran's use or nonuse of terror, at least historically. If, as the Bush administration's indictment asserts, the Al-Khobar incident shows that Iranian intelligence services maintained active contacts with radical Islamist elements opposed to the United States, that should not come as a great surprise. If, however, the Iranian government deliberately orchestrated an attack on U.S. installations and personnel as a means, for example, of driving Americans out of the Gulf region, that would be evidence of a significant shift in Iranian policy toward the United States and Saudi Arabia. Only the year before, Iran had offered a major offshore development contract to a U.S. company as a signal of interest in improved relations and was engaged in a major strategic effort to develop closer relations with Saudi Arabia.

It is impossible to conclude on the basis of the Bush administration's indictment that the Al-Khobar attack constituted a major shift in Iran's willingness to use terror against Saudi Arabia and the United States . As former U.S. national security adviser under the Clinton administration Sandy Berger described the Al-Khobar investigation: "We know it was done by the Saudi Hizballah. We know that they were trained in Iran by Iranians. We know there was Iranian involvement. What has yet to be established is how substantial the Iranian involvement was."[8]

Khatami and the New Iranian Diplomacy

With Khatami's landslide election in 1997, Iran's official foreign policy focused more intently on integrating Iran into the international community and on presenting a visage of Iran quite different from the scowling fanaticism of the earliest days of the revolution. Khatami's determination to change Iran's image became clear in January 1998, early in his first term, when he used the occasion of a CNN interview with correspondent Christiane Amanpour to deliver a message to the people of the United States. In carefully prepared remarks, he addressed all the outstanding issues between the United States and Iran, including terrorism:

> We believe in the holy Quran that says: slaying of one innocent person is tantamount to the slaying of all humanity. How could such a religion,

> and those who claim to be its followers, get involved in the assassination of innocent individuals and the slaughter of innocent human beings? We categorically reject all these allegations.... Terrorism should be condemned in all its forms and manifestations; assassins must be condemned. Terrorism is useless anyway and we condemn it categorically.... At the same time, supporting peoples who fight for the liberation of their land is not, in my opinion, supporting terrorism. It is, in fact, supporting those who are engaged in combating state terrorism.[9]

When further asked, "Regardless of the motive, do you believe that killing innocent women and children is terrorism, as for instance what happens on the streets of Israel?" Khatami replied, "It is definitely so. Any form of killing of innocent men and women who are not involved in confrontations is terrorism; it must be condemned, and we, in our term, condemn every form of it in the world."

This statement was and remains the most complete and authoritative to date regarding Iran's formal government policy on terrorism. Khatami's subsequent handling of the "serial murders" of Iranian intellectuals lent some credibility to his statement. At least four intellectuals were brutally murdered in quick succession in November and December 1998 in what may have been an effort to destabilize the Khatami government. Khatami conducted an investigation, and his government arrested a group of ultraconservative officials, headed by Deputy Director Saeed Emami, in the Ministry of Intelligence. These men were hired originally by Ali Fallahian, the former minister of intelligence, and their arrest was widely seen as a public rebuke to the conservatives as well as a rare case of transparency in the security services. Before the case came up for trial, however, Emami reportedly killed himself in prison by ingesting a toxic powder normally used for hair removal.

When Khatami first took office, he had wanted to remove Emami and his associates from the Intelligence Ministry but had not succeeded in overcoming conservative objections. After Emami's arrest, Khatami was able to replace many of Fallahian's people in the ministry and to install an intelligence minister of his choosing. The unprecedented revelations of rogue operations in the security services, including widespread allegations that Emami was killed to prevent him from implicating other ultraconservative figures at the very highest levels of the clerical leadership, created a public sensation and seemed to indicate that unauthorized terrorist operations might become subject to internal and perhaps even public scrutiny and control. Such a hope was unduly optimistic as no further examples have followed, but Emami's arrest and death did confirm widespread suspicions that pockets of extremists inside and outside the revolutionary structure were operating without the review or approval of the elected government.

September 11 and the Iranian Response

After the September 11 attacks, in sharp contrast to much of the Arab world's scarcely concealed glee that the United States had gotten a taste of its own medicine, Iran responded with official statements of condolences and unofficial candlelight vigils in support of the American people. Although Iran officially opposed the subsequent U.S. attack on Afghanistan, it made no effort to interfere and even cooperated quietly on issues such as humanitarian relief, search and rescue, and other practical matters. After the Taliban government was deposed, Iran participated positively and creatively in the Bonn talks to establish a new interim government in Afghanistan, drawing rare praise from U.S. officials.[10] At the Tokyo donors conference in January 2002, Iran pledged a total of $560 million for the reconstruction of Afghanistan—the largest donation of any developing country. Speculation emerged among pundits that this would be the beginning of a new U.S.-Iranian relationship. Then, in his 2002 State of the Union address, President George W. Bush identified Iran as the third member of an "axis of evil," along with Iraq and North Korea, stating that terrorism was a major concern:

> Iran aggressively pursues these weapons [of mass destruction] and exports terror, while an unelected few repress the Iranian people's hope for freedom.... They could provide these arms to terrorists, giving them the means to match their hatred.... The United States of America will not permit the world's most dangerous regimes to threaten us with the world's most destructive weapons.[11]

Why did the Bush administration go from praising to excoriating Iran in only six weeks? One likely reason was the Israeli intercept and capture in January 2002 of the *Karine-A*, a ship secretly purchased by the Palestinian Authority (PA) that was allegedly carrying some 50 tons of weapons and explosives from Iran's Kish Island to Palestine. Israel arrested the ship's captain, Omar Akawi, who later spoke to the press from his prison cell and identified himself as a member of Arafat's Fatah movement and a lieutenant colonel in the PA's naval police.[12] The Palestinians and Iranians denounced the event as an Israeli setup intended to influence U.S. policy. If so, it worked perfectly. A senior administration official told *The New York Times* that the incident "was a sign to the president that the Iranians weren't serious." [13]

Ties with Al Qaeda?

The United States also began asserting publicly that members of Al Qaeda were taking refuge in Iran across the border from western Afghanistan. Zalmay Khalilzad, the administration's special envoy to Afghanistan, put the U.S. case succinctly: "Hard-line, unaccountable elements of the Iranian regime facilitated the movement of Al Qaeda terrorists escaping from Afghanistan."[14] The government in Tehran initially denied that any Al Qaeda partisans were in Iran. The very lengthy border between Iran and Afghanistan and Iran and Pakistan is riddled with drug smuggling routes and is far from secure, however, and after some weeks, Iran announced that it had located Taliban and Al Qaeda supporters within its borders and that they were being returned to their countries of origin. Over the following year, the Iranian government detained and extradited more than 500 fugitives, largely volunteers from various Muslim countries who had gone to Afghanistan to join the jihad against the West.

> Claims of an alliance between Al Qaeda and some elements in Iran strain credulity.

Why would members of the Iranian security services look the other way or perhaps even facilitate the passage of these fugitives? No doubt money was the primary reason. Besides money, however, some hard-line elements may have also seen an opportunity to recruit agents or to incorporate some militant Afghan cadres into their own operations. One can only speculate, though, because neither Washington nor Tehran disclosed the identity of these individuals nor suggested their possible motives.

Some reports, usually ascribed to anonymous intelligence sources, have mentioned a connection between Al Qaeda and some elements in Iran, possibly via Hizballah.[15] Those allegations strained credulity, however, given Iran's vigorous opposition to the Taliban government in Afghanistan and its Al Qaeda supporters. Al Qaeda is a Sunni Muslim group that espouses the views of the most extreme proponents of the Salafi (often called Wahhabi) school of Islamic thought, which regards Shi'ism, the religion practiced most in Iran and by Hizballah in Lebanon, as heretical. One can imagine some low-level tactical contact between the two groups, particularly in view of their shared opposition to the Western presence in the Gulf region. Claims of an alliance, however, lack evidence and logic.

The issue of potential Iranian ties with Al Qaeda took on much greater significance in May 2003 when three suicide car bombs exploded almost simultaneously in Riyadh, Saudi Arabia. Thirty-five people, including nine bombers, died in the explosions, which targeted housing compounds for Americans and other Westerners living and working in the Saudi kingdom. The attack was carried out by a group of Saudi militants, who had previously been identified by Saudi security forces and were on the run, operating under Al Qaeda's direction. Many of the perpetrators were arrested in the following weeks, but the United States released unconfirmed intelligence reports that Iran was sheltering some senior Al Qaeda operatives who may have been involved in planning the attack. Iran denied involvement, then announced that it had several Al Qaeda members in custody, reportedly including some very senior individuals.

The United States responded quite sharply, calling the action taken by the Iranian government insufficient and suspending the potentially significant informal talks that had begun to take place on a regular basis between U.S. and Iranian officials. These talks had been warily resumed in Geneva, technically under the aegis of an informal UN committee created to deal with Afghanistan after the Afghan and the Iraq wars had underscored the mutual interests of the United States and Iran on a number of practical issues, such as preparing for refugee movements and search and rescue missions as well as maintaining stability after war had ended. The discussions were reportedly businesslike and many observers saw them as a precursor to a possible improvement in U.S.-Iranian relations, despite the two countries' many differences and the sour taste left by the axis of evil speech. As had happened in the past, U.S. charges of Iranian association with terrorist activities brought potentially constructive contacts to a halt.

Has Khatami Ended Support for Terrorism?

Iran has clearly changed its policies substantially over time. The hostage-taking and regional destabilization campaigns of the early days of the Iranian revolution that were so immensely costly to Iran's image and that continue to plague its international relations have vanished. As Khatami delicately put it in his CNN interview, there is no longer any need for such "unconventional methods."[16] Assassinating enemies of the Islamic Republic in Europe ended in 1994. Later killings outside Europe focused primarily on members of the Mujahideen-e Khalq, but those have also largely ceased in recent years and may have been rendered pointless by U.S. occupation of Mujahideen-e Khalq camps in Iraq and severe crackdowns on the organization in France and elsewhere.

As far as we can tell, Iranian direct involvement in terrorist activities in the past—kidnappings, maritime attacks, assassinations—seems to have given way in recent years almost entirely to proxy support for non-Iranian organizations. If so, this may be attributable simply to the realization that these actions were doing immense harm to Iran's broader national objectives and that their cost far outweighed whatever perceived benefits may have been gained. Iran may have taken a very long time to reach what might appear a fairly obvious conclusion, but it suggests at a minimum a capacity to modify its policies in the face of persistent pressure and experience.

> Hostage-taking and regional destabilization campaigns have vanished.

The most substantial changes in Iran's apparent policies and behavior have come with Khatami's election. Although Khatami has been largely unsuccessful in his attempt to move the ruling clerical elite toward his vision of greater political liberty, civil society, and rule of law, he has changed the political discourse in Iran. His housecleaning of the Intelligence Ministry—one of the few genuine achievements to come out of his many confrontations with the conservative power structure—may have significantly curtailed Iran's earlier tendency toward interventionism and feckless adventurism.

At the same time, Iran undoubtedly continues to consort with and provide support to organizations that are committed to the destruction of Israel. The list begins with Lebanese Hizballah and extends to include Hamas, the Palestinian Islamic Jihad, and the Popular Front for the Liberation of Palestine-General Command. Virtually all elements of the Iranian leadership do not deny this association; they actually take pride in it. Members of these and other militant organizations are brought to Iran repeatedly for various conferences and meetings; their leaders meet openly with Iran's top leaders, including Khatami and his foreign minister; other Iranian officials meet with them on trips to Lebanon and Syria; and Iran provides material support. Iran regards this as legitimate activity in support of resistance movements fighting against illegal occupation of their land. Although Khatami, as indicated earlier, asserts that bombings of innocent people are prohibited in Islam and are opposed by Iran, many other Iranians, including very senior clerics and officials, maintain that such acts are legitimate and may well be prepared to countenance or encourage violence.

The United States and much of the West regard these organizations as terrorists. Iran's more tolerant view, however, is not that different from popular Islamic opinion (and some official opinion, whether public or private). Iran envisions itself as the true world leader of political Islam, and fierce opposition to Israeli occupation is a touchstone of that core belief. Despite its own strong views, Iran has stated repeatedly that it would accept any settlement that is satisfactory to the Palestinians and that it will not try to impose its views by force. Judging from the fiery anti-Israeli rhetoric of many Iranian leaders and their failure to criticize or condemn even the most extreme actions or claims of its friends in the Palestinian-Israeli arena, including repeated suicide bombings by organizations such as Hamas and Islamic Jihad, Israel and the West have every reason to be skeptical of those assurances.

The alleged sheltering in Iran of Al Qaeda members and other fugitives, such as the Al-Ansar group in Iraq, is a different problem that is less obvious than it may appear. Even without porous borders and isolated, lawless regions, the apprehension of Al Qaeda operatives is not a simple matter, as evident elsewhere. Osama bin Laden and some of his contingent reportedly move back and forth across the Afghan-Pakistani border almost at will, despite the best efforts of both the United States and the Pakistani government to locate and intercept them. The United States itself has repeatedly discovered cells of Al Qaeda operatives within its own borders, including some members who had recently arrived and were reportedly conducting training operations not far from the nation's capital. Washington is quick to assume the worst with Iran, especially in light of Iran's lack of transparency concerning issues of intelligence and national security. Nevertheless, after massive misjudgments of intelligence concerning Iraq, the United States might be well advised to regard its present intelligence reports on Iran with a bit more caution.

Policy Options

The United States faces two severe problems in dealing with Iran and terrorism. The first is the difficulty of dealing with the legacy of the past. Terrorist acts in which Iran may have had direct or indirect involvement have seriously harmed many U.S. citizens (and others). The U.S. Congress has attempted to address this by passing legislation permitting victims to bring cases to U.S. courts, with awards granted on the basis of uncontested evidence because Iran refuses to appear. The awards are supposed to be paid from Iranian assets, but that would set a precedent that could harm U.S. interests around the world; so, large awards are paid to these plaintiffs from the U.S. Treasury on the presumption that eventually they will be recovered from Iran. The Bush administration fiercely opposes efforts to prosecute U.S. officials or military personnel for possible violations of international law in the courts of other countries or at the International Criminal Court. Yet, U.S. courts are now routinely prosecuting Iranians and others for alleged support of terrorist actions by Hizballah and other militant organizations, mocking judicial due process. The past must be dealt with, but the present remedy will only complicate future efforts to settle past grievances.

The more immediate problem for the United States and the international community is how to deal with Iran's proxy support for pro-Palestinian groups that oppose Israel and the peace process and who resort to terrorist attacks against civilian targets. At least since Khatami's election seven years ago, this proxy support has been the focus of virtually all accusations about Iran's role as a state sponsor of terrorism.

Iran undoubtedly continues to support organizations committed to destroying Israel.

Resolving the Israeli-Palestinian dispute would, among other benefits, remove the *raison d'être* of these violent factions and eliminate Iran's rationale for providing political and financial support. Iranian involvement is, of course, not the primary concern of those involved in the peace process. Nevertheless, as the heat of the intifada increased, with resultant devastating pictures on regional television, so too did Iran's rhetoric and its presumed material support to the extremist opposition. Iran insists that its support of the "forces of national liberation" is not terrorism, but its fervor rises and falls with the intensity of the Israeli-Palestinian conflict. Because of its distance from the conflict, Iran can adopt an irresponsible rhetorical stance that is "more Palestinian than the Palestinians" if only because it sounds appropriately revolutionary in speeches and distracts from the many domestic failures of the Iranian leadership. This is not a factional issue in Iranian politics; reformers and conservatives tend to sound very much alike. Pressure tactics and sanctions have been totally ineffective in changing Iran's behavior on this issue in the past, and there is no reason to believe that the future will be any different. Among the side benefits of progress in the peace process almost surely would be a cooling of Iranian rhetoric, a reduction in Iranian temptation to meddle in Palestine, and a corresponding improvement in U.S.-Iranian relations.

The most complex element of Iran's involvement in terrorist activities is the fact that Iran has two different ruling structures. As Khalilzad has noted, Iran's worst behavior often originates with "hard-line, unaccountable elements of the Iranian regime." How can the United States deal with that reality of Iranian politics? The short answer is regime change. The longer and more thoughtful answer is regime change that grows out of Iranian domestic needs and demands, not imposed by an external power.

One of the few unquestioned positive achievements of the 1979 revolution was its lesson to the Iranian people that they were in charge of their own destiny, rather than blaming every political development on foreign hands. Losing that would be a huge setback. Iran has been in a century-long struggle for freedom that started with the Constitutional Revolution of the early twentieth century. It has not been an easy or linear process, and the outcome is far from certain. Any attempt to short-circuit the process by sticking a U.S. finger in the Iranian pie, however, is a formula for disaster. Success in prompting a revolt would bring a crushing response from the conservative forces that would at least temporarily halt the democratization movement. Even if U.S. calls for revolution went unheeded, they might taint those seeking change as lackeys of a foreign power.

> How can the United States deal with the unaccountable reality of Iranian politics?

During nearly a quarter century of Islamic revolutionary rule, Iran has changed and continues to change. This is as true of the country's involvement in terrorist activities as it is in any other aspect of its political life. Iran's early ventures into hostage-taking, bombings, and subversion gave way to terror at sea during the long war with Iraq and then to a vicious vendetta of assassination against its perceived political enemies. Increasingly, Iran has shifted its focus to financing, training, and supporting proxy organizations whose actions provided some measure of deniability for Iran but could not overcome suspicion of Iranian involvement, if not actual control. Over the past seven years, the focus of this proxy relationship has been on the Israeli-Palestinian conflict.

Throughout much of this history, there has been a gap between Iran's declaratory policy and the actions of malevolent forces embedded in Iran's security services. Khatami has been successful in weeding out some of these individuals, but the job is far from complete. The magnitude of the problem that remains may be reflected in alleged Iranian support for arms shipments to Palestine and providing refuge to Al Qaeda fugitives. Iran's denial of involvement is insufficient. For the sake of its credibility, Iran must demonstrate a genuine determination to investigate such charges and to remedy any abuses. Its extradition of hundreds of Al Qaeda fighters was a step in the right direction, but Iran needs to clean its house of all known terrorists, including Lebanese and Palestinian figures with long histories of involvement in bombings and assassinations.

Confronting the hard-line elements that distort its foreign and domestic policies goes far beyond allegations of international terrorism. That struggle lies at the heart of Iran's political identity and will determine the course of its future. The United States and the international community can keep the spotlight on Iran's abuses and press hard for change. If the pressure for change is applied fairly and if Washington acknowledges Iran's accomplishments as well as its failures, the world will be assured of staunch allies within Iran. Change is a slow and often uncertain process, but it is something that can be done only by Iran itself.

Endnotes

1. There is no generally accepted definition of terrorism. For a discussion of the definitional problems, see A. William Samii, "Tehran, Washington, and Terror: No Agreement to Differ," *Middle East Review of International Affairs* 6, no. 3 (September 2002), http://meria.idc.ac.il/journal/2002/issue3/jv6n3a5.html (accessed July 23, 2003).
2. For details on this episode, see Robin Wright, "A Reporter at Large: Teheran Summer," *New Yorker*, September 5, 1988, pp. 32–72.
3. See Daniel Sobelman, "Hizbollah Two Years after the Withdrawal: A Compromise between Ideology, Interests, and Exigencies," *Strategic Assessment* 5, no. 2 (August 2002).
4. See Gary Sick, "Slouching Toward Settlement: The Internationalization of the Iran-Iraq War, 1987-88," in *Neither East Nor West: Iran, the Soviet Union, and the United States*, eds. Nikki Keddie and Mark Gasiorowski (New Haven: Yale University Press, 1990), pp. 219–246.
5. The indictment is located at `www.usdoj.gov:80/opa/pr/2001/June/khobarindictment.wpd` (accessed July 23, 2003).
6. For a thorough review of the charges and countercharges, see Elsa Walsh, "Annals of Politics: Louis Freeh's Last Case," *New Yorker*, May 14, 2001, p. 68.
7. See statement by Ashcroft released by the Department of Justice on June 21, 2001.
8. Walsh, "Annals of Politics."

9. "Transcript of Interview with Iranian president Mohammad Khatami," CNN.com, January 7, 1998, www.cnn.com/WORLD/9801/07/iran/interview.html (accessed July 23, 2003) (hereinafter Khatami interview).
10. Richard Haas, the U.S. special coordinator for Afghanistan, complimented Iran's "constructive role" in talks on the future of Afghanistan and providing the Afghan people with humanitarian aid. Agence France Press, December 6, 2001.
11. George W. Bush, State of the Union address, Washington, D.C., January 29, 2002, www.whitehouse.gov/news/releases/2002/01/20020129-11.html (accessed July 23, 2003).
12. For a detailed review of reportage on the incident from an Israeli perspective, see "Operation Noah's Ark," Ha'aretz.com, www.haaretzdaily.com/hasen/pages/ShArt.jhtml?itemNo=114367&contrassID=3&subContrassID=0&sbSubContrassID=0 (accessed July 23, 2003).
13. David E. Sanger, "Bush Aides Say Tough Tone Put Foes on Notice," *New York Times*, January 31, 2002.
14. Zalmay Khalilzad, speech to the American-Iranian Council, Washington, D.C., March 13, 2002.
15. For a detailed examination of the facts and allegations concerning Iranian terrorist activities for the first year after the September 11 attacks, see Samii, "Tehran, Washington, and Terror."
16. Khatami interview

Gary Sick is director of the Middle East Institute at Columbia University.

The Growing Syrian Missile Threat: *Syria after Lebanon*

Lee Kass

Even though international pressure succeeded in forcing Damascus to withdraw its troops from Lebanon, the Syrian regime remains in the cross hairs of U.S. defense and intelligence concern about four other Syrian activities. First, the Syrian regime has continued its attempts to acquire sophisticated surface-to-surface missiles. Second, U.S. intelligence officials remain concerned that the Syrian government has become custodian to Iraq's biological and chemical weapons. Third, questions remain about whether Damascus benefited from the network of Abdul Qadir Khan, the Pakistani nuclear scientist who sold nuclear secrets to a number of rogue regimes. Lastly, Bashar al-Assad continues to flirt with international terrorism. The young president shows no inclination to cease the behavior that has for more than a quarter century led the U.S. government to designate Syria a state-sponsor of terrorism.

Left unresolved, such questions about Syrian proliferation ambitions, coupled with the regime's demonstrated willingness to use terrorism to advance its goals, will make any rapprochement between Washington and Damascus impossible.

A SYRIAN BALLISTIC MISSILE?

Much of Syria's arsenal consists of Cold War remnants received from the Soviet Union. The Syrian military has already begun upgrading its tanks, acquiring the faster, tougher T-72s from a cash-starved Russian military industry.[1] Analysts believe that Damascus acquired the tanks for their speed—to maneuver and advance more effectively on the Golan Heights. The Syrian regime has also sought to upgrade its air force. While much of the fleet is old, the Syrian military still has enough planes to saturate Israeli air defenses and conduct a significant strike against the Jewish state. Still, the Israeli air force remains far superior, and because Syrian air defenses are old and lack complete interoperability,[2] Jerusalem still maintains a large advantage.

Perhaps to compensate for this weakness, the Syrian regime has sought to upgrade its weapons capability. When Israeli warplanes struck a Palestinian Islamic Jihad base ten miles northwest of Damascus in October 2003 following the terrorist group's suicide bomb attack in a Haifa restaurant, Iraqis who were in Damascus at the time said Syrian air defense did not react.

The Syrian regime's efforts to upgrade its missile capability threaten U.S., Israeli, and Turkish interests. With a stronger Syrian missile capability, the Assad regime could launch either a preemptive strike or, more likely, feel itself secure enough in its deterrent capability to encourage terrorism without fear of consequence.

Launched from Damascus, the Iskander-E could reach Tel Aviv in less than three minutes.

Syrian officials have sought to obtain the advanced SS-X-26 surface-to-surface missiles, also known as Iskander-Es, from Russia, but Russian president Vladimir Putin cancelled the deal after learning from his experts that Israel would not have a capability to intercept the missiles.[3] With a range of 174 miles (280 kilometers), the Iskander-E could have hit cities such as Tel Aviv, Jerusalem, and

Haifa. While a significant threat due to the proximity of Israeli population centers, the missiles fall under the 186 mile (300 kilometer) range subject to the Missile Technology Control Regime to which Russia, the United States, and thirty-two other countries are subject. It is unclear from unclassified sources whether countries that obtain Iskander-Es can extend the missiles' range, but if so, they would pose an enhanced threat to Turkey, Jordan, and Iraq as well.[4] Regardless, the chance that the Syrian government might provide the missile to terrorists or other rogue states undermines both the spirit and the effectiveness of the Missile Technology Control Regime and other nonproliferation agreements.

The Iskander-E would be a particularly dangerous upgrade. Unlike Scuds, Iskander-Es have solid fuel propellants. Solid propellants are less complicated because the fuel and oxidizer do not need to travel through a labyrinth of pumps, pipes, valves, and turbo-pumps to ignite the engines. Instead, when a solid propellant is lit, it burns from the center outward, significantly reducing launch preparation time.

Immediately after launch, Iskander-Es perform maneuvers that prevent opponents from tracking and destroying the launchers. Once in flight, the Iskander-Es can deploy decoys and execute unpredictable flight paths to confuse missile defense systems.[5] Moreover, they are fast. According to Uzi Rubin, former head of Israel's Arrow-Homa missile defense program, the Iskander can fly at 1,500 meters per second, equivalent to 3,355 miles (5,370 kilometers) per hour,[6] Launched from Damascus, the Iskander-E could reach Tel Aviv in less than three minutes, sooner if the Iskanders' mobile launchers were moved closer to the border. This capability might prevent Israel's multi-tiered missile defense shield from adequately protecting the country.

Even though Iskander-Es lack the range to hit many strategic targets, their accuracy and varied warhead types make them an adaptable military system. The missile was intended to obliterate both stationary and mobile targets, particularly short-range missile launchers, ports, command and control facilities, factories, and hardened structures. Such flexibility would allow Syria to destroy an enemy's existing military capabilities and its ability to wage a future war.[7]

These concerns have led both the U.S. and Israeli governments to criticize the Syrian regime's attempts to acquire the new technology. One U.S. official stated, "We don't think that state sponsors of terrorism should be sold weapons of any kind.[8] Israel's government is focused on the possibility that Palestinian terrorists might obtain the equipment.[9] According to the State Department's *Patterns of Global Terrorism*, Syria supports or provides safe-haven to a number of terrorist groups, including Islamic Jiliad. Hamas, and the Popular Front for the Liberation of Palestine-General Command.[10]

Russian defense minister Sergey Ivanov acknowledged such concerns when he announced, at least temporarily, that Moscow would halt export of the missile to Syria.[11] At an April 2005 meeting with senior Israeli officials, Russian president Vladimir Putin confirmed that he cancelled the Syrian Iskander contract because Israel lacks the ability to intercept those missiles.[12]

Instead, Putin said that the Russian government would only authorize sale of Strelet surface-to-air systems that are unable to penetrate Israel.[13] While a nominal downgrade, even with a range of just three miles (five kilometers),[14] the system can pose a significant threat to Israel. These missiles can proliferate to Hezbollah and other terrorist organizations supported by the Syrian regime. In such hands, the Strelets could endanger passenger planes on descent to Ben Gurion International Airport, outside Tel Aviv and just four miles from the West Bank.[15] Russian officials say they will only sell Damascus the vehicle-mounted version and not the shoulder-held type, but Western defense officials say operators can easily dismantle Strelets to make them transportable.[16]

Augmenting concern was the Israeli disclosure of a Syrian launch of three Scud missiles on May 27, 2005.[17] The tests were the first since 2001 and represented a significant milestone in the country's missile program—the three carried airburst warheads. This capability reinforced Israeli concerns that Syria could use the Scuds to deliver chemical weapons. One of the missiles launched was an older Scud B, with a range of about 185 miles (300 kilometers), while the remaining two were newer Scud-Ds with a range of approximately 435 miles (700 kilometers).[18] The greater range not only gives Syria greater reach but also allows launches from deeper within Syrian territory, making it more difficult to undertake a preemptive aerial attack on the launchers.

Questions about a possible transfer of Iraqi weapons to Syria remain unanswered.

U.S., Israeli, and other Western governments' concerns over Russian missile sales to Syria will likely go unheeded. After all, international security concerns have not stopped Russian support for the Iranian nuclear program.[19] Sergey Kazannov, head of the Russian Academy of Sciences World Economics and International Relations Institutes' Geopolitics Division, said that in Soviet times, political reasons and the need to maintain the Soviet defense industry motivated Moscow's arms sales.[20] The post-Cold War climate undercut opportunities for the Russian defense industry. He elaborated, "Seventy percent of our defense complex's output goes for export. And depriving ourselves of that factor under our unenviable conditions is almost tantamount to death." He also added that the missile sales allow Moscow, Damascus, and other regional actors the independence to develop policies without regard to U.S. pressure.[21] As relations between Putin and the West worsen, such political calculations might re-enable the Iskander-E sale.

IRAQI WEAPONS IN SYRIA?

White Western governments were able to pressure Moscow to alter its weapons shipments, Bashar al-Assad may not have limited himself to over-the-counter weapons purchases. The Syrian military's unconventional weapons arsenal already has a significant stockpile of sarin. The Syrian regime has also attempted to produce other toxic agents in order to advance its inventory of biological weapons.[22]

Several different intelligence sources raised red flags about suspicious truck convoys from Iraq to Syria in the days, weeks, and months prior to the March 2003 invasion of Iraq.[23]

These concerns first became public when, on December 23, 2002, Ariel Sharon stated on Israeli television, "Chemical and biological weapons which Saddam is endeavoring to conceal have been moved from Iraq to Syria."[24] About three weeks later, Israel's foreign minister repeated the accusation.[25] The U.S., British, and Australian governments issued similar statements.[26]

The Syrian foreign minister dismissed such charges as a U.S. attempt to divert attention from its problems in Iraq.[27] But even if the Syrian regime were sincere, Bashar al-Assad's previous statement—"I don't do everything in this country,"[28]—suggested that Iraqi chemical or biological weapons could cross the Syrian frontier without regime consent. Rather than exculpate the Syrian regime, such a scenario makes the presence of Iraqi weapons in Syria more worrisome, for it suggests that Assad might either eschew responsibility for their ultimate custody or may not actually be able to prevent their transfer to terrorist groups that enjoy close relations with officials in his regime.

Two former United Nations weapon inspectors in Iraq reinforced concerns about illicit transfer of weapon components into Syria in the wake of Saddam Hussein's fall. Richard Butler viewed overhead imagery and other intelligence suggesting that Iraqis transported some weapons components into Syria. Butler did not think "the Iraqis wanted to give them to Syria, but ... just wanted to get them out of the territory, out of the range of our inspections. Syria was prepared to be the custodian of them."[29] Former Iraq Survey Group head David Kay obtained corroborating information from the interrogation of former Iraqi officials. He said that the missing components were small in quantity, but he, nevertheless, felt that U.S. intelligence officials needed to determine what reached Syria.[30]

Baghdad and Damascus may have long been rivals, but there was precedent for such Iraqi cooperation with regional competitors when faced with an outside threat. In the run-up to the 1991 Operation Desert Storm and the liberation of Kuwait, the Iraqi regime flew many of its jets to Iran, with which, just three years previous, it had been engaged in bitter trench warfare.[31]

Subsequent reports by the Iraq Survey Group at first glance threw cold water on some speculation about the fate of missing Iraqi weapons, but a closer read suggests that questions about a possible transfer to Syria remain open. The September 30, 2004 Duelfer report,[32] while inconclusive, left open such a possibility. While Duelfer dismissed reports of official transfer of weapons material from Iraq into Syria, the Iraq Survey Group was not able to discount the unofficial movement of limited material. Duelfer described weapons smuggling between both countries prior to Saddam's ouster.[33] In one incident detailed by a leading British newspaper, intelligence sources assigned to monitor Baghdad's air traffic raised suspicions that Iraqi authorities had smuggled centrifuge components out of Syria in June 2002. The parts were initially stored in the Syrian port of Tartus before being transported to Damascus International Airport. The transfer allegedly occurred when Iraqi authorities sent twenty-four planes with humanitarian assistance into Syria after a dam collapsed in June 2002, killing twenty people and leaving some 30,000 others homeless.[34] Intelligence officials do not believe these planes returned to Iraq empty. Regardless of the merits of this one particular episode, it is well documented that Syria became the main conduit in Saddam Hussein's attempt to rebuild his military under the 1990-2003 United Nations sanctions,[35] and so the necessary contacts between regimes and along the border would already have been in place. Indeed, according to U.S. Defense Department sources, the weapons smuggling held such importance for the Syrian regime that the trade included Assad's older sister and his brother-in-law. Assaf Shawqat, deputy chief of Syria's military intelligence organization. Numerous reports also implicate Shawqat's two brothers who participated in the Syrian-Iraqi trade during the two years before Saddam's ouster.[36]

While the Duelfer report was inconclusive, part of its failure to tic up all loose ends was due to declining security conditions in Iraq, which forced the Iraq Survey Group to curtail its operations.[37] The cloud of suspicion over the Syrian regime's role in smuggling Iraq's weapons—and speculation as to the nature of those weapons—will not dissipate until Damascus reveals the contents of truck convoys spotted entering Syria from Iraq in the run-up to the March 2003 U.S.-led invasion of Iraq.[38] U.S. intelligence officials and policymakers also will not be able to end speculation until Bashar al-Assad completely and unconditionally allows international inspectors to search suspected depots and interview key participants in the Syrian-Iraqi weapons trade. Four repositories in Syria remain under suspicion. Anonymous U.S. sources have suggested that some components may have been kept in an ammunition facility adjacent to a military base close to Khan Abu Shamat, 30 miles (50 kilometers) west of Damascus.[39] In addition, three sites in the western part of central Syria, an area where support for the Assad regime is strong, are reputed to house suspicious weapons components. These sites include an air force factory in the village of Tall as-Sinan; a mountainous tunnel near Al-Baydah, less than five miles from Al-Masyaf (Masyat); and another location near Shanshar.[40]

While the Western media often focus on the fate of Iraqi weapons components, just as important to Syrian proliferation efforts has been the influx of Iraqi weapons scientists. *The Daily Telegraph* reported prior to the 2003 Iraq war that Iraq's former special security organization and Shawqat arranged for the transfer into Syria of twelve mid-level Iraqi weapons specialists, along with their families and compact disks full of research material on their country's nuclear initiatives. According to unnamed Western intelligence officials cited in the report, Assad turned around and offered to relocate the scientists to Iran, on the condition that Tehran would share the fruits of their research with Damascus.[41]

THE WEAPONS PROLIFERATION HYDRA

The Iraqi government may not have been Bashar al-Assad's only source of advanced weapons technology. Following his January 29, 2002, State of the Union speech. Bush launched the Proliferation Security Initiative.[42] Participation grew quickly to include over sixty countries. Participants seek to deter rogue states and non-state actors from obtaining material for weapons of mass destruction and ballistic missile initiatives through various activities—interdiction of suspicious shipments, streamlined procedures to analyze and disseminate information, and strengthened national and international laws and regulations. Liberia and Panama's participation marked a key development because vessels registered from both countries account for approximately 50 percent of the world's total shipping.[43]

The Syrian government remains convinced that U.S. efforts to isolate it will fail.

In 2003, cooperation between U.S. and British intelligence and coordination with their militaries led to the seizure of a Libya-bound ship that carried material for its nuclear weapons program. The capture partly led to Tripoli's agreement to dismantle and destroy its weapons of mass destruction capabilities.[44] Additionally, it was the seizure of this ship that unraveled Pakistani nuclear scientist Abdul Qadir Khan's clandestine nuclear proliferation network. While the exposure of the network drew international attention, the limelight did not eradicate the program. As one former aid of Khan's acknowledged, "The hardware is still available, and the network hasn't stopped."[45]

Khan visited various countries throughout Europe, Africa, and Asia. While no credible evidence yet links Khan's network to Damascus, Western diplomats said that he gave numerous lectures on nuclear issues in late 1997 and early 1998 in Damascus. According to sources, starting in 2001, meetings with the Syrians were held in Iran to avoid any possible linkages between Damascus and Khan's nuclear network. One senior U.S. official stated that an experimental electronic monitor recognized the unique patterns of operational centrifuges in Syria in early to mid-2004. The source reaffirmed Washington's suspicion that the technology originated from Khan's nexus.[46]

The Pakistani government has been unwilling to cooperate fully in the investigation of Khan's activities. As Pakistani president Pervez Musharraf explained, "This man is a hero for the Pakistanis, and there is a sensitivity that maybe the world wants to intervene in our nuclear program, which nobody wants ... It is a pride of the nation."[47] In addition to safeguarding the nuclear weapons program, some analysts note that Islamabad fears Khan might disclose the extent of support he obtained from the Pakistani military. Further complicating efforts to determine what assistance, if any, the Khan network provided Syria. Pakistan's interior ministry denied exit visas to over a dozen technicians who worked in the country's nuclear weapons program. The officials were also barred from meeting or exchanging information with any foreigner.[48] Such unknowns about the extent of weapons know-how and material acquired from Iraq and Pakistan may not equate to proof, but they raise serious concerns about Syrian intentions, all the more so because Damascus has not been forthcoming with explanations and simultaneously has worked to acquire potential delivery systems from Russian firms.

ASSAD'S TERRORIST GAME

U.S. concerns about Syrian weapons ambitions are magnified by the Syrian regime's flirtation with terrorism as a method to advance policy. According to the 2004 *Patterns of Global Terrorism* report, the Syrian regime provides Hezbollah, Hamas, Islamic Jihad, and other groups both logistical and financial assistance.[49]

The Syrian government denies harboring terrorists although much of this denial is based on unwillingness to recognize terrorist groups as such. Damascus views

Syrian willingness to encourage terrorism, not only against Israel but also against other neighbors, is well documented.

many terrorists as soldiers in its war against Israel. Syrian-backed terrorists have attacked Israel, often from Syrian-occupied Bekaa Valley in neighboring Lebanon.[50] Even though the Syrian military has officially ended its occupation of Lebanon under terms of U.N. Security Council Resolution 1559,[51] the Syrian intelligence presence remains significant.[52]

Syrian willingness to encourage terrorism, not only against Israel but also against other neighbors, is well documented. Until 1999, the Syrian regime provided Kurdistan Workers Party (PKK) terrorists safe-haven from which to strike at Turkey.[53] Syrian intelligence or its proxies remain the chief suspect in the February 14, 2005 assassination of former Lebanese prime minister Rafik Hariri.[54]

The Syrian regime has also played a double game with regard to Iraq. General John Abizaid, the commander of U.S. forces in the Middle East, commented that although Damascus made some progress in the curtailment of insurgents entering Iraq, "I don't regard this effort as being good enough ... I cannot tell you that the level of infiltration has decreased."[55] CIA director Peter Goss concurred. In March 2005, he told the U.S. Senate Armed Services Committee, "Despite a lot of very well-intentioned and persistent efforts to try and get more cooperation from the Syrian regime, we have not had the success I wish I could report."[56] Syrian support for terrorism combined with its lack of support for the new Iraqi government make more troubling the possibility that the Syrian regime became custodian to Iraq's chemical and biological weapons capability in the final days of Saddam Hussein's regime.

The confluence of weapons of mass destruction ambitions and Syrian willingness to sponsor terrorism make Syrian ambitions particularly dangerous. In April 2004, for example, Jordanian authorities intercepted, arrested, or killed several Al-Qaeda-sponsored terrorists who planned to attack the U.S. embassy and Jordanian targets in Amman with chemical weapons. The terrorists gathered their materials in Syria and used that country as a base from which to infiltrate Jordan.[57] While the Syrian government denies any role, the implication that it participated in such a potentially catastrophic tragedy underlines Damascus's opposition to the war on terrorism.

The Syrian government may feel that it can ameliorate or outlast U.S. concerns about its flirtation with terrorist groups. In the aftermath of 9-11, Syrian officials detained some alleged Al-Qaeda operatives, but they allowed U.S. officials only to submit questions in writing, not to interrogate the suspects directly.[58] Realists within the Bush administration did not sanction such a la carte for the war on terrorism. Deputy Secretary of State Richard Armitage, for example, remarked. "If you oppose terrorism, you oppose all terrorism."[59] Secretary of State Condoleezza Rice's new lineup at the State Department shows no sign of deviating from such positions.

The Syrian government may also believe that Washington is not able to back its rhetoric against Syria with action. With more than 100,000 U.S. troops committed in Iraq, looming crises over the Iranian and North Korean nuclear programs, and European Union cynicism, the Syrian government remains convinced that U.S. efforts to isolate Damascus will fail. As Assad recently told the Italian newspaper *La Repubblica*, "Sooner or later [the Americans] will realize that we are the key to the solution. We are essential for the peace process, for Iraq. Look, perhaps one day the Americans will come and knock on our door."[60]

But, Assad's belief that Washington needs his cooperation may be a significant misread of U.S. policy. Partly in response to Damascus's refusal to cooperate completely in the war on terrorism. President Bush signed the "Syria Accountability and Lebanese Sovereignty Restoration Act of 2003."[61] Under terms of the act, American firms cannot export any products to Syria beyond food and medicine. The president can wave this provision for an unspecified duration provided he determines that it would further U.S. national security and he submits justification for the waiver to the appropriate Congressional committees.[62] However, Bush increasingly shows little inclination to waive such provisions. In late February 2005, some U.S. government officials suggested that the Bush administration was exploring additional measures against Syria. Under the Syrian Accountability Act, Bush could cut off Syrian access to U.S. banks, limit the travel of Syrian diplomats within the United States, and freeze Syrian assets.[63] Other provisions call for the secretary of state to submit to Congress an annual report on provisions relating to the prevention of dual-use technologies that Damascus could use to advance its ballistic missiles and weapons of mass destruction projects; such reports will also prevent questions over Syrian compliance to fade from policymakers' attention.[64]

FUTURE POLICY

In a recent interview. Bashar al-Assad stated, "I am not Saddam Hussein. I want to cooperate."[65] Evidence indicates otherwise. Syrian attempts to obtain a sophisticated Russian ballistic missile undermine Washington's ability to prevent terrorist sponsors from advancing their military capabilities. Damascus's failure to come clean regarding prewar Iraqi convoys and immigration of Iraqi weapons personnel, as well as its flirtation with Abdul Qadir Khan, raise questions about Assad's sincerity.

The Syrian withdrawal from Lebanon has neither changed basic Syrian behavior nor altered its regional ambitions.

The unknowns regarding Syria's weapons programs are especially worrisome given Assad's continued rejection of international norms of behavior. Syrian obstructionism and attempts to augment its weapons of mass destruction stock make expansion and enforcement of the Proliferation Security Initiative imperative, a strategy supported by U.S. defense secretary Donald Rumsfeld.[66] Offering economic or political incentives to Yemen, Turkey, Egypt, and other countries which retain close relationships with Syria might help shut down avenues which the Syrian regime uses to advance its weapons projects although the damage to counter-proliferation efforts caused by Abdul Qadir Khan's network suggests that there should be a verification mechanism beyond simple diplomatic assurance.

Failure to counter Syrian weapons ambitions could undercut U.S. democracy and antiterror initiatives. The Syrian withdrawal from Lebanon has neither changed basic Syrian behavior nor altered its regional ambitions. The combination of a ballistic missile capability, chemical and biological weapons, and a willingness to arm or turn a blind-eye to terrorists—including those targeting the U.S. presence in Iraq—might lead to bolder terror initiatives, like the attempt in Amman in April 2004, as well as embolden rejectionism by a Syrian regime feeling its arsenal sufficient to deter a U.S. response. Only with sustained pressure can Washington prevent the Syrian regime from such a miscalculation.

Notes

1. *Yedi'ot Ahronot* (Tel Aviv), Sept. 16. 1994.
2. *Syria Primer*, Virtual Information Center, Apr. 24, 2003, p. 38, 47.
3. Associated Press, Apr. 28, 2005.
4. "The SS-26," The Claremont Institute, accessed June 8, 2005.
5. Ibid.
6. *Ha'aretz* (Tel Aviv), Jan. 13, 2005.
7. "The SS-26," The Claremont Institute.
8. Agence France-Presse, Feb. 16, 2005.
9. Radio Free Europe/Radio Liberty, Jan. 13, 2005.
10. "Overview of State-Sponsored Terrorism," *Patterns of Global Terrorism, 2004* (Washington, D.C.: U.S. Department of State, Apr. 2005).
11. *Agenlstvo Voyennykh Novostey* (Moscow), Mar. 25, 2005.
12. Associated Press, Apr. 28, 2005.
13. Associated Press, Apr. 21, 2005.
14. Associated Press, Apr. 26, 2005.
15. Yaakov Amidror, "Israel's Requirements for Defensible Borders," *Defensible Borders for a Lasting Peace* (Jerusalem: Jerusalem Center for Public Affairs), p. 33.
16. Reuters, Apr. 21, 2005.
17. *The Jerusalem Post*, June 5, 2005.
18. *The New York Times*, June 4, 2005.
19. *BBC News*, May 21, 2005.
20. *Potitkum.ru* (Moscow), Feb. 21, 2005.
21. Ibid.
22. "Unclassified Report to Congress on the Acquisition of Technology Relating to Weapons of Mass Destruction and Advanced Conventional Munitions, 1 July through 31 Dec. 2003," CIA, Nov. 2004, p. 6.
23. *The Washington Times*, Oct. 28, 2004.
24. Israel's Channel 2, Dec. 23, 2002.
25. *Petah Tiqva*, Yoman Shevu'i supplement (Tel Aviv), Feb. 21, 2003.
26. "Syria's Weapons of Mass Destruction and Missile Development Program," testimony of John R. Bolton. U.S. undersecretary of arms control and international security, before the House International Relations Committee, Subcommittee on the Middle East and Central Asia, Sept. 16. 2003; *BBC News*, Apr. 14, 2003; Alexander Downer, Australian minister of foreign affairs, news conference, Canberra, June 5, 2003.
27. Agence France-Presse, Apr. 17, 2003.
28. *Time*, Mar. 14, 2005.
29. Agence France-Presse, Apr. 15, 2003.
30. *Sunday Telegraph* (London), Jan. 25, 2005.
31. *Los Angeles Times*, Oct. 8, 1991.
32. Complied by Charles Duelfer, special advisor for strategy to the director of Central Intelligence.
33. *Comprehensive Report of the Special Advisori- to the DCI on Iraq's WMD*, vol. 1 (Washington, D.C.: CIA, Sept. 30, 2004), hereafter, Duelfer report, p. 104.
34. *The Times* (London), June 17, 2002.
35. Duelfer report, p. 239.
36. Dueller report, p. 104.
37. *Addendums to Comprehensive Report of the Special Advisor to the DCI on WMD,* Mar. 2005, accessed on June 8, 2005.
38. *The Washington Times*, Oct. 28. 2004.
39. *Petah Tiqva*, Yoman Shevu'i supplement, Feb. 21, 2003.
40. *De Telegraaf* (Amsterdam), Jan. 5, 2004.
41. *The Daily Telegraph* (London), Sept. 26, 2004.
42. State of the Union Address, Jan. 29, 2002.
43. "The Proliferation Security Initiative," U.S. Department of State, Bureau of Nonproliferation, July 28, 2004.
44. *The Washington Times*, Dec. 23, 2004.
45. *Time*, Feb. 14, 2005.
46. *Los Angeles Times*, June 25, 2004.
47. *Los Angeles Times*, Dec. 6, 2004.
48. *The News* (Islamabad), Jan. 5, 2005.
49. *Pattern of Global Terrorism*, 2004, p. 93.
50. *Patterns of Global Terrorism*, 2003 (Washington, D.C.: U.S. Department of State, Apr. 2004), p. 93.
51. U.N. Security Council Resolution 1559, S/RES/1559 (2004), Sept. 2, 2004.
52. Reuters, May 20, 2005.
53. Ben Thein, "Is Israel's Security Barrier Unique?" *Middle East Quarterly*, Fall 2004, p. 29.
54. Agence France-Presse, Mar. 25, 2005.
55. Associated Press, Mar. 1, 2005.
56. Reuters, Mar. 17, 2005.
57. *CNN News*, Apr. 26, 2004.
58. *The Washington Post*, June 19, 2002.
59. U.S. Embassy news release, Sept. 10, 2004.
60. *BBC NEWS*, Feb. 28, 2005.

61. "Fact Sheet: Implementing the Syria Accountability and Lebanese Sovereignty Restoration Act of 2003," White House news release, May 11, 2004.
62. Ibid.
63. *The Washington Post*, Feb. 17, 2005.
64. "Fact Sheet," May 11, 2004.
65. *Time*, Mar. 14, 2005.
66. *The Washington Times*, Dec. 23, 2004.

***Lee Kass** is an analyst in the research and analysis division of Science Applications International Corporation (SAIC). The views expressed in this article are his own.*

Terrorists Don't Need States

Fareed Zakaria

STEPPING AWAY FROM THE PARTISAN screaming going on these days, the 9/11 commission hearings and—far more revealing—the panel's staff reports paint a fascinating picture of the rise of a new phenomenon in global politics: terrorism that is not state-sponsored but society-sponsored. Few in the American government fully grasped that a group of people without a state's support could pose a mortal threat. The mistake looks obvious in hindsight, but was, sadly, understandable at the time of 9/11. What is less understandable is that this same error persists even today.

Before the mid-1990s, almost all terrorism against the United States had been backed by a state. The Soviet Union had financed and trained terror groups around the world. Syria, Iran, Iraq and Libya had all sponsored terrorism. The most dramatic attacks on Americans—the Beirut Marine-barracks bombing in 1983, and Pan Am 103 in 1988—had both been encouraged if not planned by governments. Even Saudi Hizbullah, the group that bombed Khobar Towers, the American barracks in Saudi Arabia, got support from Iran.

Around 1997, members of the intelligence community—and others, like Richard Clarke—began focusing on a Saudi man, Osama bin Laden, who they realized was the financier and leader of a new group, Al Qaeda. Few in government shared their concern. In 1997 Al Qaeda was not confirmed to have executed a single terrorist attack against Americans. "Employees in the government told us that they felt their zeal attracted ridicule from their peers," the commission's report on intelligence says.

In due course, some senior officials in the Clinton administration awakened to the threat: CIA Director George Tenet, national-security adviser Sandy Berger and Clinton himself. But they never proposed a full-fledged assault on it. Their one dramatic attack—bombing the Afghan terror camps and Sudanese factory in 1998—proved unsuccessful and led to domestic criticism, and they did not think they could do something more ambitious. The Pentagon, which comes off poorly in the commission reports, was stubbornly unwilling to provide aggressive and creative options.

The Bush team, distrustful of anything Clinton's people said, did not see Al Qaeda as an urgent threat. They held few meetings on it and in other ways were inattentive to it. One example from the panel's report: the senior Pentagon official responsible for counterterrorism is the assistant secretary for special operations and low-intensity conflict. Even by September 11, 2001, no one had been appointed to that post. The Bush administration came to office with different concerns. During the 1990s conservative intellectuals and policy wonks sounded the alarm about China, North Korea, Cuba, Iran and Iraq, but not about terror. Real men dealt with states.

The danger is less that a state will sponsor a terror group and more that a terror group will sponsor a state—as happened in Afghanistan

Even after 9/11, many in the administration wanted to focus on states. Bush spoke out against countries that "harbor" terrorists. Two days after the attacks, Paul Wolfowitz proposed "ending states that sponsor terrorism." Beyond Iraq, conservative intellectuals like Richard Perle and Michael Ledeen insist that the real source of terror remains the "terror masters," meaning states like Iran and Syria.

I asked an American official closely involved with counterterrorism about state sponsorship. He replied, "Well, all that's left is Iran and to a lesser extent Syria, and it's mostly directed against Israel. States have been getting out of the terror business since the late 1980s. We have kept many governments on the list of state sponsors for political reasons. The reality is that the terror we face is mostly unconnected to states." Today's terrorists are harbored in countries like Spain and Germany—entirely unintentionally. They draw on support not from states but private individuals—Saudi millionaires, Egyptian radicals, Yemenite preachers.

Afghanistan housed Al Qaeda, and thus it was crucial to attack the country. But that was less a case of a state's sponsoring a terror group and more one of a terror group's sponsoring a state. Consider the situation today. Al Qaeda has lost its base in Afghanistan, two thirds of its leaders have been captured or killed, its funds are being frozen. And yet terror attacks mount from Indonesia to Casablanca to Spain. "These attacks are not being directed by Al Qaeda. They are being inspired by it," the official told me. "I'm not even sure it makes sense to speak of Al Qaeda because it conveys the image of a single, if decentralized, group. In fact, these are all different, local groups that have in common only ideology and enemies."

This is the new face of terror: dozens of local groups across the world connected by a global ideology. Next week I will explain how best to tackle this threat. But first we need to see it for what it is.

Write the author at comments@fareedzakaria.com.

Guerrilla Nation

The arrest of FARC terrorist Ricardo Granda sheds new light on Hugo Chavez's ongoing support of terrorism.

Thor Halvorssen

SIMON TRINIDAD is the *nom de guerre* of Ricardo Palmera, a high-ranking terrorist of the Fuerzas Armadas Revolucionarias de Colombia (FARC), the deadliest and largest terrorist organization in the world. Thanks to Colombia's president, Alvaro Uribe, Trinidad was extradited to the United States last month. He now awaits trial for a lengthy list of crimes involving the recent kidnapping and murder of American citizens in Colombia. Trinidad's capture was a victory in the fight against global terror (see **Note**), but it is unlikely that the FARC terrorists will be defeated as long as Venezuelan president Hugo Chavez continues to use his government to harbor, equip, and protect them.

Since assuming the presidency of Venezuela in 1999, Lieutenant Colonel Hugo Chavez has often sympathized with global terrorism. Not only has he proclaimed his "brotherhood" with Saddam Hussein and bestowed kind words on the Taliban, but he also maintains close economic and diplomatic ties with the leaders of Iran and Libya. Moreover, President Chavez is increasingly identified with the FARC terrorists. Although the full extent of Chavez's involvement with FARC is unknown, he has been accused of everything from sympathizing with the group to providing it with weapons and monetary support. The allegations against Chavez are numerous and it is likely that some of them are either exaggerated or untrue. Even so, President Chavez's activities reveal a consistent pattern of sympathy for terrorists.

The FARC terrorist group has been fighting the democratic government of Colombia for almost 40 years. Founded as the armed wing of the Colombian Communist party, this 16,000-strong terrorist force recruits children and funds its activities with billions of dollars collected as taxes on the cocaine trade. The group's explicit objective is to take Colombia by force. In pursuing its mission, FARC terrorists have kidnapped, extorted, and executed thousands of innocent civilians, bombed buildings, assassinated hundreds of political leaders, and, with two other local terrorist organizations, have turned Colombia into one of the most violent and dangerous countries in the world. All in all, FARC has caused the deaths of more than 100,000 people.

The U.S. Department of State has designated FARC a Foreign Terrorist Organization—yet FARC leaders are welcomed in Venezuela and treated as heads of state. The prominent FARC leader Olga Marin, for example, spoke on the floor of Venezuela's National Assembly in the summer of 2000, praising Hugo Chavez as a hero of the rebel movement and thanking the Venezuelan government for its "support." Weeks later, the Colombian government announced that it had confiscated from terrorists more than 400 rifles and machine guns bearing the insignia of the Venezuelan armed forces. Although President Chavez claimed this was a smear campaign against him and that many of those weapons could have come into terrorist hands as a result of border skirmishes with Venezuelan armed forces, his explanation was less than plausible, since some of the guns had sequential serial numbers and were therefore likely part of a unified arms shipment.

In February 2001, months after the Chavez government denied supporting FARC, the capture of a Colombian terrorist revived the debate. Jose Maria Ballestas, a leader of Colombia's other left-wing terrorist organization, the National Liberation Army (ELN), was captured in Venezuela's capital by Interpol operatives working in conjunction with the Colombian police. Although Ballestas was wanted for a 1999 commercial airliner hijacking, he was immediately released from custody by order of the Chavez government. As the Colombian media cried foul, Chavez officials denied that Ballestas had ever been arrested and claimed that "news" of his arrest was actually a story concocted by enemies of the Chavez government. When Colombian officials responded by releasing a video of the arrest, the Chavez government tried to claim that Ballestas was seeking asylum from political persecution in Colombia. As diplomatic tension reached a fever-pitch, Venezuela re-arrested Ballestas and grudgingly extradited him to Colombia.

Seeking to repair relations with Colombia's president, President Chavez paid a state visit to Colombia in May 2001. While there, he allowed a FARC associate, Diego Serna, to serve as his personal bodyguard. Serna was arrested months later and told the magazine Cambio (published by Nobel laureate Gabriel Garcia Marquez) that President Chavez was in constant and secret touch with the FARC leadership. Serna remarked that in Colombian television broadcasts of the presidential summit "you can see not only our closeness, but also the confidence and the comments he made to me on various occasions." Indeed, the footage shows Chavez laughing, jostling, and whispering in Serna's ear.

Three months after diplomatic tension over the Serna incident died down, the Chavez-FARC connection surfaced again when Venezuela's intelligence chief, Jesus Urdaneta, publicly denounced Chavez for supporting FARC. A lifelong friend and military colleague of President Chavez, Urdaneta publicized documents showing that the Chavez government offered fuel, money, and other support to the terrorists. The documents included signed letters from a Chavez aide detailing an agreement to provide support for FARC. That aide later became Chavez's minister of justice, a position which gave him oversight of the entire Venezuelan security apparatus.

Less than a week after Urdaneta went public, a group of female journalists released a video showing meetings between Venezuelan military leaders and FARC guerilla commanders. The next day, hundreds of miles away, the Colombian Air Force captured a Venezuelan plane loaded with ammunition. Colombian intelligence established that the supplies were meant for the FARC terrorists.

THE COLOMBIAN GOVERNMENT is currently embroiled in the most momentous FARC-related matter since Simon Trinidad's extradition. On December 14, 2004, Ricardo Granda, widely known as FARC's "foreign secretary," was arrested on the Colombian border. One of the most senior, well connected, and highly skilled political strategists in FARC's history, Granda had been living in Venezuela's capital.

In Caracas Granda enjoyed Venezuelan citizenship (granted by government decree), took advantage of state-supplied protection, and even, on December 8, participated in a government-sponsored networking conference attended by Chavez, Daniel Ortega, and other revolutionary socialists. Today, Chavez expresses fury that Granda was captured, lamenting that Granda was apprehended in Caracas, stuffed in the trunk of a car, and driven to Colombia where he was then given to Colombian authorities by junior Venezuelan military and police officers working for cash rewards. The Venezuelan government has announced it will issue arrest warrants for the Colombian Defense secretary and for the Colombian attorney general, who are to be charged with "kidnapping."

THE COLOMBIAN GOVERNMENT has understandably become exasperated by the impunity with which Chavez has permitted terrorists to use Venezuela as a safe haven and justifies its actions by claiming that the United Nations forbids members to harbor terrorists in either an "active or passive" manner. Last week the Colombian foreign secretary went public with a list of senior FARC terrorists living in Venezuela.

Thus far, the U.S. State Department has been exceedingly tame with the Venezuelan government. Perhaps the Granda case will spur the new secretary of state to focus more on terrorist threats plaguing our own hemisphere. Should she do so, she will effect a necessary and long overdue shift in U.S.-Venezuela relations.

Thor Halvorssen, *a human rights and civil liberties advocate, is First Amendment Scholar at the Commonwealth Foundation. He lives in New York.*

NOTE: FARC terrorist Simon Trinidad's indictment last month includes information about the murder and kidnapping of American citizens in Colombia last year. Trinidad's actions were not exceptional; killing Americans is routine for FARC. For example, in 1999 FARC terrorists killed three American activists who were in Colombia on a humanitarian mission. They were Terence Freitas, 24; Ingrid Washinowatok, 41; and Lahe'ena'e Gay, 39.

Apprehended after attending a religious ceremony on an Indian reservation, Freitas, Washinowatok, and Gay were initially held for ransom but were later taken into Venezuela and executed in cold blood. Washinowatok, a New Yorker, was the head of the Fund for Four Directions, a Rockefeller-supported charity which helps indigenous peoples. Lahe'ena'e Gay was an award-winning Hawaiian photographer. Terry Freitas was an environmental activist from California. All three progressive activists had colorful life stories. Washinowatok, for example, was a Menominee Indian from Minnesota, daughter of a tribal chieftain, and personal friend of Nobel Peace Prize laureate Rigoberta Menchu. She studied in Havana and is described by her friends as a champion of the oppressed. Her lifeless body, found just inside the Venezuelan border, was impossible to identify since her face had been destroyed by gunshot. The autopsy revealed that she had been forced to march barefoot through the jungle for several days despite having been bitten by a poisonous spider. She was only identified when her foundation's American Express card was found hidden in her clothing. Washinowatok and her friends were executed for one chilling reason: They were Americans.

UNIT 4

International Terrorism

Unit Selections

Key Points to Consider

- What conditions contributed to the emergence of extremist groups in Egypt? What lessons can be learned from the Egyptian example?
- What are the causes of the current crisis in Colombia? What impact have U.S. policies had on Colombia's problems?
- How can the Russian government end Chechen terrorism?
- Will Batasuna's political proposals end the ongoing conflict in Spain? Why or why not?

Student Website

www.mhcls.com/online

Internet References

Further information regarding these websites may be found in this book's preface or online.

Arab.Net Contents
http://www.arab.net/sections/contents.html

International Association for Counterterrorism and Security Professionals
http://www.iacsp.com/index.html

The International Policy Institute for Counter-Terrorism
http://www.ict.org.il

International Rescue Committee
http://www.intrescom.org

United Nations Website on Terrorism
http://www.un.org/terrorism/

United States Institute of Peace
http://www.usip.org/library/topics/terrorism.html

International terrorism has changed significantly. Simply said, it has become more complex. Increased organizational complexity, improved communications, and an increased willingness to cause mass casualties pose new challenges for the international community.

Individuals and small groups dominated international terrorism in the 1970s. Larger groups and organizations played a critical role in international terrorism in the 1980s. More complex multinational terrorist networks emerged in the 1990s. More recently, small independent groups of individuals—of local origin—have emerged to carry out attacks in their home countries in support of broad global movements. As we enter the 21st century all four generations and levels of organizational structure appear to exist. Sometimes terrorists act locally or regionally to pursue independent agendas. At other times they take advantage of cross-national links to obtain greater access to weapons, training, or financial resources. On occasion, they may even temporarily set aside local interests and objectives, to cooperate within loosely connected international networks to pursue broader ideological agendas. At a given point in time international terrorists may appear to be engaged in activities at all levels posing unique challenges to those engaged in the study of international terrorism.

Modern communications technologies have changed the way international terrorists operate. The cellphone and the laptop computer have become as important as the bomb and the AK-47 in the terrorist arsenal. The Internet has provided terrorists with instant access to global communications. It enhances their ability to exchange information and provides them with an effective vehicle to rally their supporters. Almost all major international terrorist organizations operate their own web sites and communicate via the Internet.

A particularly disturbing trend in contemporary international terrorism is the increased willingness, by some terrorists, to cause mass casualties. While the potential causes of this trend are subject to debate, this trend has again elevated terrorism to the top of the international agenda. While over the past several decades, the number of international terrorist incidents has declined, the casualties caused by international terrorism have steadily increased. More importantly this trend has focused international attention on terrorist methods deemed unlikely only a few years ago. Potential threats posed by biological, chemical, or radiological weapons are again at the forefront of international concern.

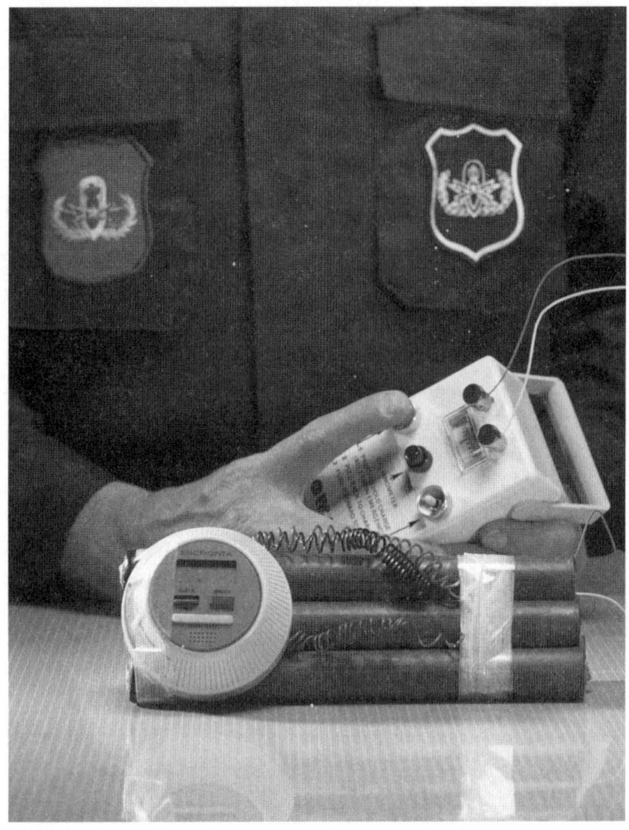

The selections in this unit reflect some of the diversity in international terrorism. The first article traces the development of Islamic extremism in Egypt from the Muslim Brotherhood to al-Qaeda. It contends that there are important lessons to be learned from the Egyptian experience. The second article focuses on the shift in U.S. policy in Colombia "from counternarcotics to counterterrorism." The article argues that U.S. policy may have made "an already grave situation" even worse. In the third article Ben-Meir argues that as long as Russia continues to ignore the root causes of Chechen terrorism, there is no chance of diminishing or eliminating it. Finally, Meredith Moore maintains that despite calls for peace, inaction on both sides will lead to a continuation of Basque violence in Spain.

Extremist Groups in Egypt

On 11 September, terrorism became a much greater reality for Americans and much of the rest of the world. Indeed, that date marks the beginning of a new era for the global community, an era that may be called the Violent New World Order, or the Age of Fear. Since the events of 11 September, the United States has been working to build a coalition against terrorism consisting of countries throughout the world. More recently, the United States launched attacks on positions inside Afghanistan in order to wipe out Osama bin Laden's al-Qaeda organization and assisted the Northern Alliance in the overthrow of the Taliban government that had protected al-Qaeda. Given the military prowess of the United States and its allies, these acts have largely achieved their short-term goals. Over the long term, however, the United States, its allies, those countries that have been breeding grounds for terrorist organizations and those whose citizens sympathize with terrorist organizations need to look deeper at the causes of terrorism. On the surface, the religious zeal associated with the most prolific terrorist organizations appears to be something with which the United States and its allies cannot negotiate. The frustrations that drive people to acts of terror, however, are often rooted in adverse socio-economic conditions as well as cultural and political tensions that need to be addressed by underdeveloped nations and the larger international community. Since the deserts of Egypt gave birth to the rise of the first Islamic militant organizations, the Egyptian experience provides a perspective.

JEFFREY A. NEDOROSCIK

In the early 1990s, Egypt was gripped by years of terrorist acts by radical Islamic groups. This violence helped cause the demise of the country's economy, which relies on tourism for hard currency, and sent the government scrambling for control. The Western press largely dismissed these acts as those of religious fanatics with whom it is impossible to reason or negotiate. Rather, according to traditional thought, they needed to be eliminated. In a project that spanned two continents, however, I joined some of my colleagues in investigating the roots of the violent internal conflict that was growing stronger in Egypt.

The foundation for our project rested on the theory that there could be lessons learned about the roots of violent internal conflict in underdeveloped nations by comparing conflicts such as those in Mexico and Egypt.[1] A workshop in Cairo in 1996 took a deeper look into these two conflicts to understand what they had in common. Three hypotheses emerged from our investigation. The first and primary hypothesis was that violent internal conflict results from socio-economic, cultural and political tensions. These tensions are exacerbated by the mobilizing efforts of anti-status quo activists, as well as by the impact and perceived impact of economic liberalization and structural adjustment policies. There are various sub-hypotheses:

a. The actors in the conflict stretch across four intricately linked levels: the locus of the conflict, the institutional system, civil society and the international environment. The locus of the conflict is the geographic area from which the conflict has erupted or a group of people who, for various reasons, are somehow separated from mainstream society and are in a less advantageous position socially and economically than their countrymen.
b. The locus is in a situation where it is being marginalized by the institutional system.
c. Civil society demonstrates sympathetic tendencies towards the anti-status quo group.
d. The international environment strongly influences the response of the government towards the conflict.

This workshop led to further investigation of violent internal conflict. Again, the efforts focused on the conflict situations in Egypt and Mexico. In Egypt, it soon became clear that the movement that had caused acts of terror throughout the country was not simply one based on religious extremism. Rather, this movement grew out of the socio-economic conditions as well as the cultural and political tensions existing for the poorest of Egypt's poor.

The militant groups carrying out these acts of terror were from Upper Egypt, and the movement was very region-biased as they viewed other Islamist groups, largely from the north, as having failed to address what they saw as the immediate issues at hand—the dire socio-economic conditions faced by their people each day. While other groups tended to look beyond Egypt's borders at issues such as Pan-Arabism and the liberation of Palestine as priorities, Upper Egyptian Islamists looked homeward at the dismal socio-economic conditions of the region and the policies of the government in power that perpetuated the status quo. This was evident in the assassination of President Anwar Sadat at the hands of Upper Egyptian Islamists as he attempted to roll back reforms that President Nasser had set in place that had benefited Upper Egyptians.

Upper Egypt has historically been isolated, geographically, politically and economically, from the rest of Egypt. Until recently, there was little effort at development in Upper Egypt, by either the Government of Egypt or the international donor community, as efforts focused on Cairo and the Nile Delta region. As a result, Upper Egypt has soaring rates of unemployment, illiteracy, malnutrition, morbidity and mortality. Upper Egypt ranks at the bottom of virtually every social and economic indicator for Egypt and has the highest percentage of ultra-poor—those who survive on a day-to-day basis. As such, it served as a perfect breeding ground for militant activity.

The rise of low-intensity conflict in Egypt culminated in November 1997, when 58 tourists were brutally killed in Luxor, Upper Egypt. This attack was attributed to the southern militant movement al-Gama'a al-Islamiyya, one of the organizations included on the US Government's list of 'terrorist organizations' later released following the 11 September attacks. In recent years, al-Gama'a al-Islamiyya is said to have increasingly close ties to Osama bin Laden as he sought to create a worldwide terrorist network.

The attack in Luxor succeeded in devastating Egypt's economy by scaring away tourists and their much-needed hard currency. This attack was the pinnacle of the movement that began its terrorist operations with vigor in 1992, when anti-status quo groups such as al-Gama'a al-Islamiyya began targeting Egypt's secular government. During the height of fundamentalist activity in Egypt (between April 1992 and October 1993), some 222 people were killed in anti-status quo violence, including 66 members of government security forces, 76 'terrorists' and 6 foreign tourists.

The Egyptian government has mostly fought violence with violence. Whereas they have succeeded in keeping the movement under control in recent years, the root causes of the Upper Egyptian movement continue to exist: the bleak socio-economic conditions characterizing the lives of most Upper Egyptians. As a result, Upper Egypt has become the breeding ground for the fundamentalist movement and is the area where the movement receives its most widespread support. Still, the media, the Egyptian government and the international community focused on the violence itself and largely ignored its root causes.

The fundamentalist groups that have grown up in Egypt trace their roots to an ideology that dates to the immediate post-independence period and in the creation of the Muslim Brotherhood in the 1920s. Although the Brotherhood has historically been a more mainstream organization that operated within the established institutional system, they have had a profound influence on the groups that are active today.

The Muslim Brotherhood in Egypt

The Muslim Brotherhood was founded in 1928 by Hasan al-Banna. Al-Banna was born in the Nile Delta in 1906, the son of an Islamic scholar. He grew up in the Delta province of al-Buhayra before attending the teacher's college in Cairo. There, al-Banna learned the concept of Islam as a self-sufficient ideology and became aware of the 'dangers' of Westernization.

A rural-to-urban migrant, al-Banna was initially shocked by life in Cairo. In his memoirs he wrote,

> Young men were lost, and the educated were in a state of doubt and confusion... I saw that the social life of the beloved Egyptian nation was oscillating between her dear and precious Islamism which she had inherited, defended, lived with and become accustomed to, and made powerful during thirteen centuries, and this severe western invasion which is armed and equipped with all destructive and degenerative influences of money, wealth, prestige, ostentation, material enjoyment, power and means of propaganda.[2]

Al-Banna began to think of all of the problems that Egypt faced as a result of the influence of the West and the straying of Egyptians from the straight path of Islam.

In 1927, al-Banna began teaching at a government primary school in Ismaliya. There, he came into daily contact with British soldiers and he found their presence to be offensive. He began to form the goal of ridding Egypt of this foreign occupation.

The following year, al-Banna established the Muslim Brotherhood, predicated on radical nationalist ideas and religious practices. At first, the group was simply an Islamic revivalist movement. Quickly, however, the group developed a political orientation that was very anti-Western and opposed to secular politics. An underground paramilitary wing was also established, primarily to fight against the British occupation of Egypt. It was also active, however, in attacking Jewish interests in Egypt as well as government figures.

The Muslim Brotherhood has often been called the first wide-ranging, organized and international Islamic movement of modern times. The group's message: struggle to rid Egypt of foreign occupation and defend and obey Islam. Its slogan: 'God is our purpose, the Prophet our leader, the Qur'an our constitution, Jihad our way and dying for God's cause our supreme objective.' Its banner: two crossing swords, a copy of the Holy Qur'an and the word 'prepare'.

The Brotherhood recognized early on the power of education in spreading its message and engaged in many educational activities. The group established primary and secondary schools for boys and girls as well as technical schools for workers.

There were also Qur'anic classes and basic skill classes for the illiterate. Many of the Brotherhood's members were recruited from among the students of these institutions. In her book, *Egypt from Independence to Revolution,* historian Selma Botman commented that,

> From the beginning of its organization, the Muslim Brotherhood took the recruitment of students seriously and considered them the organization's 'striking force'. Student members were organized in 'families' which met regularly every week in the house of one of them to study an Islamic educative syllabus specially designed to suit their age group and educational level.[3]

The Brotherhood has also been active in encouraging 'appropriate' social and economic reform in Egypt. As part of their programme, they set up urban projects that provided jobs for the unemployed and poor and set up industrial and commercial enterprises that could compete with Egypt's non-Islamic entrepreneurs.

The Brotherhood drew much of its support from recent rural-to-urban migrants who had found little satisfaction in Cairo. To them, the Brotherhood offered associations that embraced each member as a family. Soon, however, support for the Brothers grew beyond students and the urban poor to the middle classes, who were equally disenchanted with the government, the economy and the continued presence of foreigners on Egypt's soil.

The Depression of 1929 and the economic distress experienced by Egyptians contributed to the growing popularity of the Brotherhood and its message. By the time of the Second World War, Brotherhood membership was estimated to be anywhere from hundreds of thousands to greater than one million members. These included government employees, students, policemen, lawyers and soldiers, as well as the urban and rural poor. John Esposito wrote that, 'The Muslim Brotherhood grew at a time when the Egyptian community was in crisis. They blamed European imperialism and Westernized Muslim leadership for many of the problems Egypt faced.'[4] He later continued, 'In the 1930s and 1940s, the Muslim Brotherhood reasserted the relevance of Islam to all areas of life, diagnosed the ills of Muslim society, and offered an Islamic activism aimed at redressing issues of religious identity and social justice.'[5]

The Brotherhood was strongest in the mid to late 1940s, when its influence on lower- and middle-class Egyptians was considerable. The Brotherhood organized numerous demonstrations, marches and protests during this period and acted as a counterweight to the Communists and Wafdists. These organizations were secular and envisioned a more democratic, constitutional and cosmopolitan society. In contrast, the Brotherhood spread a message of Islam as a self-sufficient, all-encompassing way of life, an alternative to Marxism and Western capitalism. John Esposito argued that, 'the message of the Brotherhood... was the conviction that Islam provided a divinely revealed and prescribed third alternative to Western capitalism and Soviet Marxism'.[6] The Brotherhood told people to turn to religion, as opposed to political parties, for answers.

The Brotherhood was frequently accused of sabotaging meetings of the Communists and Wafdists, precipitating clashes in public, damaging property and carrying out political assassinations. According to Botman, 'the Brotherhood advocated a militancy that went far beyond the imaginations of Egypt's established political leaders'.[7]

It is clear that the Brotherhood was hostile to the West. They saw Egypt as dependent on the West, politically ineffectual and socially and culturally weak. They viewed Westernization as threatening not only independence, but the identity and way of life of Muslims. Still, the Brothers made a distinction between Westernization and modernization. Esposito explains, 'They engaged in modern organization and institution building, provided educational and social welfare services, and used modern technology and mass communications to spread their message and mobilize popular support.'[8]

At the same time that the Brotherhood condemned the West, they realized that the predicament that Egypt was in was foremost a Muslim problem, that is, the result of Muslims failing adequately to observe their religion. Indeed, 'Rebuilding the community and redressing the balance of power between Islam and the West must begin with a call or invitation (*dawa*) to all Muslims to return to and appropriate their faith in its fullness—to be born again in the straight path of God.'[9]

After the Second World War, the Brotherhood continued to be active in education and enjoyed enormous popularity on university campuses. The group continued to recruit heavily among students, who joined their ranks in large numbers.

At the same time that students were being recruited, the Brotherhood was also recruiting workers. Al-Banna's interest here was twofold: to save all Muslims and to protect Egyptians from the dangers of having foreigners controlling Egypt's economy. To help carry out its interest in this area, the Brotherhood formed labour unions among workers in many different trades. In its union activities, the Brotherhood spread its fundamentalist ideas among workers. The Brotherhood also hoped that these activities would weaken the force of the Communists and Wafdists in labour affairs and would demonstrate their commitment to the average wage-labourer.

The Brothers also looked beyond Egypt's borders and became deeply concerned with and involved in the struggle in Palestine. The group collected money and arms, trained volunteers and sent troops to Palestine in 1948 to fight alongside the Palestinians and other Arab soldiers. By most accounts, the participation of many Brothers in the effort to liberate Palestine was an event that led to the strengthening of a growing paramilitary wing of the Brotherhood. In this respect, Palestine served as a training ground for Brotherhood militants.

The Palestine conflict made the growing power and popularity of the Brotherhood all too obvious. The Egyptian government began to fear the growing strength of the Brotherhood and recognized the danger the group posed to state security. During the political tension of the Palestine conflict, Egyptian Prime Minister Mahmoud al-Nuqrashi imposed military law and outlawed the Muslim Brotherhood in December 1948. He claimed that the group had plotted revolution against the government and had repeatedly carried out terrorist attacks on individuals.

In retaliation to the government's actions, an angry member of the Brotherhood assassinated al-Nuqrashi. The government did not sit by idly, however. Revenge was sought after: as the 'Supreme Guide' of the Brotherhood, al-Banna was also murdered in 1949, presumably by a member of the government's police forces.

The Brotherhood movement was forced underground as it tried to regroup after the loss of its founder and leader. After the Wafd regained power in 1951, the Brotherhood was temporarily allowed to resume activities. They were restricted, however, to cultural, social and spiritual services. The group soon resumed banned activities, however, and was an active presence in the guerrilla war being fought against British occupation in the Suez Canal area.

Sharing many of the same goals, the Muslim Brotherhood gave their support to the Free Officers' movement in the early 1950s in its attempt to overthrow the government. The Officers succeeded with the Brothers' help and set up a government under the leadership of Gamal Abdul Nasser. In doing so, however, the revolutionaries cut the Brotherhood out of any leadership role. Afaf Lutfi Marsot wrote that,

> They [the Muslim Brothers] had helped the Free Officers come to power and had expected a share in the government of the country. But once in power, the Officers saw no need to associate the Brethren with their government, especially since the Brethren possessed a massive, popular power base, and the officers had no power base yet.[10]

In October 1954, an assassination attempt was carried out against Nasser. Popular thought was that the Brotherhood was responsible. As a result, the Brotherhood was banned by the government, and many of the Brothers were arrested, imprisoned and tortured. The remaining members of the organization were forced underground again.

The late 1950s and the 1960s saw an increased militancy among the Brotherhood as a consequence of the group's hostility toward the Egyptian government. This new trend of radical Islam was led by Muslim Brotherhood member Sayyid Qutb (1906–66). Qutb took the ideological writings and beliefs of al-Banna and built them into a revolutionary call to *jihad*. Esposito attested that, during the 1950s,

> Qutb emerged as a major voice of the Muslim Brotherhood and its most influential ideologue. His commitment, intelligence, militancy, and literary style made him especially effective within the context of the growing confrontation between a repressive regime and the Brotherhood. Government harassment of the Brotherhood and Qutb's imprisonment and torture in 1954 for alleged involvement in an attempt to assassinate Nasser only increased his radicalization and confrontational worldview.[11]

Qutb's major work, *Signposts on the Road,* is said to mark the beginnings of Islamic fundamentalism. Qutb, an Upper Egyptian, wrote about the concept of *jahiliyya.* He wrote that there exist two kinds of societies: Muslim and *jahiliyya.* In Muslim societies, Islam is fully applied across the board. In *jahiliyya* societies, it is not applied. He went on to say that even if its members proclaim themselves as Muslim, they are not if their legislation does not have divine law as a basis. This statement was made in order to point out that the government of Nasser was *jahiliyya.* Qutb then went on to describe how the *jahiliyya* government must be destroyed so that an Islamic state could be established in its place. Qutb called for a revolution under the vanguard of the *umma* (the community of believers). Qutb called for a *jihad* and said that this holy war could not be waged by words alone. Thus, it was under Qutb that the Muslim Brotherhood began to splinter into moderate and radical factions. Both factions retained the same objective (that is, the creation of an Islamic system of government in Egypt) but disagreed on the means of attaining this objective. More moderate and traditional members of the Brotherhood felt that change could be brought about by working within the existing political system. Radical members, however, felt that attempts at change within the system were futile (given the authoritarian nature of the existing government). These Brothers, inspired by the writings and leadership of Qutb, felt that the only option was armed struggle.

The two different paths taken by the moderate and radical factions—evolutionary and revolutionary—are described as thus,

> Qutb's formulation became the starting point for many radical groups. The two options—evolution, a process which emphasizes revolutionary change from below, and revolution, the violent overthrow of established (un-Islamic) systems of government—have remained the twin paths of contemporary Islamic movement.[12]

In 1965, the government clamped down on the Muslim Brotherhood once again. Many Brotherhood leaders, including Qutb, were arrested and executed. Thousands of other Brotherhood members were arrested and tortured. Those who escaped the grasp of the government went underground or left the country. Those who went underground were again faced with the task of rebuilding the organization while consciously avoiding confrontation with the government.

The Brotherhood was seen by many as a broken and harmless organization in the early years of the Sadat presidency. In fact, Sadat released the Brotherhood's Supreme Guide, Omar Tilmassani, from prison and formed a somewhat co-operative relationship with the group. Taking advantage of this relative freedom, the Brotherhood was soon preaching and publishing its message once again, as well as setting up institutions for its social welfare and financial activities.

At the same time, more radical factions had gone underground and were planning violent activities aimed at overthrowing the government. They criticized the Brotherhood as a 'has been' organization that had allowed itself to be co-opted by the Egyptian government. These factions received great inspiration in the 1970s from the Islamic Revolution in Iran. Sociologist Saad Eddin Ibrahim (recently jailed for criticizing the

Egyptian government on human rights practices) interviewed a number of jailed militants during this period. Although they demonstrated a respect for al-Banna and other Brotherhood leaders, they criticized the state of the post-1965 Brothers. He wrote,

> But these militants took some exception to the current practices of surviving members of the Brotherhood. They consider some of the surviving members as weak and 'burned out' or bought off. Some of the early members of the Military Academy group reported having gone to visit older members of the Brotherhood, to seek advice and offer support. They were advised to mind their individual businesses, to stay out of trouble, and to worship God. Quite disillusioned, the youngsters then decided to form their own organization.[13]

Still, the Brotherhood continued to attract members from the lower and middle classes and began to rebuild its power base. They quietly criticized the government but strongly and consistently denounced violence and criticized the armed struggle of the radical Islamic groups.

> It [the Brotherhood] clearly opted for socio-political change through a policy of moderation and gradualism which accepted political pluralism and parliamentary democracy, entering into political alliances with secular political parties and organizations as well as acknowledging the rights of Coptic Christians... Though sympathetic to many of the concerns of extremist groups, it remained steadfast in its rejection of violence and terrorism and stayed scrupulously within the limits of Egyptian law.[14]

Following this trend, the 1980s and 1990s have witnessed the Muslim Brotherhood's return as a major force in Egyptian society and politics, having rebuilt its wealth and power. The group gathered such institutions among its assets as banks, schools, factories and mass communication organizations. In addition, the Brotherhood became a major force in professional organizations. In the 1990s, they gained control of the powerful doctors', lawyers' and engineers' syndicates. They also returned as a force in university politics.

At the national level, although the Brotherhood could not compete in elections as a political party, it joined forces in 1984 with the Wafd party. In 1987, it formed an 'Islamic Alliance' with the Labour Party and won 17 per cent of the general vote, making the coalition the major opposition to President Hosni Mubarak's National Democratic Party.

Since 1993, Mubarak's government has become increasingly aware of the Brotherhood's growing power base and has sought to break its hold over professional syndicates and university politics. Law 100 of 1993 made syndicate elections valid only if winners received greater than a 50 per cent quorum of syndicate member voters, a plateau that is rarely achieved. In the event of receiving less than 50 per cent, the government reserves for itself the right to appoint people to the most powerful positions in the syndicate's legislative bodies. The government has also undertaken audits of Brotherhood-controlled syndicates, accusing the Brotherhood leadership of financial irregularities. Likewise, in university elections, the names of Islamist candidates have been removed from election lists in an effort to break their hold over student unions.

General elections were held in November 1995. Unlike previous elections, candidates were allowed to run for office as independents. In the months leading up to the election, the Brotherhood announced that it would field some 150 candidates. Many of these candidates and their supporters, however, ended up watching election results from behind bars, as the government once again clamped down on the Brotherhood in three sets of arrests carried out over six weeks in September and October. Seventy-nine members of the Brotherhood were ordered to stand trial in military court for 'increasing their activity in the Muslim Brotherhood organization and provoking citizens against the government, inciting hatred for it and claiming that it strives to strike at the Islamist movement in the country on the advice of foreign powers'.[15] Human rights activist Hesham Kassem denounced the government's actions, saying, 'it is a direct reaction to the declaration by the Muslim Brotherhood to field 150 candidates in the elections'.[16]

Even with many of its leaders and its most efficient organizers in prison, the Brotherhood still succeeded in fielding 150 candidates in the elections, competing in 17 governorates. This marked the first time in the organization's history that they had such a large number of candidates and covered such a large geographical area. Hisham Mubarak, of the Centre for Human Rights Legal Aid, commented on the Brotherhood's presence in the election. He said, 'It proves that it [the Brotherhood] is a big political force, that it can stand the blows from security, and that it has not lost its political viability.'[17]

1996 brought other changes and challenges for the Muslim Brotherhood as well. In January, the group's Supreme Guide, Mohamed Hamed Aboul Nasser, died after a long battle with illness. In a swift transfer of power, Mustafa Mashour was named as Aboul Nasser's successor on the same day of his death. Mashour has effectively exercised leadership of the organization for the previous two years while Aboul Nasser's health was fading. Therefore, radical changes in the organization's leadership and activities did not occur. The new leader reiterated the Brotherhood's commitment to 'moderate, non-violent, non-extremist' activities. He stated in an interview, 'We will not turn to violence and we hope that God will allow an opening between us and the government to put an end to injustice of which we have been victims.'[18]

Another challenge for the Brotherhood came with the announcement that a group of younger Brotherhood members had officially requested to establish a new political party called al-Wasat (Centre Party). The group, which also includes women and Christians on its slate, was led by the deputy secretary-general of the engineers' syndicate, Abul Ella Madi. Although the request received limited attention in the mass media, one journalist claimed that it 'represents the first open move by the young generation of syndicate activists in the Muslim Brotherhood to break with the aging leadership of the organization who have opted to lie low while the state rages against them'.[19]

Islamic Militant Groups in Egypt

As demonstrated above, armed Islamic militant groups in Egypt grew out of the Muslim Brotherhood and were a response to the Brotherhood's increasing moderation in the face of confrontation with the state. Former Egyptian minister of the interior General Hasan al-Alfi (fired after the 1997 massacre in Luxor) has stated that,

> If we go back to the 1940s we realize that the Muslim Brothers are the root group to all these [militant] organizations. If we look at the leaders of the radical terrorists groups—they are all defected Muslim Brotherhood members, there is a link—not a direct one but an indirect one.[20]

Both the Muslim Brotherhood and militant groups share a common vision for Egypt: the establishment of an Islamic state based on the principles of *shari'a.* They disagree, however, on the means of achieving this goal. Still, their common vision and historical ties leave them intricately connected and sympathetic to each other's causes.

The Muslim Brotherhood had an underground paramilitary group as early on as the late 1930s. It was particularly active in the 1940s, carrying out major violent activities aimed at British forces, Jewish businesses and government officials. Amira Howeidy wrote that,

> In just one week in 1946, four attacks, in which guns and explosives were used, were directed at British occupation forces, wounding 128 people. A group of Brotherhood figures were put on trial and found guilty by judge Ahmed El-Khazindar. Eight months later, the judge was assassinated by two Brotherhood members.[21]

The paramilitary group also bombed several Jewish-owned businesses in Cairo in 1947 and 1948.

After the Brotherhood was outlawed in 1948 by Prime Minister Mahmoud El-Naqrashi, the paramilitary wing responded by assassinating the prime minister inside the interior ministry building. At this point, Hassan al-Banna appeared to realize that the paramilitary wing was raging out of control. He denounced the assassination and declared that the members who had carried it out were 'neither Brothers nor Muslims'.[22]

Al-Banna was assassinated in 1949. Without his voice of moderation, militancy increased among the Brotherhood's ranks as it began its split into competing factions. Esposito explained that,

> Though ostensibly one organization, the Muslim Brothers had in fact split into several factors. After al-Banna's death, no single leader enjoyed his authority. While radical rhetoric appealed to a more militant wing, especially disaffected youth, many other Brothers, including their leader or Supreme Guide, were wary of direct confrontation with the regime. The political strategy of the Brotherhood remained unresolved, with two competing models, evolution and revolution. While some Brothers conspired to overthrow the government, many of the older guard, fully aware of the power of the state, preferred to pursue change through preaching and social activism.[23]

After government crackdowns on the Brotherhood in the 1950s and 1960s, it became evident that the two Brotherhood factions could no longer live together in the same group. Whereas the older and more traditional members continued to moderate their stand, younger members chose a more militant route. Among them were Shukri Mustafa, founder of Takfir wal-Hijra, and Salih Siriya, founder of the Military Academy group. These leaders drew heavily on the ideology of Hassan al-Banna and Sayyid Qutb, taking their call for Islamic revolution and pushing it to its logical conclusion.

When Egyptian sociologist Saad Ibrahim interviewed jailed militants in the late 1970s, he was quick to recognize that the militants were intricately connected with the ideological view of the Brotherhood and that they were building on the foundation that the Brotherhood had established in Egypt. He wrote, 'In terms of religious dimensions of ideology, their reading of history and their overall vision for the future, the militants expressed no differences with the Muslim Brotherhood. In fact, they consider themselves a natural continuation of the Brotherhood.'[24]

Militant Islamic groups have continued to orchestrate campaigns of terror in Egypt in their violent attempt to overthrow the government. At the same time, the Muslim Brotherhood has been consolidating its power base in professional organizations, universities and on the national political scene. They represent two groups that share a common ideological basis, a common history and a common vision for the future. The difference between the two lies in the lengths they feel their religion allows them to go to in order to achieve that vision. El-Sayid Yassan, former director for the al-Ahram Centre for Political and Strategic Studies, argued that these commonalities create an underlying bond between the militant groups and the Brotherhood: 'They both have the same project, the establishment of an Islamic state. It is true that one of them believes in violence while the other denounces it, but the unity of goals definitely creates sympathy.'[25]

There are many well-known northern Islamic militancy groups that have been active in Egypt for years. These include the Islamic Liberation Organization (also known as 'Shabab Mohamed' or Mohamed's youth as well as Salvation from Hell). Most of the recent violence, however, has been carried out by militants from Upper Egypt. Indeed, most of the acts of terrorism that have occurred since 1992 have been attributed to al-Gama'a al-Islamiyya, the main violent anti-status quo group from Egypt's south. In the last decade, al-Gama'a has stepped up its activities and has been blamed for attacks on the Egyptian minister of information, Safwat Sharif, as well as an attack against General Hasan al-Alfi, the former Egyptian minister of the interior. The group has also been blamed for the murder of secular commentator Farag Foda, the 1994 assassination attempt on nobel laureate Naguib Mahfouz, the fatal shooting of 18 Greek tourists in Cairo in 1996, the bombing of the Egyptian Embassy in Islamabad, Pakistan in 1995, the failed attack on the American Embassy in Albania in 1998, as well as numerous other bombings, assassination attempts and attacks on foreign tourists, including the massacre in Luxor in 1997. Since 1992, more than 1,200 people have been killed in violence attributed to Islamic fundamentalists in Egypt. More recently, al-Gama'a appears to have become closely associated with Osama bin Laden's terrorist network. Whereas some of al-Gama'a's leadership formally renounced acts of terrorism, others joined the ranks of al-Qaeda.

Another Egyptian fundamentalist group was the Jamaat al-Muslimin, or Society of Muslims (also known as Takfir wal-Hijra, or Excommunication and Emigration—this was the popular name given to the organization by the Egyptian media). This group was founded by Shukri Mustafa, an Upper Egyptian from Asyut. Shukri's group earned its name because it chose to withdraw (*hijra*) from the *jahilyya* society governing Egypt by hiding in caves in the desert outside of Asyut. In 1977, this group succeeded in kidnapping Muhammad al-Dhahabi, the former Egyptian minister of *waqfs* (property whose funds are used to support religious foundations). The group killed al-Dhahabi when their demands were not met, causing the government to hunt down the members of the organization, including Shukri Mustafa, imprison and execute them.

Jamaat al-Jihad, another Egyptian militant group, gained notoriety when it orchestrated the assassination of Egyptian President Anwar Sadat in 1981. Al-Jihad was formed by survivors of another group, Muhammad's Youth, which had staged an abortive coup in 1974. The ideology of al-Jihad was put forward in *The Neglected Obligation,* by Muhammad al-Farag. Building upon al-Banna's writings as well as those of Sayyid Qutb, al-Farag called true believers to rise up and fight a holy war against Egypt's un-Islamic government. Al-Farag saw *jihad* as the only effective way to fight for the establishment of an Islamic state, because all other means had been tried before and failed. Only through *jihad* could the glory of Islam be restored. Khalid al-Islambuli was the al-Jihad member who killed President Sadat in 1981. Upon firing the bullets that killed Sadat, he reportedly exclaimed, 'I have killed Pharoah'. Al-Islambuli was incited into action when his brother Mohammad, leader of the al-Gama'a organization at the university in Asyut, was taken out of bed in the middle of the night and arrested by Egyptian police forces during Sadat's crackdown on Islamists. Al-Jihad was led by the infamous Sheikh Omar Abdel Rahman. The group drew further worldwide attention after claiming responsibility for the February 1993 bombing of the World Trade Center in New York (for which Abdel Rahman is serving a life sentence in the United States).

Another prolific militant group that became active in Egypt in the 1990s and a sibling of al-Jihad is Jihad al-Gadid, or the New Holy War. This group has been claiming responsibility for acts of violence since 1993. Still, many observers do not feel that this group is new at all. Rather, it is believed that the group is made up of veterans of the Afghanistan war splintered off from al-Gama'a. Indeed, the level of success of its acts of terrorism (27 people were killed in 3 attacks in 1993 and 1994) indicates that the group is familiar with sophisticated weaponry and explosives, most likely a result of training in al-Qaeda camps in Afghanistan.

Militant groups are said to receive support from abroad. This is becoming evident as the United States and its allies try to break the flow of finances to terrorist groups as part of their current campaign. In the past, the Sudan, Iran and Saudi Arabia were often listed as the chief external financiers of the movement in Egypt (as argued by the Egyptian government). Egyptian government officials have also accused Western governments such as Britain of harboring so-called 'terrorists' who influence the activities of militants in Egypt.[26] Today, we know that terrorist cells existed throughout Europe and the United States. In the month following the 11 September attacks, some 225 people in a dozen countries outside of the United States were rounded up after intelligence indicated that they were involved in plotting or assisting terrorism.

The Afghanistan Connection

Al-Gama'a al-Islamiyya has historically been made up of two types of members, one of which has a strong connection to Afghanistan and the al-Qaeda network. The first type consists of Egyptians recruited locally from Upper Egypt, including those who rose and who continue to rise through the ranks of the universities. A second and important type of member includes those Upper Egyptians who fought in the war of independence in Afghanistan and then returned home to their native country. Indeed, as the Islamic Revolution in Iran fuelled Islamic extremism in Egypt in the 1970s, the Soviet withdrawal from Afghanistan in 1989 and the subsequent formation of an Islamic state boosted the fundamentalist movement in Egypt in the 1990s. Alongside the Afghan *mujahideen* who were fighting against the Soviets were between 5,000 and 10,000 zealous youths from around the Muslim world, including many from Egypt. These youths were trained in military camps set up along the Pakistan–Afghanistan border and sent to the front lines. These camps continued to operate after the rise to power of the Taliban, and continued to be a training ground for youths from around the Muslim world who prescribed to a more militant form of Islam.[27]

After the Soviets withdrew from Afghanistan, these young fighters, fresh from the battle, looked throughout the Muslim world for places where they could try to replicate their victory. Many sided with al-Gama'a al-Islamiyya in its fight against the Egyptian government. The subsequent rise in terrorist attacks in the 1990s in Egypt is largely a result of the return home of fighters from Afghanistan's war of independence. The presence of these members in al-Gama'a has made the continuation of ties with Afghanistan and Osama bin Laden's al-Qaeda organization a natural consequence. Al-Qaeda continued to nurture these ties, recruit from al-Gama'a, and provide training and other support to Egypt's militants. After the *mujahideen* succeeded in ridding Afghanistan of Soviet occupation, al-Qaeda, originally set up to provide logistical, financial and military support for the fight for independence, decided to stay intact. The group, led by bin Laden, chose to use the skills, resources and experience that it had acquired to fight for Islam on a more global scale. Originally, the targets for this fight were the governments of Egypt and Saudi Arabia. As a result, al-Qaeda provided significant support to al-Gama'a al-Islamiyya. Indeed, al-Gama'a had its foreign headquarters in Peshawar, Pakistan (the origin of many of the faxes that have been sent out by the group), allowing the group to be in close contact and co-ordination with al-Qaeda operatives. During the crackdown of the Egyptian government on fundamentalist groups, some al-Gama'a members fled Egypt and took up posts within al-Qaeda in Afghanistan and Pakistan. As the focus of al-Qaeda began to shift to the United States, al-Gama'a soon began to join al-Qaeda in its attacks on the US and its allies around the world. The connection between Egyptian militants and al-Qaeda has remained strong. Today, bin Laden's lieutenant is Ayman al-Zawahiri, an Egyptian physician who headed the Egyptian group al-Jihad until its effective merger with al-Qaeda in 1998. Al-Qaeda's military commander, Mohamed Atef, is a former Egyptian policeman and member of al-Jihad. He recently married off his daughter to bin Laden's son Mohamed. Suspected bin Laden finance operative, Mustafa Ahmed, has typically worked out of Egypt and the United Arab Emirates.

Why Upper Egypt?

It is clear that the militant activity that grew up in Upper Egypt is rooted in the experience and ideology of the Muslim Brotherhood. Militants in Upper Egypt, however, more than in any other part of the country, have come to view the Muslim Brotherhood and other fundamentalist groups as having failed to address the immediate issues at hand. This sense of failure has caused a new Upper Egyptian militant movement to grow that is very regional in outlook. Whereas the Brotherhood and other groups have tended to look beyond Egypt's borders as international issues, such as the promotion of pan-Arabism and the liberation of Palestine as major priorities, Upper Egyptian Islamists, until their recent union with al-Qaeda, have looked homeward, focusing on the dismal socio-economic conditions in Upper Egypt created by policies of the Sadat and Mubarak regimes. Islam as the solution to poverty and injustice in Egypt, and in particular in Upper Egypt, has been their overriding conviction. As the Muslim Brotherhood grew in force during the distress of the Depression and as a reaction to the influence of Western powers in Egypt, the Upper Egyptian militant movement similarly grew as a result of a regime in power that has been perceived as corrupt, and as a reaction to the implementation of Western policies of economic liberalization.

The Upper Egyptian movement, as an entity separate from other Islamic organizations, originated in the 1970s at Asyut University and its various regional campuses across southern Egypt. The movement's leadership consisted of university students. This was in sharp contrast to the fundamentalist movements in northern Egypt that were made up of doctors, engineers and other professionals for the bourgeoisie. Unlike their northern brothers, these southern students who sought solace in Islam were mainly from lower middle- and working-class families. Many were the sons and daughters of *fallaheen,* Egypt's peasant farmers, who were able to get an education thanks to the reforms of President Nasser. Even though their education was free, the conditions in which these students had to study were often destitute, especially in Asyut. Gilles Kepel described the scene,

> In Asyut in particular there has grown up around the university what can only be called a 'belt', al-Hamra. An entire universe of poverty-stricken students is packed into it, cut off from their family milieu and highly receptive to any voices that manage to make themselves heard and promise an improvement in their conditions.[28]

The students were united in the southern universities, as the education code in Egypt prevents a student from studying outside of the region from which they originate, unless the subject that they wish to study is not offered in the given region. The fact that the students were all from the south helped to solidify the growing Islamist movement's southern focus.

The difference in the backgrounds of the southern and northern Islamists contributed to them having significantly different agendas. Whereas both groups agreed that Islam is the solution, they differed in answering the question: *a solution to what?* Upper Egyptians agree with northern Islamic revolutionaries that the system must be changed and that it cannot be changed from within the current legal system. The Upper Egyptian militants, however, have a much more regional focus given their particular situation. This became starkly clear when President Sadat began to roll back many of Nasser's reforms. Nasser's form of socialism had attempted to be an equalizer among Egyptians. His many programmes included providing free university education, giving government jobs to graduates and instituting a program of land reform. Whereas many of Nasser's programmes can be seen as having failed, they were a source of hope and promise for Egypt's poor. Through his de-Nasserization program, Sadat was destroying the hopes of millions of Egypt's less fortunate and restoring the bond between the government and Egypt's wealthy, land-owning elite. The Muslim Brotherhood and its Islamist colleagues in the north (largely from the bourgeoisie) supported the de-Nasserization programme. The southern Islamist movement, formed by university students—the sons and daughters of peasant farmers who had been given hope and opportunity by the very programmes that Sadat sought to recall—stood firmly against the de-Nasserization programme; a programme that they viewed as an attack against the lower classes. They began to view the northern government as greedy, corrupt and un-Islamic.

On 6 October 1981, President Anwar Sadat was shot and killed by Lieutenant Khalid al-Islambuli, an Upper Egyptian from the town of Minia. Out of the 280 people implicated in the conspiracy surrounding the death of President Sadat, 183 were Upper Egyptians. Another 73 were from neighbourhoods in Cairo where a significant portion of the population consists of rural-to-urban migrants from Upper Egypt.[29] Hence, the Upper Egyptian Islamic militant movement had risen from secluded university classrooms in Asyut and Minia to having forced a major change in Egyptian politics, shocking all of Egypt and the world in the process.

The assassination of Sadat was followed by massive sweeps and arrests of suspected Islamists. Those who were not taken into custody were forced underground. President Hosni Mubarak, Sadat's successor, has largely continued the political and economic policies that Sadat had set in motion. Given the crackdown, the Islamists' resources and strength were largely depleted. Still, their grievances against a government that they continued to see as un-Islamic and as oppressive to Upper Egypt's disenfranchised remained. Given time to rebuild and regroup, the southern Islamists would begin a new attack aimed at destabilizing Egypt's government in the 1990s, as Egypt's accelerated programme of structural adjustment, as recommended by Western institutions such as the International Monetary Fund (IMF) and the World Bank, began to squeeze the majority of poor Egyptians, in particular Upper Egyptians, who felt even further ostracized from Egypt's developmental goals. This attack against the Government of Egypt culminated in the massacre in Luxor in November 1997.[30]

As stated earlier, it is very easy to look at the conflict between Upper Egyptian Islamic fundamentalists and the Egyptian government and security forces as a confrontation based simply in religion. It has been demonstrated, however, that the roots of the conflict in Upper Egypt lie much deeper, and the Upper Egyptian movement, although it has a basis in the ideological thought of the Muslim Brotherhood, stands in stark contrast to northern Islamist movements. This unique movement has grown out of the experience of poverty, discrimination and a history of neglect that the majority of Upper Egyptians have felt throughout Egypt's modern period. This argument was put forward by political scientist Mamoun Fandy. Mamoun stated that the targets of these militants offer proof that this co-called holy war is not merely based in religion. If this were true, he argued, the militants would be targeting Sufi mystics and Muslims who combine pagan and Christian rituals into their religion. For instance, many Muslim Upper Egyptian peasants still practice annual celebrations that date to pharaonic times and have strong pagan characteristics. However, Fandy wrote,

> there is no single report of the Islamists objecting to these 'un-Islamic' practices of the common folk. The difference is that these are the indigenous customs of the poor and powerless, whereas the practices of the government are part of the corrupt, pseudo-Westernized upper classes' monopoly of political and economic power.[31]

Upper Egypt has been left behind in many aspects of Egypt's development so that today, not only does it have the worst socio-economic indicators of any area in Egypt, its inhabitants also have a culture and a legal system that sets itself apart from the rest of the country.

In Egypt, the Arabic word used for Cairo is *Misr*—the same name in Arabic for Egypt. In many ways, Cairo is Egypt, that is, Cairo is where the government is centred and where the government spends the vast majority of its resources. Cairo is the unchallenged economic, political and cultural centre of Egypt and dominates all other areas of the country, the south in particular. The south does not even have its own media. Dr Abdel-Mo'ati Shaarawi, a professor of sociology at Cairo University, summed up the feelings of Upper Egyptians:

> They do not feel they belong to the country because in the south, justice is denied, poverty is enforced, ignorance prevails, and people feel that society is conspiring to oppress, rob, and degrade them, so neither person nor property will be safe.[32]

Upper Egypt has historically been isolated geographically, politically and economically from the rest of the country. Up until the last two decades, travel between Middle and Upper Egypt was difficult. Even the British never truly colonized Upper Egypt, failing to maintain an outpost there. In the post-revolutionary period, Egypt's central government has also left Upper Egypt largely to itself, relying on local notables to enforce order. These notables have allowed the traditional customs and traditional structure of society to govern, and Upper Egypt has largely remained divided along tribal lines. Egyptian civil law has been widely ignored in Upper Egypt. Political thought and economic growth has been concentrated in the north. Until recently, there was little effort at development in Upper Egypt, with the notable exception of the Aswan Dam and a few other agricultural and manufacturing projects. Still, these projects were controlled by the north and the north has reaped their benefits. The donor community in Egypt, including the United States Agency for International Development (USAID), was also late in realizing the great developmental needs of Upper Egypt and how this area was falling behind the rest of the country. Therefore, traveling between the north and south of Egypt often seems like travelling between two different worlds: one desperately trying to modernize and the other left far behind and losing ground.

Upper Egyptians are typically classified as three different groups of people. These groupings have structured the hierarchy of Upper Egyptian society in the following way: the *ashraf* claim descent from the Prophet Muhammed and see themselves as superior to all other Upper Egyptians. The Arabs claim lineage from central Arabia. They are inferior to the *ashraf* but superior to the majority of Upper Egyptians who are the *falaheen,* or the non-Arab Egyptians. The *falaheen* consider themselves the direct descendants of the pharaohs. But the *ashraf* and the Arabs see the *falaheen* as those who converted to Islam under threat of the sword and therefore as inferior. The *ashraf* and the Arabs have typically filled any positions of power and influence and tend to own land, be wealthier, and to have better access to social services such as health care and education than the *falaheen,* who are typically peasant farmers.

The above structure dates back to the Arab conquest of Egypt (AD 622) and has remained the norm throughout Upper Egypt's modern history as well. The *ashraf* and the Arabs have typically used interpretations of Islam that appear to accept inequality in society to solidify their hold on society and their dominance over the *falaheen.*

The post-revolutionary reforms of President Nasser posed the first serious challenge to the social structure of Upper Egypt. Under Nasser, for the first time ever, there was talk of equality among Egyptians, and the sons and daughters of *falaheen* were allowed to receive a free government education and to qualify for government jobs. Educated, many of these children of farmers began to challenge the local hierarchy and to prescribe to other interpretations of the Qur'an that preached that there is no difference between the Arab and non-Arab Muslims. Still, very little else changed. The economic situation for the majority of the *falaheen* remained dismal and the status of the *falaheen* as third-class citizens was upheld by the *ashraf* and Arabs, who continued to be favoured by Cairo. In addition, the local administrators were largely able to avoid Nasser's attempts at land reform as well.

President Sadat offered hope for the *falaheen* at the start of his presidency through his use of Islamic rhetoric. His de-Nasserization programme, however, left these poor Upper Egyptians disillusioned, as they became afraid of losing even the small gains that they had received during Nasser's legacy. Sadat's economic programme clearly favoured the southern families that had historically held wealth and power. The *falaheen* began to feel that Egypt was going back in time, given that some of the few plots of land that had been redistributed under Nasser were even taken back by the government. This was a source of enormous frustration for the *falaheen.* As stated earlier, it is from the ranks of these newly educated sons and daughters of the *falaheen* that Upper Egypt's unique Islamic movement rose. Mamoun Fandy noted, 'the *falaheen* took refuge in Islam and Sai'di [Upper Egyptian] regionalism… The Islamists' reform and revolt in the south is informed by the *falaheen's* desire to rearrange the rules of southern social structure and centre-periphery relations'.[33]

As Fandy also noted, another major force in the growth of the Islamist movement in Upper Egypt was the influx of oil money from the Gulf Arab states. At the same time that Sadat was introducing his economic liberalization programme, the Gulf states were experiencing an oil boom and needed migrant workers from surrounding poorer countries. Many of the sons of Upper Egyptian *falaheen,* eager to do hard labour for cash, travelled to the Gulf. Many later returned with significant savings that were used to buy land and start businesses. Many of these returning migrant labourers also used their newly found wealth to influence society and challenge the traditional hierarchy by building mosques and setting up social services for poor Upper Egyptian communities. These mosques preached a message that was non-traditional to Upper Egypt: that all Muslims, Arab and non-Arab, are equal. Differences should be recognized in terms of one's piety, not one's ancestral background. As the number of mosques grew, this message was spread across Upper Egypt and the new Islamist force grew.[34] Fandy claimed that, 'The power of the Islamic network in Upper Egypt rivals that of the government and other social organizations and has emerged as an alternative to the old *'umdas* [governors] in mediating local disputes.'

Dr Mohamed Abul-Issad, a Professor of History at Minia University in Upper Egypt, once stated that,

> The deadliest disease is despair. Poverty is only a symptom. The lack of any governmental attention to development has created a vast underclass which has no stake in the society or government.

Development Indicators in Upper Egypt

A visitor to rural areas in Upper Egypt quickly becomes aware of the desperate situation that many of the Upper Egyptian *falaheen* live in on a daily basis. According to World Bank senior economist Marcelo Giugale, Upper Egypt has the highest percentage of Egypt's so-called 'ultra poor', those surviving on a day-to-day basis.[35] With an annual per capita income of just over $300, and soaring rates of unemployment, illiteracy, malnutrition, morbidity and mortality, Upper Egypt ranks at the bottom of virtually every social and economic indicator for Egypt. In 1995, during the height of Islamist activity, the Institute of National Planning published a human development report that looked at various social and economic indicators in the numerous governorates of Egypt. Each governorate was then ranked as low, medium or high in terms of human development. Only Giza and Aswan in Upper Egypt achieved the rank of medium. The remaining six governorates were all ranked at the bottom of the human development survey.[36] A project paper by USAID's Cairo Mission, completed during the same period, reiterated these facts by stating that, 'according to every indicator, [developmental] progress has been slowest in Upper Egypt, especially rural Upper Egypt'.[37] Given that the 1999 United Nations development index ranked Egypt as a whole as 105th out of 162 countries, Upper Egyptians face conditions of dire poverty.

A look at a few of the key development indicators paints a very bleak picture:

a. *Health.* Upper Egypt suffers from important health problems and lags behind the progress made in other parts of the country. There is a general lack of health facilities. Clinics are rare, and physicians can often not be found for miles. This is a considerable journey when your only means of transportation may be on the back of a donkey. Even if healthcare is available, it is mostly unaffordable for rural Upper Egyptians, who will attempt to cure illnesses with traditional homemade remedies in order to avoid doctors' fees. Most babies are born in the home without the assistance of a medical professional. Almost every mother has a story of a stillborn birth or a child who died before the age of five. Many of these deaths were from diarrhoea or respiratory infections that could have been cured given the proper care. The rate of immunization coverage is much lower in Upper Egypt than in other parts of the country, and many *fallaheen* have never heard of vaccinations that are standard practice in other parts of the country. Knowledge of proper hygiene practices is also minimal in many areas. It is important to point out that not only does Upper Egypt lag far behind Cairo and other urban areas in Egypt, but it lags far behind Lower Egypt as well. A USAID Cairo report pointed out that, 'The gaping discrepancies between Lower and Upper Egypt and between rural and urban areas according to all human development indicators cannot be ignored.'[38]

b. *Population.* In the area of population, Upper Egyptians are less educated in family planning methods and less likely to use family planning. As a result, families in Upper Egypt tend to be much larger than in Egypt's urban governorates and in Lower Egypt. Women in Upper Egypt get married at the average age of 17.2 years. The median age for the first time to give birth is 19.3 years and the average number of births is six. This figure is some 2 births higher than in rural Lower Egypt. This is due in part to the fact that the rate of usage of modern family planning methods in Lower Egypt is some 20 per center higher than in Upper Egypt, where 19 per cent of households expressed disapproval of family planning and 24 per cent claimed that it is against their religious practices.[39]

c. *Education.* As noted earlier, Nasser's education reforms gave many of the sons and daughters of Upper Egyptian peasants the opportunity to receive an education. Still, Upper Egypt trails the rest of the country in terms of educational facilities and educational attainment. Many villages have no educational facilities at all. As a result, children have to walk for miles to get to the nearest school. Many of the educational facilities that do exist are overcrowded and poor. After a few years of schooling, a child may know little more than how to write his or her name. Given this, parents often feel that it is more practical for a child to be assisting in harvesting or other household and manual tasks. In addition, there still exists a strong unwillingness among a large part of the rural Upper Egyptian population to educate girls. For Upper Egyptian males, the average number of years spent in school is 4.2. Approximately 44 per cent of men have no education at all. Only 30 per cent of women have had any formal schooling. The adult literacy rate in Upper Egypt is 46.7 per cent.

d. *Infrastructure.* Many of the villages of Upper Egypt have no electricity or piped-in water. Few have sanitation facilities beyond primitive drainage tanks attached to a homemade latrine. Access to villages is oftentimes difficult. Some villages can only be accessed by boat or on foot. Homes are typically simple mud huts with dirt floors that are poorly constructed. Huts that collapse during rainstorms or for various other reasons are commonplace. Villages are often centred along a canal from the Nile, with stagnant and unclean water that is used to launder clothes, wash dishes, as well as for bathing and drinking. Water buffalo, donkeys, geese, ducks and other creatures are scattered about making use of the same water source. Only around 38 per cent of households in Upper Egypt have piped-in water. Around 32 per cent of households do not have any toilet facilities. Nearly three-quarters of households in Upper Egypt have sand or dirt floors and the number of people sleeping per room averages 3.6.

The above statistics and description of life demonstrate that rural Upper Egyptians are more likely than their fellow countrymen to live in ill-constructed dwellings that are overcrowded and that lack clean, piped-in water, sanitation facilities, electricity and durable goods. In addition, their access to health care and education is limited and, due to a lack of knowledge about family planning as well as a lack of modern family planning products, large families are the norm in Upper Egypt.

The militant groups that have formed in Upper Egypt, with their unique characteristics that have set them apart from other Egyptian Islamist groups, have risen from the experience described above; an experience that is seen as being perpetuated by the current regime in power and by the spread of globalization. In this sense, it becomes clearer that the rise of militant activity in Upper Egypt is not only a call for a more Islamic government, but a call for a more just government as well as a protest against the dismal status quo. The Islamists view the government as a puppet of the West that has bought into a globalization and an Americanization that is leaving most of Egypt's people behind and given them little voice in the future. Desperate for change, they have resorted to acts of violence to bring attention to the plight of Upper Egyptians and to the disparity that exists in the country.

Responses to the Development Challenge and Terrorism

As stated earlier, the Egyptian government has responded to the militant's threats and activities in Upper Egypt with force. Police and security battles with militants have been frequent. The government has used massive sweeps through the countryside to gather up suspected militants, who may be detained without trial, tortured into confessing crimes that they may or may not have committed, and tried at a military court. They have also offered incentives for people who have information on suspected militants or militant activities and have burned thousands of square kilometers of sugarcane fields—traditional hideouts for the militants.

After the violence in Luxor in November 1997, President Mubarak fired many of the security personnel in that district and forced the resignation of General Hassan al-Alfi, the then minister of the interior. These moves, however, were largely symbolic gestures meant to assure the international community that he was acting to improve security.

The government's anti-terrorism policies have been met with mixed reactions. Cairenes and other northerners have generally supported the government's heavy-handed actions and have seen them as justified to stop the growing militancy movement. Still, there is evidence of much support for the militants in Upper Egypt. In addition, there are countless others who, although they may not support the militants' activities, are sympathetic to the militants' cause. Many Upper Egyptians who claimed not to be supporters of the militancy movement declared that, nonetheless, they fear and do not trust the police and the security forces as much as they fear and do not trust the militants. One Western diplomat living in Cairo in the mid-1990s commented that, 'The populace [of Upper Egypt], for the most part, stands aside or sympathizes with the militants. The Government is losing the war down there.'[40]

Egypt's central government has also acknowledged that the development of Upper Egypt will play a role in the fight to cease the violence of the region's Islamic militants (although it has not been widely publicized). In 1996, the government announced the formulation of an Upper Egyptian development campaign. The campaign, to be financed 25 per cent by the Egyptian government and 75 per cent from private sources, aims at establishing sustainable development programmes in the region over 22 years. Still, the government appears to have realized the fact that development is a major component needed in the fight against militants only years after the violence actually began.

The government's plan for Upper Egypt is to be undertaken in four stages and will cost between \$60bn and \$100bn. The plan is expected to produce some 2.8 million new jobs in the agricultural and industrial sectors. Still, this plan is seen by many as constituting too little too late. In autumn 1996, after the plan was announced, journalist Ahmed Ragab wrote,

> We have to stretch our hands, like good neighbours do, to the state of Upper Egypt which is south of Cairo and belongs to the Fourth World. We have to change the concept that Upper Egypt is the exile of bad employees. If the government is taking care of the slums in the city, why doesn't it take care of the slum south of Cairo which suffers from poverty, unemployment and terrorism.[41]

The rise of terrorism in Egypt did not occur in a vacuum and is not a problem specific to Egypt. Rather, the story of Upper Egypt and the rise of militant factions has been duplicated in other countries around the world as the people react to the adverse socio-economic conditions and neglect that they face in their everyday lives, with little hope for the future. For some, religious extremism and desperate militant acts become a favourable alternative.

Osama bin Laden has succeeded in creating a worldwide network among various loosely knit terrorist cells and organizations, including some of Egypt's militant groups, around the world. Most likely, there is no single master plan that exists that dictates the actions of terrorists worldwide. Rather, terrorist organizations are united in their vision as well as their admiration for bin Laden and his cause. They have received necessary training, financial and logistical support from bin Laden and al-Qaeda. It is largely believed that within al-Qaeda exists a *shura* council that acts as a board of directors and includes representatives from various terrorist groups and cells from across the globe. This group meets in Afghanistan and reviews and approves terrorist operations proposed by members of the network. As a result of this networking and of his efforts to build an umbrella structure that unites so many groups, bin Laden has succeeded in streamlining the activities of these groups into actions that serve the purpose of a single, larger Islamic crusade against the West and those governments viewed as puppets of the West. At their heart, however, these individual militant/terrorist movements have homegrown objectives that have arisen out of their individual experiences—experiences often rooted in poverty, lack of opportunity, neglect and discrimination. Osama bin Laden has succeeded in moving these groups beyond their regional causes and in convincing them that American-driven globalization is the larger force that keeps the governments that they are fighting against in power and that perpetuates the dismal living conditions of the people they claim to represent. Indeed, bin Laden has been very good at calling on sympathizers of causes such as the Palestinian and Iraqi causes, as well as militants from Algeria, Egypt and numerous other nations, and focusing their energies and directing their anger towards the United States and its allies. For the network that he has created, globalization is a continuation of the imperialism, colonialism and neo-colonialism that has kept the Muslim world from advancing since the seventeenth century.

Today, the world is in a battle against the forces of terror. In this fight, President Bush has stated that the nation states of the world are either against the US or with it. It is countries such as Egypt, however, for which this choice is very difficult to make. Certainly, the Egyptian government would like to pledge their fully-fledged support to the US-led war. At the same time, this could provoke another rise of militant activity in the country as a significant portion of the population sympathizes with the fundamentalist cause. Whereas many may not support the specific actions that took place on 11 September, they feel that the United States is guilty of worsening and perpetuating a situation of social and economic injustice around the world, particularly against Muslims.

What Can the United States and the Coalition Learn from the Egyptian Experience?

The Egyptian experience demonstrates that the argument that religious fundamentalism cannot be dealt with logically and cannot be negotiated with is largely untrue. Many of these groups have been able to rise in force and in numbers because they offer an 'Islamic' alternative to governments that are seen as corrupt and not representing the common person. They represent hope to people living in dire socio-economic conditions with no political voice. In countries such as Egypt, there is a large, dissatisfied, underemployed or unemployed segment of the population that is easily susceptible to ideas put forth by Islamic fundamentalist groups. These groups, in addition to establishing charities, businesses, schools and hospitals for those people whom the official governments have ignored, have conducted acts of terror in an attempt to make their cause known and to destabilize the governments that they are fighting against. Osama bin Laden was able to take some of these groups 'global' by convincing them that the system that has created and perpetuated the conditions that they are fighting to escape is driven by American-led globalization. The ties that have brought these groups together were forged in the military training camps in Afghanistan. As a result of these alliances, terrorist attacks have escalated beyond the countries that gave birth to these groups and to a global level that culminated in the 11 September attacks. Today, it is believed that bin Laden's group controls 3–5,000 terrorists worldwide in a loose organization.

In the war against terrorism, the United States and its allies cannot just focus on the fight at the global level to which Osama bin Laden and his al-Qaeda organization has elevated terrorism. Certainly, the military strikes in Afghanistan aimed at wiping out al-Qaeda's base may succeed in destroying the network of communication, logistical and financial support among terrorist organizations that bin Laden has achieved. As demonstrated by the Egyptian experience, however, this is a short-term solution to a much larger and widespread problem. Indeed, Egyptian authorities may have been overly successful in their violent crackdown on terrorists. As a result of the crackdown, many of Egypt's militants fled the country only to set up terrorist cells elsewhere, including the United States. If al-Qaeda is destroyed, individual terrorist and militant organizations will continue to exist. If governments are able to use this environment to crack down on these groups, others will surely rise in their place. This will continue to happen until the root causes of their existence—so often the adverse socio-economic and political conditions existing in the loci of their countries—are cured. Egyptian sheikh Yusef al-Qaradawi has stated on Egyptian television that fighting terrorism by merely waging a huge war would be using the same logic as the terrorists. He argued instead that true Muslims need to advocate for the middle path of Islam, and that the United States and its allies need to make efforts to understand the psychology of the terrorists.[42] Similarly, Algerian President Abdelaziz Bouteflika recently stated that the international coalition against terrorism would gain more support if it included efforts to 'settle problems and injustices which fanaticism exploits to feed despair that nurtures terrorism'. He continued, 'We should all work together to correct the flagrant injustices of the world today, which globalization only exacerbates.'[43] The Egyptian government realized this quite late in the game in its fight against terrorism. It is hoped that the coalition that the United States has built will not. Secretary of State Colin L. Powell recently stated that international development programmes are 'at the core of our engagement with the world... over the long-term, our foreign assistance programs are among the most powerful national security tools'.[44] Military and police measures alone will not eradicate the forces of terrorism. Americans must learn that globalization does not just facilitate the movement of products or capital around the world, it also brings with it the risk of facilitating the movement of terrorism grown in another country into its own backyard. A long-term battle aimed at ridding the world of large-scale terrorism must be comprehensive in scope. The battle must include an intensified fight against the poverty existing in the world that drives individuals with no hope to such extreme acts of violence and terrorism as well as facilitating greater political participation for those people who are under represented. Such an approach will also benefit non-Islamic countries, such as Mexico, that have witnessed the rise of militant anti-status quo groups. If the military campaign does not go hand-in-hand with a campaign for international development and social justice, the United States and its allies risk creating more enemies than the ones that they will eliminate. Only by working to create a world where social justice is distributed more equitably can the ideological roots of terrorism, as demonstrated in the rise of the Muslim Brotherhood and subsequent militant organizations in Egypt, be eradicated.

The views expressed in this article are those of the author and not necessarily those of USAID or the US Government.

NOTES

1. This concept originated with Dr Dan Tschirgi of the Political Science Department of the American University in Cairo.
2. See Selma Botman, *Egypt from Independence to Revolution* (Syracuse: Syracuse University Press 1991) p. 120.
3. Ibid. p. 94.
4. John Esposito, *Islam: The Straight Path* (Oxford: Oxford University Press 1998) p. 120.
5. Ibid.
6. John Esposito, *The Islamic Threat* (Oxford: Oxford University Press 1992) p. 123.
7. Botman (note 2) p. 122.
8. Esposito (note 6) p. 122.
9. Ibid. p. 124.
10. Afaf Lutfi al-Sayyid Marsot, *A Short History of Modern Egypt* (Cambridge: Cambridge University Press 1992) p. 109.
11. Esposito (note 6) p. 128.
12. Ibid. p. 129.
13. Saad Eddin Ibrahim, 'Egypt's Islamic Militants', in Nicholas Hopkins and Saad Eddin Ibrahim (ed.), *Arab Society* (Cairo: American University in Cairo Press 1998) p. 501.
14. Esposito (note 6) p. 132.
15. See Andrew Hammond, 'State Defies Support for Brothers', *Middle East Times,* 22–28 Oct. 1995, p. 1.
16. Ibid. p. 20.
17. See Fatemeh Farag and Steve Negus, 'Brothers Form Electoral Pact with Liberals and Labour', *Middle East Times,* 15–21 Oct. 1995, p. 1.
18. See Fatemeh Farag, 'New Guide for the Brotherhood', *Middle East Times,* 21 Jan–3 Feb. 1996, p. 1.
19. See Andrew Hammond, 'New Centre Party Takes State by Surprise', *Middle East Times,* 21–27 Jan. 1996, p. 1.
20. See Miriam Shahin, 'Egypt Cracks Down on Terrorism', *Middle East,* May 1996, p. 15.
21. Amira Howeidy, 'Politics in God's Name', *al-Ahram Weekly,* 16–22 Nov. 1995, p. 3.
22. Ibid. p. 3.
23. Esposito (note 6) p. 130.
24. Ibrahim (note 13) p. 500.
25. Amira Howeidy, 'Debating Democratic Credentials', *al-Ahram Weekly,* 2–8 Feb. 1995, p. 2.
26. For example, London is home to Yasir al-Siri, who has claimed asylum in Britain from Egypt, where he was sentenced to death for the attempted murder of the prime minister in 1993.

27. Note that, in 1992, the Egyptian government passed a law stating that anyone having received military training in a foreign country could face the death penalty. This law was aimed at Egyptians who had received training in camps in Afghanistan, Pakistan and the Sudan.
28. Gilles Kepel, *Muslim Extremism in Egypt* (Berkeley, CA: University of California Press 1993) p. 137.
29. See Mamoun Fandy, 'Egypt's Islamic Group: Regional Revenge', *Middle East Journal* 48/4 (Autumn 1994) p. 607.
30. Note that only one of the six attackers in the Luxor incident was identified. The identified attacker, Medhat Abdel Rahman, had been trained in military camps in Afghanistan.
31. See Mamoun Fandy, 'The Tensions Behind the Violence in Egypt', *Middle East Policy* 2/1 (1993) pp. 25–34.
32. These comments were made in an interview with Omayma Abdel-Latif, *al-Ahram Weekly,* 29 Aug.–4 Sept. 1996, p. 15.
33. See Fandy (note 29) p. 614.
34. In 1992, the Egyptian government passed a law stating that all mosques, including those that were built privately, would come under the control of the government. As a result, all sermons being delivered at mosques had to be approved by government officials.
35. See Omayma Abdel-Latif, 'Making the Future into a Site of Hope', *al-Ahram Weekly,* 29 Aug.–4 Sept. 1996, p. 14.
36. See Egypt Human Development Survey, the Institute of National Planning, Nasr City, Cairo, Egypt, 1995.
37. See USAID/Cairo, 'Healthy Mother/Healthy Child', project paper, 8 June 1995, p. 1.
38. Ibid.
39. The statistics are taken from the Egypt Demographic and Health Survey, The Population Council, Cairo 1993.
40. See Chris Hedges, 'Egypt Loses Ground, to Muslim Militants and Fear', *New York Times,* 5 Feb. 1994.
41. Quoted in Abdel-Latif (note 35) p. 14.
42. Al-Qaradawi's comments were cited in the special report on fighting terrorism, *Economist,* 20 Oct. 2001.
43. Comments reported in 'Algeria says U.S. Should Address Arab Discontent', CNN.com, 5 Nov. 2001.
44. Secretary of State Colin Powell, speech given at USAID's 40th anniversary celebration.

Colombia and the United States: From Counternarcotics to Counterterrorism

"The worldview that has molded Washington's twin wars on drugs and terrorism constitutes an extremely narrow framework through which to address the complex problems Colombia faces. National security, defined exclusively in military terms, has taken precedence over equally significant political, economic, and social considerations."

ARLENE B. TICKNER

During the past several years, United States foreign policy toward Colombia has undergone significant transformations. Long considered a faithful ally in the fight against drugs, as well as showcasing Washington's achievements in this camp, Colombia became widely identified as an international pariah in the mid-1990s during the administration of Ernesto Samper because of the scandal surrounding the president's electoral campaign, which was said to have been funded by drug money. Although the inauguration in 1998 of President Andrés Pastrana—a man untainted by drugs—marked the official return to friendly relations with the United States, Colombia came to be viewed as a problem nation in which the spillover effects of the country's guerrilla war threatened regional stability. The events of September 11, combined with the definitive rupture of the Colombian government's peace process with the rebels in February 2002, have converted this country into the primary theater of United States counterterrorist operations in the Western Hemisphere today.

THE PERVERSE EFFECTS OF THE "WAR ON DRUGS"

Any discussion of United States policy in Colombia must begin with drugs. Since the mid-1980s, when illicit narcotics were declared a lethal threat to America's national security, the drug issue has been central to relations with Colombia. Washington's counternarcotics policies have been based on repressive, prohibitionist, and hard-line language and on strategies that have changed little in the last few decades. The manner in which Colombia itself has addressed the drug problem derives substantially from the United States approach, with most of Bogotá's measures to fight drug trade the result of bilateral agreements or the unilateral imposition of specific strategies designed in Washington.[1] These American-guided efforts to combat illegal drugs "at the source" have produced countless negative consequences for Colombia, aggravating the armed conflict that continues to consume the country and forcing urgent national problems such as the strengthening of democracy, the defense of human rights, the reduction of poverty, and the preservation of the environment to become secondary to countering the drug trade.

Perhaps the most perverse result of the United States–led "war on drugs" is that it has failed to reduce the production, trafficking, and consumption of illicit substances. Between 1996 and 2001, United States military aid to Colombia increased fifteenfold, from $67 million to $1 billion.[2] During this same period, data from the United States State Department's annual *International Narcotics*

THE "REALIST" APPROACH TO DRUGS AND TERRORISM

WITH THE END of the cold war the United States lost its most significant "other," the Soviet Union; it also lost a clear sense of the national security interests of the United States. Drugs, long considered a threat to United States values and society, became an obvious target. Viewed in this light, the "threat" represented by illegal drugs in the United States is not an objective condition; rather, narcotics constitute one of the "cognitive enemies" against which United States national identity attempted to rebuild, albeit only partially, until September 11. In this sense, drugs are seen as "endangering" the American way of life and social fabric, much like the challenge posed by the communist threat to America's values during the bipolar conflict.

Given the sense of moral superiority that has traditionally characterized United States relations with the rest of the world, drug consumption is understood as being prompted by the availability of illegal drugs, which are concentrated, unsurprisingly, in the countries of the periphery; it is not seen as a problem originating in the demand for drugs in the United States or in the prohibitionist strategies that have traditionally characterized America's handling of this issue. While this rationale clearly runs contrary to commonsense economic rules of supply and demand, it tends to reinforce the underlying assumption of moral purity on which America's sense of self is partly based.

The terrorist attacks of September 11, 2001 and the United States–led retaliation mirror this perspective on drugs. Just as the drug issue fails to conform to typical notions of security and threat from a realist perspective, so September 11 challenges traditional views of international relations. The attacks came from within America's borders, not without, and were perpetrated by nonstate actors with little or no military power. But terrorism, rather than being seen as a diffuse, nonterritorial problem, has been associated by the Bush administration with state-based territories—Afghanistan and the entire "axis of evil"—and personified in figures such as Osama bin Laden and Saddam Hussein. The exercise of military power in countries threatening "freedom" and "justice" in the world constitutes the cornerstone of the United States strategy. And the zealous language accompanying the fight against terrorism—"those who are not with us are against us"—eerily recalls the cold war period.

The similarities between the wars on drugs and terrorism and the war on communism notwithstanding, a crucial difference exists: the enemies of these new wars are not readily identifiable, making victory nearly impossible. Hence, any explanation of the role of drugs and terrorism in United States domestic and international politics must necessarily return to the concepts of danger and threat. Although the policies implemented by the United States have failed in reducing the availability of illegal substances—and will most likely be unsuccessful in erasing terrorism from the globe—drugs and, more important, terrorism occupy a crucial discursive function in support of American identities and values. Both are considered lethal threats to United States security—and the political costs associated with directly challenging existing policies in Washington are extremely high. At the same time, the need to persevere in the war on drugs has received an additional push from the war on terrorism; the financing of terrorist activities with drug money has received much greater attention in United States policymaking circles in the aftermath of September 11.

A.B.T.

Drug Control Strategy report show that coca cultivations in Colombia grew 150 percent, from 67,200 to 169,800 hectares (1 hectare = 2.471 acres). Clearly, the high levels of military assistance received by Colombia have had little effect on illicit crop cultivation in the country.

Efforts to eradicate coca cultivation, primarily through aerial spraying, have also increased progressively in Colombia. In 1998, for example, 50 percent more hectares were fumigated than in 1997; in 2001 the Colombian National Police fumigated nearly two times more coca than in 2000. In both instances, fumigation had no effect or even an inverse effect on the total number of hectares cultivated.

Intensive aerial fumigation—particularly in southern Colombia, where Plan Colombia efforts are concentrated—has created public health problems and led to the destruction of licit crops. According to exhaustive studies conducted by Colombia's national human rights ombudsman in 2001 and 2002, aerial spraying with glyphosate has not only killed the legal crops of many communities in southern Colombia but has also caused health problems associated with the inhalation of the pesticide and contact with human skin.[3] On two separate occasions, the ombudsman called for a halt to aerial fumigation until its harmful effects could be mitigated. Echoing similar concerns, in late 2001 the United States Congress, as a precondition for disbursing the aerial-fumigation portion of the 2002 aid bill to the Andean region, requested the State Department to certify that drug-eradication strategies currently employed in Colombia do not pose significant public health risks. On September 4, 2002 the State Department issued its report, arguing that no adverse effects had been found. Members of the scientific community and environmental nongovernmental organizations in the United States and Europe criticized the report, primarily on methodological grounds.

Eradication efforts also have not affected the costs to users: in November 2001 the United States Office of National

Drug Control Policy acknowledged that the price of cocaine in principal American cities has remained stable during the past several years. Yet Washington and Bogotá continue to insist that the war on drugs can be won simply by intensifying and expanding current strategies.

The United States war on drugs is nearly inseparable from counterinsurgency efforts in Colombia.

THE "WAR ON DRUGS" AND COUNTERINSURGENCY

The cold war's end saw drugs replace communism as the primary threat to United States national security in the Western Hemisphere. Military assistance to Latin America became concentrated in the "source" countries, particularly Colombia. At the same time, the definition of "low-intensity conflict"—the term used to describe the political situation in Central America during the 1980s—was expanded to include those countries in which drug-trafficking organizations threatened the stability of the state. And the strategies applied in the 1980s to confront low-intensity conflict in the region were subsequently adjusted in the 1990s to address the new regional threat: drugs.

In Colombia this view of the drug problem, and of the strategies needed to combat it, is especially troublesome, given that illegal armed actors, especially the leftist Revolutionary Armed Forces of Colombia (FARC) and the paramilitary United Self-Defense Force of Colombia (AUC), maintain complex linkages with the drug trade. At conceptual and practical levels, the United States war on drugs is nearly inseparable from counterinsurgency efforts in Colombia.[4]

The conflation of low-intensity counterinsurgency tactics with counter-narcotics strategies was facilitated initially through the "narcoguerrilla theory" (a term first made popular in the 1980s by former United States Ambassador to Colombia Lewis Tambs, who accused the FARC of sustaining direct links with drug traffickers). However, the fact that paramilitary organizations, most notably MAS (Muerte a Secuestradores, or Death to Kidnappers), were created in the early 1980s and financed by drug traffickers in retaliation for guerrilla kidnappings, seemed to belie the theory's validity. Yet by the mid-1990s, references to the "narcoguerrilla" slowly began to find their way into the official jargon of certain sectors of the United States and Colombian political and military establishment. Robert Gelbard, United States assistant secretary of state for international narcotics and law enforcement, referred to the FARC as Colombia's third-largest drug cartel in 1996. During his administration, President Ernesto Samper himself began to use the narcoguerrilla label domestically in an attempt to discredit the FARC, given the group's unwillingness to negotiate with a political figure that the guerrilla organization considered illegitimate.

Ironically, when the Colombian military during the Samper administration tried to convince Washington that the symbiosis between guerrillas and drug-trafficking organizations was real, and that counternarcotics strategies needed to take this relationship into consideration, the United States argued against the idea that the guerrillas were involved in the drug traffic. Indeed, although Tambs and others had made the accusation, the United States had never categorically associated Colombian guerrilla organizations with the latter stages of the drug-trafficking process. Only in November 2000 did the State Department accuse the FARC of maintaining relations with Mexico's Arellano-Félix Organization, one of the most powerful drug cartels in that country; it also argued that "since late 1999 the FARC has sought to establish a monopoly position over the commercialization of cocaine base across much of southern Colombia." One week later, United States Ambassador to Colombia Anne Patterson affirmed that both the FARC and the paramilitaries had "control of the entire export process and the routes for sending drugs abroad" and were operating as drug cartels in the country.

In principle, the "narcoguerrilla theory," as employed in Colombia, argues that: 1) the FARC controls most aspects of the drug trade, given the demise of the major drug cartels in the mid-1990s; 2) the Colombian state is too weak to confront this threat, primarily due to the inefficacy of the country's armed forces; and 3) United States military support is warranted in wresting drug-producing regions from guerrilla control.

The events of September 11 and America's war on terrorism have introduced an additional ingredient to United States policy in Colombia: counterterrorism.

Although bearing a certain degree of truth, this description grossly oversimplifies the Colombian situation. For example, while a general consensus exists that the FARC derives a considerable portion of its income from the taxation of coca crops and coca paste and that members of this organization have participated in drugs for arms transactions, the involvement of the FARC in the transportation and distribution of narcotics internationally is still uncertain. (Contrary to the claims made by the United States State Department and its representative in Colombia, for example, the Drug Enforcement Agency has never directly accused the FARC of operating as an international drug cartel.)

The involvement of paramilitaries in drug-related activities clouds this picture even further. According to some sources, paramilitary expansion in southern Colombia during late 2000, in particular in the Putumayo region, was largely financed by drug-trafficking organizations in response to the FARC-imposed increases in the price and taxation of coca paste. This is not surprising, since the leader of the AUC, Carlos Castaño, has personally acknowledged since March 2000 that a large percentage of this organization's revenues, especially in the departments of Antioquia and Córdoba, are derived from participation in the drug trade.

Yet even with evidence that the "narcoguerrilla theory" is simpleminded, it seems to have informed many United States and Colombian political and military actors in the search for policy options in the country, while also lending credence to those who argue that counterinsurgency techniques used in other low-intensity conflicts can be applied successfully in Colombia.

SEPTEMBER 11 AND COUNTERTERRORISM

The events of September 11 and America's war on terrorism have introduced an additional ingredient to United States policy in Colombia: counterterrorism. On the day of the attacks, United States Secretary of State Colin Powell was to have visited Bogotá on official business. Although Washington's concern about the FARC's abuse of a swath of Colombia designated as a demilitarized zone created to facilitate peace talks was clear (the FARC was accused of using the zone to cultivate coca, hold kidnapping victims, and meet with members of the Irish Republican Army, allegedly to receive training in urban military tactics), some members of the American government were beginning to express reservations about the depth and nature of United States involvement in Colombia and the effectiveness of counternarcotics strategies in the country. To a large degree, the incidents of the day facilitated shifts in United States policy that had begun taking shape much earlier.

Colombia's insertion into the global antiterrorist dynamic leaves scant room for autonomous decision-making by the new president.

In a congressional hearing held on October 10, 2001, Francis Taylor, the State Department's coordinator for counterterrorism, stated that the "most dangerous international terrorist group based in this hemisphere is the Revolutionary Armed Forces of Colombia." Both Secretary of State Colin Powell and United States Ambassador to Colombia Patterson also began to refer to Colombian armed actors, in particular the FARC, as terrorist organizations that threaten regional stability.[5] Given that the global war on terrorism has targeted the links that exist among terrorism, arms, and drugs, a new term was coined, "narcoterrorism," to describe actors such as the FARC and the AUC that fund terrorist-related activities with drug money.

The Colombian government's termination of the peace process with the FARC on February 20, 2002 placed Colombia squarely within Washington's new counterterrorist efforts. Until that day, the government of President Andrés Pastrana had never publicly referred to the guerrillas as terrorists. In a televised speech announcing his decision to call off the peace talks, however, Pastrana made this association explicit. Echoing this change, the presidential electoral battle of 2002 centered on the issues of counterterrorism and war, and led to the election of hard-liner Álvaro Uribe on May 26.

Colombia's insertion into the global war on terrorism has been reflected in concrete policy measures in the United States. In simple terms, Colombia is now viewed through the lens of counterterrorism. Public officials from both countries must frame Colombia's problems along antiterrorist lines to assure continued United States support. This shift in terminology has led to the complete erasure of differences between counternarcotics, counterinsurgency, and counterterrorist activities that formerly constituted the rhetorical backbone of United States policy in Colombia. For many years, Washington stressed the idea that its "war" in Colombia was against drug trafficking and not against the armed insurgents. As was noted, some began to openly advocate reconsideration of this policy as early as November 2000. Tellingly, United States Representative Benjamin Gilman (R., N.Y.), in a letter written that month to drug czar Barry McCaffrey that criticized the militarization of counternarcotics activities in Colombia, suggested the need for public debate concerning counterinsurgency aid to the country. A RAND report published in March 2001 also affirmed that Washington should reorient its strategy in Colombia toward counterinsurgency to help the local government regain control of the national territory.[6]

On March 21, 2002 President George W. Bush presented a supplemental budgetary request to the United States Congress totaling $27 billion for the war on terrorism and the defense of national security. The request solicited additional funding for Colombia as well as authorization to use counternarcotics assistance already disbursed to the country. The antiterrorist package finally approved by Congress in July contains an additional $35 million for counterinsurgency activities in Colombia as well as authority to use United States military assistance for purposes other than counternarcotics—namely, counterinsurgency and counterterrorism.

In its 2003 budget proposal submitted to Congress on February 4, 2002, the Bush administration also requested, for the first time, funding for activities unrelated to the drug war in Colombia. The aid package, which totals over $500 million, includes a request for approximately $100 million to train and equip two new Colombian army brigades to protect the Caño Limón-Coveñas oil pipeline, in which the American firm Occidental Petroleum is a large shareholder.

Militarization and human rights

One of the most severe challenges to United States policy derives from the human rights situation in Colombia. According to the United States State Department *Report on Human Rights* for 2001, political and extrajudicial actions involving government security forces, paramilitary groups, and members of the guerrilla forces resulted in the deaths of 3,700 civilians; paramilitary forces were responsible for approximately 70 percent of these. During the first 10 months of 2001, 161 massacres occurred in which an estimated 1,021 people were killed. Between 275,000 and 347,000 people were forced to leave their homes, while the total number of Colombians displaced by rural violence in the country during only the last five years grew to approximately 1 million. More than 25,000 homicides were committed, one of the highest global figures per capita, and approximately 3,041 civilians were kidnapped (a slight decline from the 3,700 abducted in 2000).

Although Colombian security forces were responsible for only 3 percent of human rights violations in 2001 (a notable improvement over the 54 percent share in 1993), the report notes that government security forces continued to commit abuses, including extrajudicial killings, and collaborated directly and indirectly with paramilitary forces. And although the government has worked to strengthen its human rights policy, the measures adopted to punish officials accused of committing violations and to prevent paramilitary attacks nationwide are considered insufficient. In the meantime, paramilitary forces have increased their social and political support among the civilian population in many parts of the country. Increasingly, Colombians sense that the paramilitaries constitute the only force capable of controlling the guerrillas' expansion. The AUC have also adopted parastate functions in those regions in which the government's presence is scarce or nonexistent.

Because of the questionable human rights record of the Colombian armed forces as well as Bogotá's unwillingness to denounce this publicly, United States military assistance to the country was severely limited during much of the 1990s. Nevertheless, the United States continued to provide the armed forces with military training, weapons, and materials. In 1994 the United States embassy in Colombia reported that counternarcotics aid had been provided in 1992 and 1993 to several units responsible for human rights violations in areas not considered to be priority drug-producing zones. As a result, beginning in 1994 the United States Congress anchored military aid in Colombia directly to antidrug activities. The Leahy Amendment of September 1996—introduced by Senator Patrick Leahy (D., Vt.)—sought to suspend military assistance to those units implicated in human rights violations that were receiving counternarcotics funding, unless the United States secretary of state certified that the government was taking measures to bring responsible military officers to trial.

The Colombian government itself began in 1994 to adopt a stronger stance on human rights and in January 1995 publicly claimed responsibility in what became known as the Trujillo massacres (committed between 1988 and 1991): more than 100 assassinations carried out by government security forces in collaboration with drug-trafficking organizations. Other measures directly sponsored by the Samper government in this area included the creation of a permanent regional office of the UN High Commissioner for Human Rights; the ratification of Protocol II of the Geneva Conventions; and the formalization of an agreement with the International Red Cross that enabled this organization to establish a presence in the country's conflict zones. Unfortunately, as the Colombian newsweekly *Semana* noted, "Little by little, the novel proposals made at the beginning of the Samper administration became relegated to a secondary status, given the government's need to maintain the support of the military in order to stay in power."

The moderate changes implemented by the Colombian government in its handling of human rights issues—combined with the intensification of the armed conflict and the military's need for greater firepower and better technology—facilitated the signing of an agreement in August 1997 in which the Colombian armed forces accepted the conditionality imposed by the Leahy Amendment. In the past, the Colombian military had repeatedly refused United States military assistance on the grounds that such unilateral impositions "violated the dignity of the army." But the marked asymmetries between United States aid earmarked for the Colombian National Police (CNP), which immediately accepted human rights conditionalities, and assistance specifically designated for the Colombian army constituted a strong incentive for the military to finally accept the conditions attached by the United States. Until the late 1990s the CNP was Washington's principal ally in the war on drugs, receiving nearly 90 percent of United States military aid given to Colombia. The 2000–2001 Plan Colombia aid package, however, reversed this trend completely: while the Colombian army received $416.9 million, primarily for the training of several counternarcotics battalions, police assistance only totaled $115.6 million.[7] In the 2002 and 2003 aid packages, the Colombian army continues to be the primary recipient of United States military assistance.

With the approval of the first Plan Colombia aid package in June 2000, the United States Congress specified that the president must certify that the Colombian armed forces are acting to suspend and prosecute those officers involved in human rights violations and to enforce civilian court jurisdiction over human rights crimes, and that concrete measures are being taken to break the links between the military and paramilitary groups. This legislation, however, gives the president the prerogative to waive this condition if it is deemed that vital United States national interests are at stake. On August 22, 2000 President Bill Clinton invoked the waiver. And although human rights organizations, the UN High Commissioner for Human Rights, and the State Department affirm that little or no improvements have been made in satisfying the human rights requirements set forth in the original legislation, President George W. Bush certified Colombia in 2002.

With the end of the peace process, human rights in Colombia have been further marginalized. (President Pastrana called off the process with the FARC on February 20, 2002, after continuous setbacks and halts in the peace talks, as well as late 2001 attempts on the part of the United Nations and several countries to serve as intermediaries and revive the process.) Several components of President Álvaro Uribe's national security strategy have caused alarm in human rights circles. Shortly after taking office on August 7, 2002, Uribe declared a state of interior commotion (*Estado de Conmoción Interio*), a constitutional mechanism that allows the executive to rule by decree. In addition to expanding the judicial powers of the police and military, plans to increase the size of the armed forces, create a network of government informants, and build peasant security forces are already under way. In a letter to the Colombian president on August 26, 2002, UN High Commissioner for Human Rights Mary Robinson expressed concern about Colombia's lack of human rights progress and suggested that some of the security measures adopted by the Uribe administration may be incompatible with international humanitarian law. In its November 2002 report on Colombia, Human Rights Watch also criticized the recent reversal of several investigations of military officers suspected of collaborating with paramilitaries.

WEAKENING THE STATE

Inherent to America's growing concern with Colombia is the perception that the state has become "weak" when it comes to confronting the domestic crisis and maintaining it within the country's national boundaries. (The new National Security Strategy of the United States, made public in September 2002, explicitly identifies weak states as a threat to global security because of their propensity to harbor terrorists.) Thus, in addition to combating drugs and terrorism and reducing human rights violations, another stated goal of United States policy is to enable the Colombian military to reestablish territorial control over the country as a necessary step toward state strengthening.

Although state weakness has been a permanent aspect of Colombian political history, during the 1990s the country's deterioration quickened—with the logic of United States "drug war" imperatives playing a direct role in this process. The expansion and consolidation of drug-trafficking organizations in Colombia during the 1980s were intimately related to increasing United States domestic consumption of illegal substances, as well as the repressive policies traditionally applied to counteract this problem. America's demand for drugs and Washington's prohibitionist strategies created permissive external conditions in which the drug business in Colombia could flourish. The appearance of these organizations coincided with unprecedented levels of corruption in the public sphere, growing violence, and decreasing levels of state monopoly over the use of force.

The dismantling in the mid-1990s of the Medellín and Cali drug cartels—the two main drug-trafficking organizations in the country—gave way to fundamentally different drug-trafficking organizations that combined greater horizontal dispersion, a low profile, and the use of a more sophisticated strategy that made them even more difficult to identify and eradicate. Part of the void created by the disappearance of these two cartels was filled by the FARC and the AUC, which became more directly involved in certain aspects of the drug business between 1994 and 1998. As a result, one might also conclude—correctly—that United States drug consumption and its counternarcotics strategies have also exacerbated the Colombian armed conflict, providing diverse armed actors with substantial sources of income without which their financial autonomy and territorial expansion might not have been as feasible.

The propensity of the United States to interpret the drug problem as a national security issue, in combination with the use of coercive diplomatic measures designed to effectively confront this threat, has forced the Colombian state to "securitize" its own antidrug strategy. One underlying assumption of this "war" is that the use of external pressure is a crucial tool by which to achieve foreign policy objectives in this area, and that United States power is an enabling condition for the success of coercive diplomacy. But realist-inspired counternarcotics efforts ignore that policy orientations in source countries must necessarily answer to domestic as well as international exigencies. If domestic pressures are ignored on a systematic basis, growing state illegitimacy and state weakness can result; in an already weak state, this strategy can accelerate processes of state collapse.

With the Samper administration, the United States drug decertifications of 1996 and 1997 and the continuous threat of economic sanctions combined with domestic pressures that originated in Samper's lack of internal legitimacy to force the government to collaborate vigorously with the United States.[8] As noted, between 1994 and 1998 the Colombian government undertook an unprecedented fumigation campaign that, while returning impressive results in terms of total coca and poppy crop eradication, saw coca cultivation itself mushroom during the same period. More significantly, the fumigation campaign had tremendous repercussions in those parts of southern Colombia where it was applied. In addition to provoking massive social protests in the departments of Putumayo, Caquetá, Cauca, and (especially) Guaviare, guerrilla involvement with drugs heightened during this period, and the FARC strengthened its social base of support among those peasants involved in coca cultivation. The absence of the Colombian state in this part of the country largely facilitated the assumption of parastate functions (administration of justice and security, among others) by the guerrillas. Paramilitary activity also increased with the explicit goal of containing the guerrillas' expansion. The result was the strengthening of armed actors and the intensification of the conflict. Although the United States was clearly not directly responsible for creating this situation, the excessive pressure placed on the Samper government to achieve United States goals did make it worse.

At the same time, Samper, because of the taint of drug money, was ostracized by the United States; increasingly, Colombia became identified as a pariah state within the international community.[9] The political costs of the country's reduced status globally were significant; during his term in office Samper received only two official state visits to Colombia, by neighboring countries Venezuela and Ecuador. On an official tour through Africa and the Middle East in May 1997, the Colombian president was greeted in South Africa by news that President Nelson Mandela had been unable to meet him. Equally considerable were the economic costs. Colombia was precluded from receiving loans from international financial institutions during the time in which the country was decertified by the United States, while United States foreign investment was dramatically reduced.

The "Renarcotization" of Relations

Confronted with growing evidence that it had aggravated Colombia's domestic crisis, Washington became increasingly sensitive to the issue of state weakness and attempted to develop a more comprehensive strategy toward the country when Andrés Pastrana was elected president in 1998. This shift in policy partly explains the initial willingness of the United States to adopt a "wait-and-see" strategy regarding the peace process Pastrana initiated with the FARC in early November 1998. Moreover, because of the marked deterioration in the political sphere, it became difficult to ignore the calls of an increasingly strong civil movement for a negotiated solution to the country's armed conflict. Thus, during the first year of his government, Pastrana was able to effectively navigate between domestic pressures for peace and United States exigencies on the drug front. But less than a year later, the assassination in early March 1999 of three United States citizens at the hands of the FARC, along with growing difficulties in the peace process itself, led to a change in both the United States and the Colombian postures and facilitated the ascendance of the drug-war logic once again.

This "renarcotization" of the bilateral agenda saw the emergence of Plan Colombia in late 1999. At home the Colombian government was able to circumvent domestic pressures by manipulating information about its intentions. This was achieved mainly through the publication of distinct versions for public consumption (in both Colombia and Europe) of arguments in which peace (and not the drug war) were adeptly presented as the centerpiece of Plan Colombia's strategy. Public statements by the government downplaying the strong emphasis the United States version of the plan placed on the drug problem reinforced this idea. When the United States Congress approved the Colombian aid package in mid-2000, sustaining this argument became increasingly difficult, primarily due to the large military component (80 percent of the total) that was designated for the drug war. Instead, the Pastrana government attempted to highlight the approximately $200 million earmarked for initiatives related to alternative development, assistance to displaced persons, human rights, and democracy, while discouraging public debate concerning the significant weight attached to the military and counternarcotics aspects of the package.

Just as war-weary Colombians welcomed Andrés Pastrana's proposal for peace in 1998, a country tired of the failed peace process overwhelmingly elected Álvaro Uribe on a national security and war platform in 2002. Uribe's plans for reestablishing state control over the national territory and for crushing militarily those armed actors unwilling to negotiate on the government's terms—goals widely supported by the Colombian population—rely heavily on United States military assistance. The use of that aid for counterinsurgency and counterterrorism is conditioned on a series of measures with which the Colombian government must comply. In addition to adopting explicit commitments in the "war on drugs," including fumigation efforts that surpass those of previous administrations, the Uribe government must implement budgetary and personnel reforms within the military and apportion additional national funding for its own war on drugs and terrorism. Some of these monies will be accrued through the creation of new taxes and reductions in the size of the state, but social spending is likely to be reduced as well. In early August 2002, Washington also requested a written statement from Bogotá conferring immunity for United States military advisers in Colombia as a precondition for the continuation of military aid.

Although at first glance Colombia and the United States share a common objective—winning the war against armed groups in the country—Colombia's insertion into the global antiterrorist dynamic leaves scant room for autonomous decision-making by the new president. In the future, the hands-on, take-charge attitude that has won Uribe a high public approval rating could be blocked by decisions made in Washington. For example, the September 2002 request for the extradition to the United States of a number of paramilitary leaders and several members of the FARC on charges of drug trafficking may work at cross-purposes with future peace talks. Although it is highly unlikely that negotiations with the FARC will resume anytime soon, on December 1, 2002 a cease-fire was declared by the paramilitaries, who have said they would like to negotiate with the government. The United States has been reluctant to state whether the extradition requests, or its classification of Colombia's armed groups as terrorists, would be revoked in the event of new peace negotiations.

The Wrong Profile

United States policy in Colombia has worked at cross-purposes in terms of reducing the availability of illegal substances, confronting human rights violations, and strengthening the state. In all these areas, United States actions may actually have made an already grave situation worse. The worldview that has molded Washington's twin wars on drugs and terrorism constitutes an extremely narrow framework through which to address the complex problems Colombia faces. National security, defined exclusively in military terms, has taken precedence over equally significant political, economic, and social considerations. Until this perspective undergoes significant change, United States policy will continue to be ill equipped to assist Colombia in addressing the root causes of its current crisis.

Notes

1. For a discussion of the role of drugs in United States–Colombian relations from 1986 to the present, see Arlene B. Tickner, "Tensiones y contradicciones en los objetivos de la política exterior de Estados Unidos en Colombia," *Colombia Internacional*, nos. 49–50, May–December 2000; and "U.S. Foreign Policy in Colombia: Bizarre Side-Effects of the 'War on Drugs,'" in Gustavo Gallón and Christopher Welna, eds., *Democracy, Human Rights, and Peace in Colombia* (Notre Dame: University of Notre Dame Press, Kellogg Series, forthcoming).
2. The first disbursement of United States aid for Plan Colombia, a multipronged strategy presented by the Pastrana administration to address problems of peace, state building, poverty, drugs, and the rule of law in the country, was made in fiscal year 2000–2001.
3. The United States and the Colombian governments argue that Roundup Ultra, which is a type of glyphosate and is used for aerial fumigation in Colombia, does not have secondary effects in human beings or surrounding plant life. But the manner in which Roundup is used in Colombia is troubling because it is applied in concentrations that exceed the technical specifications established by the manufacturer and sprayed from planes at a great distance as a defensive measure against ground fire; moreover, an additive mixed with the glyphosate to make it better stick to coca leaves also causes it to adhere to human skin and other plants.
4. In the mid-1990s, before United States military assistance to Colombia began to increase, government officials often admitted that, for Colombia, counternarcotics and counterinsurgency were essentially the same. In a 1996 interview conducted by Human Rights Watch with Barry McCaffrey, then head of the United States Southern Command, McCaffrey conceded that these facets constituted "two sides of the same coin."
5. All three of Colombia's largest armed groups, FARC, the leftist National Liberation Army (ELN), and the AUC, are classified by the United States State Department as terrorist organizations.
6. Angel Rabasa and Peter Chalk, *Colombian Labyrinth: The Synergy of Drugs and Insurgency and Its Implications for Regional Stability* (Santa Monica, Calif.: RAND, 2001).
7. An additional $330 million in police and military aid was provided through the counternarcotics budgets of the State and Defense Departments.
8. The decertifications occurred because every March the president of the United States is required to present a report to Congress certifying whether a country involved in the drug trade is in compliance with United States counternarcotics efforts. Colombia was found to be in noncompliance and thus "decertified."
9. On June 20, 1994, one day after Samper won the second round of the presidential elections, Andrés Pastrana, the conservative party candidate, released an audiotape in which Cali cartel leaders Gilberto and Miguel Rodríguez Orejuela were overheard offering several million dollars to the Samper campaign. A series of accusations and denials concerning this allegation, labeled "Proceso 8,000," ensued.

Arlene B. Tickner *is the director of the Center for International Studies and professor of international relations at the Universidad de los Andes, Bogotá, Colombia. She is also a professor of international relations at the Universidad Nacional de Colombia.*

Root Causes of Chechen Terror

Alon Ben-Meir

Although terrorism is without exception reprehensible, as long as the United States and other powers, including Russia, continue to ignore its root causes, the prospects for diminishing and eventually eliminating it will remain practically nonexistent.

Intelligence estimates originating in the United States, Israel, and Europe, as well as experiences on the ground have shown that in the past three years the ranks of terrorist organizations have dramatically swelled. That is, for every terrorist killed or captured, two or more are joining one terrorist group or another.

The stubborn refusal by the U.S. government, and now Russia's, to acknowledge that the use of force to combat terrorism, albeit necessary at times, will neither reduce or eradicate the scourge of terrorism only adds to the problem. For anything real to be accomplished, far greater attention must be focused on the social, economic, political, and ethnic/religious conflicts and grievances that create the environment for and the motivation to commit acts of terror.

The terrorism in Russia, the direct result of the Chechens' struggle, offers a stark example of how a terribly misguided policy leads to increasingly tragic consequences when the root causes of the struggle are ignored. Chechens, recognized as a distinct people since the seventeenth century, bitterly opposed Russia's conquest of the Caucasus, which began in 1818 and was completed in 1917. They were and are of an entirely different ethnicity than the rest of Russia, with a separate culture, religion, and historic background.

After Soviet rule was reestablished in 1921, the autonomous region of Chechen was created in 1922. In 1934 it became part of the Chechen-Ingush region and was made into a republic in 1936. The Chechen collaboration with the Germans in World War II prompted the Soviets to deport many Chechens to Central Asia. After the war, most of the deportees were gradually repatriated, and in 1956, the republic of Chechnya was reestablished.

During the Soviet regime, the Chechens suffered greatly from discrimination, cruelty, and institutionalized abuse. With the collapse of the Soviet Union in 1991, the Chechen Parliament seized the opportunity and declared the republic's independence. The tensions between Russia and Chechen President Dzhokhar Dudayev escalated into warfare in 1994. Grozny, the capital was totally devastated by the Russian army and tens of thousands of Chechens killed. This national tragedy affected every man, woman, and child in Chechnya. For much of the past 10 years, Russian military and security forces have continued to persecute the Chechens. According to several major human rights watch groups, abuses in Chechnya are as rampant as they were under Stalin, with disappearances, rape, imprisonments, abductions, and other severe violations commonplace.

The terrorist attacks that recently indiscriminately killed innocent children at a Beslan school, and elsewhere in Russia, airplane passengers, subway commuters, and theater audiences are abhorrent acts that must be condemned in the strongest terms. But Chechen terrorism, however abhorrent, must be seen through the prism of what has befallen the Chechen people. To view them as pure wanton acts of terrorism rather than in their historical context defies logic and will only contribute to even greater tragedies in the future.

It is true that Chechen militants are influenced by Wahhabism, a strict form of Sunni Islam practiced in Saudi Arabia, and are aided by Islamist terrorist groups, especially al-Qaida. This, however, should neither make the war against Chechnya a war against terrorism, nor blind the Russian government to its own role in what has happened and to accepting its responsibility to deal with the Chechens' legitimate grievances.

Chechen terrorism will not end because of more repression and preemptive strikes against the Chechen people. It will end when the Russian government admits it made terrible mistakes in the past, rectifies these injustices by recognizing the rights of the Chechen people, and reaches a political settlement through negotiation. A settlement leading to Chechen self-rule while at the same time safeguarding Russia's national security and economic interests (oil and gas) is not impossible. In fact, the two go hand-in-hand. They can be realized only if the Russian government and the Chechen rebels recognize each other's legitimate requirements and national interests.

Russian President Vladimir Putin's fears that allowing the Chechens self-rule will have a domino effect throughout the

Caucasus, are legitimate only to the extent that Chechnya becomes completely independent and ceases to be a part of the Russian federation. As reported by the New York Times, the Chechen rebels presented extreme demands during their seizure of the school in Beslan, including the withdrawal of Russian troops from Chechnya, the inclusion of Chechnya as a separate state within the commonwealth of the former Soviet states, and the restoration of order in the region. But the Times also reported that these demands, according to Ruslan Aushev (president of Ingushetia from 1993 to 200), who was sent by the Kremlin to Beslan to negotiate with the rebels, could have formed the basis for a negotiated settlement. And only negotiations that address the Chechen grievances, he added, can prevent future war.

Putin does not have only two choices: that of giving Chechnya complete independence or of permanently subjugating its people. (In fact, there is nothing in the history of this conflict, which spans more than two centuries, to indicate that the Chechens will ever submit to Russian domination.) Rather, given the historical reality and present situation, the only realistic solution is for Chechnya to remain part of the Russian Federation yet be permitted to run its own internal affairs as it sees fit.

For Putin to equate negotiating with the Chechen rebels to negotiating with al-Qaida, as he did recently, is both disingenuous and dangerously misleading. He may wish to find a common cause with President Bush by looking at the terrorism phenomenon in black-and-white terms. That perspective, however, will not solve the problem of the terrorism he faces any more than it has solved it for the United States. Putin will sooner than later have to answer to the Russian people about how many more of them will die before he recognizes that his strategy has failed. He should learn from the Bush administration's failure in confusing Saddam Hussein with al-Qaida and from the Israeli-Palestinian conflicts, which have claimed the lives of many thousands with no end in sight, that root causes cannot be ignored. By amassing more power under the pretext of fighting terrorism and trampling on democracy in Russia, Putin will not solve the Chechen problem or lessen the Russian people's pain over their terrible losses now or in the future.

Only a negotiated settlement with the Chechens will stop the vicious cycle and prevent this war from spreading into other republics in the region, a situation that could set the south of Russia and perhaps the entire Caucasus on fire.

Alon Ben-Meir *is the Middle East project director at the World Policy Institute and a professor of international relations at the Center for Global Studies at New York University.*

End of Terrorism?

ETA and the Efforts for Peace

MEREDITH MOORE

A nationalist hard-line party of the Basque region, which consists of northern Spain and parts of southwestern France, has asserted Basque independence for the past 40 years. This party, known as Batasuna or *Soziidistii Abvnzjileak*, has been fighting for the autonomy of three of northern Spain's Basque provinces, but it has been declining in power since being formally banned by the conservative People's Party government in 2003. The organization that is commonly assumed to be its military wing, *Eitskadi Ta Askatasuna* (ETA), has been responsible for the deaths of over 800 people since 1968. After decades of struggle, top Batasuna officials appear to want to achieve their goals through more political means, yet the new Socialist government of Spain under Prime Minister Jose Luis Rodriguez Zapatero remains skeptical. The question is whether Batasuna's proposed political methods will bring peace to this troubled region and whether the Spanish government, which has long been debating the issue of Basque self-determination, will ever believe Batasuna's claims.

Due to the increased cooperation of the Spanish and French police forces, several key Batasuna and ETA figures have been arrested. The two states have made a concerted effort to combine their forces and intelligence in an effort to disrupt potential violent attacks. As a result, an open letter from several jailed Batasuna leaders in August 2004 encouraged ETA to use diplomacy as a method to pursue its goals, since terrorism, according to them, "was not serving any purpose." In October 2004, a major raid netted Mikel "Antza" Albizu Iriarte, one of the alleged leaders of ETA, and he and other top officials have also begun to urge the party to end its violent tactics.

These numerous arrests (over 400 ETA members are currently in prison) seem to have greatly shaken the internal structure of Batasuna, and the October raid was a very serious blow for the separatist party. There have been no deaths attributed to ETA since 2003, and some government officials have taken this fact as a sign that Batasuna is losing ground and becoming increasingly weaker due to party in-fighting and conflicting goals.

Nevertheless, there have been several bombings in public places, sueh as outside office buildings and in public parks, causing material damage; these disturbances show that there is still some terrorist activity. In November 2004, Batasuna released a proposal to negotiate peace with the Spanish government that involved demilitarization. Arnaldo Otegi, the leader of Batasuna, has hinted that ETA intends to "let the weapons fall silent" and is "seeking a definitive peace scenario." In February 2005 Batasuna even sent an open letter to French President Jacques Chirac asking him to speak with ETA and commence negotiations, but the appeal has not been answered.

The People's Party, which was in power until March 2004, also refused negotiations with ETA due to its unwillingness to renounce the use of violence. They banned Batasuna in 2003 due to its ties to the terrorist group, and although support for Batasuna and ETA remained strong in the Basque region, the government claimed to have scored a significant victory. Indeed, there have been no deaths attributed to ETA since the outlawing of its supposed political arm. The People's Party, however, did not completely ignore Basque complaints. It strove to find a solution to the question of Basque sovereignty and proposed giving the region some autonomous powers, while keeping it tied to Spain. These concessions were not enough for the Basque people, and the problem remains unsolved.

Although Spain elected to replace the People's Party with the Socialists in March 2004, largely due to the premature and false accusation made by the People's Party that a terrorist attack on several Madrid trains was committed by ETA, the current Socialist administration also refuses to negotiate with ETA, and the government's policy towards and dealings with the terrorist group did not change with the election. Justice Minister Fernando Lopez Aguilar stated, "We do not want a single word with

ETA or anything that moves in its entourage." Despite the group's attempts at peace, the Socialists remain skeptical, especially since these peaceful overtures have been disagreeable to the younger members, led by Garikoitz Aspiazu, of the active forces of ETA. This disagreement is an ominous foreshadowing of continual terrorism and militancy in the region. Moreover, ETA has continued to plant minor bombs around the country', as the November proposals also did not overtly declare that ETA would end its violent tactics, an omission that the Socialist Spanish government lamented. Zapatero remarked this past December, "ETA knows that is has only one destiny and that is to end violence and throw down its weapons." Thus, terrorism and the conflict between ETA and Spanish authorities will not end until ETA itself formally renounces using violence to achieve its goals. Even a majority of Basque inhabitants told pollsters in February that they prefer that the government begin talks with ETA only after it makes this formal renunciation.

The lack of clarity in Batasuna's overtures and the internal squabbles have definitely weakened its credibility. Thus the only apparent path toward peace remains with ETA; until it pledges to put down its weapons, the Spanish government will continue to doubt ETA's supposed wish for peace and Batasuna's claim that it will end ETA's terrorist ways. For Spain, the lack of action on both sides unfortunately might mean continued violence in the region and throughout the country.

UNIT 5

Terrorism in America

Unit Selections

15. **Homegrown Terror**, Michael Reynolds
16. **Speaking for the Animals, or the Terrorists?**, Scott Smallwood
17. **Women and Organized Racial Terrorism in the United States**, Kathleen M. Blee
18. **José Padilla and the War on Rights**, Jenny S. Martinez

Key Points to Consider

- Why does "homegrown" terrorism receive less attention than international terrorism?
- Should spokespersons or defenders of terrorist groups be considered terrorists?
- Why are women in the United States more likely to be involved in terrorism against minorities rather than the government?
- Can civil rights be protected in the age of terrorism?

Student Website

www.mhcls.com/online

Internet References

Further information regarding these websites may be found in this book's preface or online.

America's War Against Terrorism
http://www.lib.umich.edu/govdocs/usterror.html

Department of Homeland Security
http://www.dhs.gov/dhspublic/index.jsp

FBI Homepage
http://www.fbi.gov

ISN International Relations and Security Network
http://www.isn.ethz.ch

The Militia Watchdog
http://www.adl.org/mwd/m1.asp

The Hate Directory
http://www.bcpl.lib.md.us/~rfrankli/hatedir.htm

Domestic terrorism remains a difficult topic for many in the United States. While Americans are all-too-willing to believe in "evil forces" with origins in other countries, many become uncomfortable at the thought of U.S. citizens, men and women, as perpetrators of political violence. Many refuse to believe that a system as free, open, and democratic as ours can spawn those who hate and wish to destroy the very system that has bestowed on them tremendous individual freedoms including the right to political dissent.

American reactions to domestic terrorists vary. While many Americans are outraged by domestic terrorism, some terrorists, like Eric Rudolph—responsible for four bombings including attacks on the Olympics in Atlanta, two women's clinics, and a bar—have achieved cult-hero status, with bumper stickers and T-shirts popularizing Rudolph's near-legendary flight from law enforcement officials. Groups like the Animal Liberation Front (ALF) and the Earth Liberation Front (ELF) continue to attract apologists, searching for ways to justify or explain the violent behavior of otherwise "good Americans." Even the case of Timothy McVeigh, who was prosecuted and executed for the Oklahoma City bombing, has attracted some that continue to believe in an international conspiracy with origins in the Middle East, despite evidence to the contrary. This apparent schizophrenia is echoed in media reporting, public opinion, and public policy.

While the media demonizes foreign terrorists, it tends to humanize American terrorists. Stories of American terrorists often emphasize a human-interest perspective. Stories about Minnesota, middle-class, soccer mom Jane Olson or the young, idealistic, obviously misguided, "American Taliban" John Walker or even the psychologically unbalanced, log-cabin-recluse Ted Kaczynski make good copy and are designed to elicit sympathy or empathy in a larger audience. In its efforts to explain how or why "good" Americans have gone "so bad," the violence and victims are often ignored.

Public opinion and public policy are also subject to this apparent dissonance. While the American public and U.S. policymakers appear to care little about the legal rights or physical detention of foreigners suspected of association with terrorist organizations, the legal rights of domestic terrorists are often the subject of intense public scrutiny and debate.

The selections in this unit look at the problem of terrorism in the United States. The first article looks at right-wing groups. Michael Reynolds argues that, preoccupied with the pursuit of Islamic terrorists, the U.S. government has downplayed the threat posed by extremists and right-wing groups. The second article examines the role of a professor at the University of Texas-El Paso, who is a spokesperson and advocate for the Animal Liberation Front (ALF). The article questions whether vocal support for a terrorist organization and its objectives constitutes terrorism. The third article examines the role of women in right-wing groups, making the case that women are more likely to participate in loosely structured groups whose direct targets are minorities. Finally, Jenny Martinez, a lawyer for José Padilla, an American citizen detained as an "enemy combatant," describes the legal hurdles Padilla's lawyers have faced while trying to ensure due process. She uses Padilla's case to illustrate how the civil liberties have been sacrificed in the name of security.

HOMEGROWN terror

A bomb is a bomb. A chemical weapon is a chemical weapon.
It won't matter to the victims whether their attacker's name is Ahmed or Bill.

Michael Reynolds

ON APRIL 10, 2003, A TEAM of federal agents armed with a search warrant entered a storage unit in a small Texas town and were stunned to find a homemade hydrogen cyanide device—a green metal military ammo box containing 800 grams of pure sodium cyanide and two glass vials of hydrochloric acid. The improvised weapon was the product of 62-year-old William Joseph Krar, an accomplished gunsmith, weapons dealer, and militia activist from New Hampshire who had moved his operations to east central Texas just 18 months earlier.

That same day the *New York Times*'s Judith Miller reported from south of Baghdad that the U.S. Army Mobile Exploitation Team had "unearthed ... precursors for a toxic agent ... banned by chemical weapons treaties." That turned out not to be the case. What the army team found was fewer than two dozen barrels of organophosphate used in pesticides.

In Chicago a month earlier, Joseph Konopka, a 26-year-old anarcho-terrorist had been sentenced on one count of possession of a chemical weapon. In March 2002, Konopka, who had appropriated an abandoned Chicago Transit Authority storage room under downtown Chicago, was found and arrested in a tunnel beneath the University of Illinois at Chicago. Konopka was a fugitive from federal charges in Wisconsin, where he had hit power substations, radio transmitters, and utility facilities in a 1999 firebombing campaign that caused 28 power outages.

An accomplished systems programmer and hacker, Konopka had assumed the online moniker of "Doc Chaos" and recruited bright teenage accomplices into a cadre he called the "Realm of Chaos." One of these accomplices was arrested with him. In a search of Konopka's subterranean outpost, authorities found nearly a pound of sodium cyanide along with substantial amounts of potassium cyanide, mercuric sulfate, and potassium chlorate.

The young man never gave a reason for why he had stockpiled the deadly chemicals, except to say they were not for "peaceful purposes."[1] He is now serving more than 21 years in federal prison for sabotage and possession of a chemical weapon.

By the time Krar pleaded guilty to one count of possession of a chemical weapon on November 11, 2003, two U.S. citizens—Krar and Konopka—were accountable for far more chemical weapons than have been found in post-war Iraq.

Chemical capers

Without diminishing the significance of Konopka's attacks on local infrastructure in Wisconsin, Krar's is the more disturbing case, given the size and capabilities of his arsenal, his history, his ideology, his discipline, and his expertise. Despite that, his case attracted little national media attention. There were no press conferences called by Attorney General John Ashcroft and FBI Director Robert Mueller, even though Krar presented the most demonstrably capable terrorist threat uncovered in the United States since September 11, 2001.

Krar's cyanide apparatus was only the most dramatic component of an extraordinary arsenal Krar and his common-law wife, Judith Bruey, had stashed in their Texas storage facility.

Along with the sodium cyanide, hydrochloric acid, acetic acid, and glacial acetic acid, Krar and Bruey's armory included nearly 100 assorted firearms, three machine guns, silencers, 500,000 rounds of ammunition, 60 functional pipe bombs, a remote-controlled briefcase device ready for explosive insertion, a homemade landmine, grenades, 67 pounds of Kinepak solid binary explosives (ammonium nitrate), 66 tubes of Kinepak binary liquid explosives (nitromethane), military detonators, trip wire, electric and non-electric blasting caps, and cases of military atropine syringes.[2]

Although Krar presented the most capable terrorist threat since 9/11, there were no press conferences called by Attorney General John Ashcroft or FBI Director Robert Mueller.

The storage unit also contained an extensive library of required reading for the serious terrorist: U.S. military and CIA field manuals for improvised munitions, weapons, and unconventional warfare; handbooks on assault rifle conversions to full-auto and manufacturing silencers; formulas for poisons and chemical and biological weapons; descriptions of safety precautions in handling; and information on means of deployment. Many of the same easily acquired, open-source materials, translated into Arabic, were found in Al Qaeda terrorist manuals recovered in Afghanistan and Europe.

As for Krar's cyanide device, according to investigators, the blueprint and formula for the weapon were in the form of a computer printout and handwritten notes that Krar either took down from the internet or obtained from another source.

Margaret Kosal, an analyst of chemical and biological weapons at Stanford University's Center for International Security and Cooperation, determined that Krar had enough sodium cyanide, combined with hydrochloric acid, to produce enough hydrogen cyanide gas to kill more than 6,000 people under optimal conditions for attack.

According to Kosal, such a device, if employed in a 9x40x40-foot conference room, would probably kill half of the room's occupants within one minute of inhalation. If the room was crowded, *immediate* fatalities could number as many as 400. More fatalities would probably follow as a result of age or ill health.

If the cyanide gas were dispersed in a larger space, say an enclosed shopping mall, hotel lobby, or school, the number of deaths would be diminished. In any case, the psychological impact on the public of a successfully deployed improvised chemical weapon in the United States would be enormous.

Kosal observed that it was not that difficult to obtain substantial amounts of sodium cyanide and acid. "While [sodium cyanide] is a DEA [Drug Enforcement Administration]-controlled compound," any notion that it "can only be acquired legally for specific agricultural or military projects is wrong," Kosal pointed out. The price of 2.5 kilograms purchased over the web "is only $105 ... without an educational discount."[3]

In statements made to the FBI after his arrest, Krar claimed he obtained his sodium cyanide and acids from a gold-plating supply house.

Found by a fluke

Krar's admission about how he acquired chemicals may be one of the few straightforward statements he has made to federal authorities since they stumbled upon him nearly two years ago.

On January 24, 2002, a UPS package was misdelivered to a family on Staten Island, New York. After inadvertently opening the packet, Michael Libecci discovered all array of identification documents with different names, all of which featured a photograph of the same man. Libecci turned over the packaging and its contents to the Middletown, New Jersey, police, who called the FBI in Newark.

The documents included a North Dakota birth certificate for "Anthony Louis Brach," a Social Security card for "Michael E. Brooks," a Vermont birth certificate for Brooks, a West Virginia birth certificate for "Joseph A. Curry," a Defense Intelligence Agency identification card, and a U.N. Multinational Force Observer identification card. The package was addressed to Edward S. Feltus in Old Bridge, New Jersey. The return address was for William J. Krar at a mailbox in Tyler, Texas. Along with the bogus IDs was a letter from Krar to Feltus.

"Hope this package gets to you O.K.," wrote Krar. "We would hate to have this fall into the wrong hands."

Seven months went by before FBI agents finally talked to Feltus, a 56-year-old employee of the Monmouth County Department of Human Services. On August 8, 2002, Feltus admitted that the forged documents were intended for him, saying he wanted "an ace in the hole" against some future "disaster" or government crackdown. The documents, he said, would allow him to travel "freely in the United States."

Feltus told the agents that he was a member of the New Jersey Militia, an anti-government right-wing paramilitary group permeated with white nationalism. FBI agents later discovered that after he requested the false IDs from Krar, Feltus had stored more than 100 rifles and pistols at a fellow militia member's residence in Vermont. Seven months after the Oklahoma City bombing, leaders of the New Jersey Militia traveled to central New Hampshire on November 22, 1995, to meet with representatives from militias in Rhode Island, Massachusetts, Maine, Connecticut, and New Hampshire, to form the New England Regional Militia. Its purpose, according to the New Jersey Militia Newsletter, was to "establish an operational framework" to "develop and implement tactical contingency plans" that would include "supply, training, public relations, and intelligence gathering."[4] A key player in the New Hampshire militia at the time was William J. Krar.

Nothing unusual?

Born in 1940, Bill Krar grew up in Connecticut, learning all about guns from his father, a gunsmith for Colt Firearms. Although he didn't serve in the military, weapons and militaria were his life's centerpiece and primary source of income. His formal education ended

Unusual suspects?

March 2000—Larry Ford, biochemist, gynecologist, and anti-government paramilitary activist, kills himself in his suburban southern California home, where police find buried caches of machine guns, assault rifles, thousands of rounds of ammunition, C-4 explosives, and canisters of ricin. In Ford's refrigerators agents discover 266 vials of assorted pathogens including salmonella, cholera, botulism, and typhoid. Ford also seemed to have some kind of working relationship with South Africa's apartheid-era bioweapons program, Project Coast.

November 2000—James Dalton Bell, anti-government militant and MIT-trained chemist, violates his parole and is charged with threatening Internal Revenue Service (IRS) agents. Bell had been convicted and sentenced to prison in 1998 on charges of attacking a Portland, Oregon, IRS office with a "stink bomb." While searching Bell's home lab, federal agents find three assault rifles, explosives, sodium cyanide, and precursor chemicals for the production of sarin nerve gas. Bell claims he had successfully manufactured a small amount of sarin. On one of Bell's computers authorities find the names and home addresses of more than 100 IRS and FBI agents along with those of local law enforcement personnel.

October 2001—Envelopes containing high-grade anthrax are mailed to a tabloid media office in Boca Raton, Florida, to major media offices in New York City, and to two Democratic senators' offices in Washington, D.C. Five die and scores are hospitalized. Although the case remains unsolved, some investigators believe the primary suspect or suspects are likely from within the American anti-government extremist movement.

March 2002—Joseph Konopka, a 25-year-old anarcho-hacker and anti-government extremist, is arrested in Chicago and charged with possession of a chemical weapon, sodium cyanide.

October 2002—Members of the Idaho Mountain Boys, an anti-government paramilitary group, are charged with possession of machine guns, plotting to kill a federal judge and a police officer, and helping fellow members escape from jail. The leader of the group, Larry Eugene Raugust, is also charged with possessing numerous bombs and booby-trap devices. Raugust is one of the leaders of the U.S. Theater Command, a nationwide militia network formed in 1997.

April 2003—William J. Krar, Judith Bruey, and Edward Feltus are arrested after Krar's weapons and chemical weapon cache are found in a Texas storage facility. Krar and Bruey are charged with possession of a chemical weapon.

October 2003—Norman Somerville, a 44-year-old anti-government militiaman is arrested. Near his rural Michigan home agents find an underground bunker stocked with 13 machine guns, thousands of rounds of ammunition, hundreds of pounds of gunpowder, and manuals on guerrilla warfare, "booby traps," and explosives. On the walls are pictures of President George W. Bush and Defense Secretary Donald Rumsfeld with the crosshairs of a rifle scope drawn over them. Somerville had also outfitted his van and Jeep Cherokee with machine guns. Somerville and his comrades had planned to use these "war wagons" in attacks on law enforcement agents. The men had been spurred by the killing of fellow militia member Scott Woodring in a shootout with police, who were attempting to arrest him for the shooting death of a state trooper. At the time of his arrest, Somerville warns of a "quiet civil war" brewing in rural Michigan. On August 10, 2004, Somerville pleads guilty to possession of machine guns and pledges "to cooperate in the hunt for shadowy rebels." Two other members of his group also enter guilty pleas to federal weapons charges.

June 2004—A Bureau of Alcohol, Tobacco, and Firearms (ATF) raid uncovers a cache of castor beans, formulas for extracting ricin from the beans, and bomb-making materials in the suburban apartment of Boston-area anti-government activist Michael Crooker. Crooker, once convicted for fraud and possession of a machine gun, had come under scrutiny by the U.S. Postal Service for shipping a silencer to a compatriot in Ohio.

July 2004—After his arrest in south Florida in November 2003, Michael Crooker. John Jordi, a Christian anti-abortion zealot and ex-U.S. Army Ranger, is sentenced to five years in prison for plotting to bomb abortion clinics, gay bars, and certain churches. U.S. District Judge James Cohn rules that Jordi is not a terrorist because federal laws require that plots have an international component to be considered terrorism.

August 2004—Two young Tennessee leaders of a dozen-member anti-government paramilitary cell called the American Independence Group (AIG) are charged with attempted bank robbery and possession of assault weapons. The AIG had intended to use the money from the bank robbery to fund their operations. According to federal agents, the AIG hated the federal government and select ethnic groups and talked of declaring war on law enforcement and killing President Bush.

Also in August, a 66-year-old convicted counterfeiter and antigovernment activist, Gale Nettles, is charged with plotting to build an ammonium nitrate/fuel oil truck bomb and use it to attack the federal courthouse in Chicago. According to federal agents, Nettles had stored 500 pounds of ammonium nitrate in a Chicago-area storage facility and was seeking more from an FBI informant. The informant also stated that Nettles was looking to make contact with either Al Qaeda or Hamas.

after a few semesters in community college. He married and had a son, but later divorced.

Exactly when Krar was drawn into the American radical-right constellation of illegal weapons dealing, shadowy paramilitaries, white nationalism, and anti-Semitic global conspiracies is unknown. According to some who knew him at the time, Krar was active in the movement by the mid-1980s. In 1984 he was dealing guns without a federal firearms license under the name of International Development Corporation (IDC) America, listed at his home address in Bedford, New Hampshire. Krar continued using IDC America as the front for his gun dealing for the next 18 years.

From 1984 to 1985, Krar was ostensibly working as a sales representative for a home-building distributorship in the nearby town of Hooksett, near Manchester. But a co-worker recalls Krar as a highly secretive man who always had a pistol at his side and stacks of *Soldier of Fortune* in his office—and who had almost no knowledge or experience of the construction business.

In an interview, this fellow employee remembered Krar and another colleague disappearing for weeks at a time, heading off to Costa Rita and other locations in Central America, even though the building supply company had no dealings beyond New England. Krar's mysterious travel activities and gun dealing occurred at the height of the

Reagan administration's "private sector" paramilitary and weapons operations in support of the *contras*.[5]

It was also in 1985 that Krar was arrested by New Hampshire state police and charged with impersonating a police officer. He entered a no-contest plea, paid a fine, and was released. Three years later, in 1988, the building supply company where Krar worked went out of business following a fire that destroyed its building. That same year, Krar stopped filing federal income taxes and effectively dropped out of the system.

In April 1995 Krar became the subject of an FBI–Bureau of Alcohol, Tobacco, and Firearms (ATF) investigation stemming from a thwarted kidnapping and bombing plot concocted by white supremacist paramilitaries in Tennessee.

Following the arrest of Timothy McVeigh, Sean Patrick Bottoms and his brother Brian became outraged by media coverage of the Oklahoma City bomber and plotted to kidnap or kill Nashville television newscaster John Siegenthaler, now with MSNBC.[6]

After an informant tipped law enforcement to the plot, the Bottoms brothers fled to east Texas, where they were arrested on April 30, 1995. During a search of the brothers' residences, FBI and ATF agents found pipe bombs, large amounts of explosives, illegal weapons, thousands of rounds of ammunition, and a business card for "William J. Kaar" of IDC America. When questioned, Sean Bottoms told agents that "William Kaar" was in fact William J. Krar. Bottoms said he had lived in Manchester in late 1994 and early 1995 and had used Krar's IDC address on his driver's license.

After his indictment on explosives charges, Bottoms said that Krar, using the alias "Bill Franco," was active in the militia movement and that Krar said he had known about the Oklahoma City bombing before it happened. Krar had also said there were more attacks to come.

In July 1995, ATF agents questioned Krar, who told them that all he had done was sell some ammo and military surplus to Bottoms. Bottoms was then given a polygraph examination, which he failed.

Krar continued his involvement with the militia movement in New England. He later told FBI agents that this was when he first obtained sodium cyanide and began working with it, though there is no evidence to support the claim.

In a separate but simultaneous FBI–ATF investigation in Boston, Krar was under scrutiny for his role in a militia with "strong/violent antigovernment views." According to an FBI affidavit, a federal law enforcement source advised that Krar was a "white supremacist due to the anti-Semitic and anti-black literature" seen at his IDC America business in Manchester, where Krar hosted militia meetings. The source went on to say that Krar was "a good source of covert weaponry for white supremacist and anti-government militia groups."

Bruey, who was president of Krar's IDC operation at the time, told an undercover federal agent of her hatred of "U.S. government policies toward its citizens" and that she believed the government was afraid "military surplus would end up in the hands of citizens rejecting their government."

Despite this report and evidence from the Bottoms case that Krar was illegally selling firearms without a federal license, Krar and Bruey were left free to soldier on until they ran afoul of federal agents in 2001.[7]

Self-store stockpiles

For a decade Krar conducted his operations out of multiple mail drops and storage units. According to investigators, he had no permanent shop, but would work in the storage units, running in electrical cords to power his tools and run lights. In June 2001 there was a fire at one of the two self-storage facilities Krar and Bruey were using in New Hampshire. Firemen discovered that Krar's unit contained thousands of rounds of ammunition and numerous firearms. ATF agents were called in and found among the weapons an assault rifle converted to full-auto. Krar said the weapons and ammo were the property of his employer, Ed Cunningham of Eagle Eye Guns, who had a federal firearms license. Krar and Bruey packed up the weapons and ammo and left, moving their stockpiles to another self-storage unit.

The manager of the new storage facility, Jennifer Gionet, recalled Krar vividly, describing him as "wicked anti-American."

According to an FBI affidavit, Gionet said that Krar told her "the U.S. government was corrupt" and that he "hated [it] and all of the cops." Krar went on to say he "hated Americans because they are 'money-hungry grubs.'" He also told Gionet he had several businesses in Costa Rita and offered to set her up with some "financial investments" down there. On September 11, 2001, Krar told Gionet that he knew the attacks in New York and Washington were going to happen and that there would be more in Los Angeles or Manchester. Gionet immediately reported this conversation to the local police, who notified the FBI.

Having drawn the ATF's attention in June, Bruey and Krar moved their operations to Flint, Texas, in October 2001. A young woman, Dawn Philbrick, who had become Krar's lover, accompanied them. Bruey rented two units at Noonday Storage, opened a mail drop at Mail Boxes Etc., and rented a secluded rural house. Krar was soon back at work in one of the storage units fabricating explosive devices and silencers and converting assault rifles to machine guns. He was also fabricating hundreds of magazines and receivers for Bushmaster Firearms in Maine, manufacturer of the civilian models of the AR-15 assault rifle, one of whose weapons was used by John Mohammed and Lee Malvo, the D.C. sniper team. Krar had done similar work for Bushmaster since at least 1998.

On his home computer, Krar was applying another of his skills—counterfeiting identification documents for his compatriots in the antigovernment paramilitary underground. He was as accomplished with counterfeiting as he was with guns and bombs. Over the years Krar used at least seven aliases with four different Social Security numbers and numerous business fronts. Some he used in the gun trade, some within the anti-government movement or offshore. According to investigators, Krar didn't sell counterfeit documents on the open market but gave them away to others in the white supremacist and anti-government movements. It is not known how many sets of documents Krar distributed.

The investigation into Krar and his bogus IDs was slow in developing. It took the FBI until November 2002, 10 months after opening the case, to begin surveillance on Krar, even though they had his address.

Busted ... for drugs

His activities were being monitored when, on January 11, 2003, Krar was arrested by a Tennessee state trooper in the course of a routine traffic stop on the outskirts of Nashville. Searching Krar's rental car, Trooper William Gregory found a plastic bag containing "seven marijuana cigarettes, one syringe of unknown substance, one white bottle with an unknown white substance, 40 wine-like bottles of unknown liquid," as well as two pistols, 16 knives, a stun gun, a smoke grenade, three military-style atropine injections, 260 rounds of ammunition, handcuffs, thumb cuffs, fuse ropes, binoculars, and "other various close hand-to-hand combat items." Gregory also found Krar's passport, a birth certificate, a California credit union card for "William Fritz Hoffner," and a Christian missionary identification card with Krar's photo and the name "W. F. Hoffner." There were also other documents, letters to IDC America, and four pages of what appeared to be a clandestine operations plan for cross-country travel and communications. Gregory busted Krar on marijuana possession, took him into custody, and impounded the car.

The Tennessee state police then called the local FBI, which in turn contacted its Tyler, Texas officer to inform him that Krar had been arrested. Nashville FBI Special Agent David McIntosh, who interviewed Krar that day in the local jail, said that Krar told the FBI that the weapons and ammo were his and that the other material was part of his stock as a gun dealer who worked gun shows. Krar said he was moving back to New Hampshire to help his girlfriend get out of a bad divorce, and that he didn't know that the bag contained marijuana—that it was something a waitress had left beside his plate that he had just stuffed into his pocket.[8]

Krar bonded out of jail the next day, leaving his property behind, and drove west out of Nashville. Trooper Gregory opened the jar of white powder, took a whiff, assumed it was cocaine, and threw it into an evidence locker. After the discovery of Krar's chemical weapon four months later, the powder was brought to the FBI lab, where it tested positive as sodium cyanide. Federal authorities have not released information as to what liquid was found in the 40 wine bottles.

Neither Krar nor Bruey gave up any information following their arrest. Krar accepted a plea agreement on possession of a chemical weapon in exchange for Bruey getting a lighter sentence—five years. Otherwise, all the leads federal agents were able to generate were through documents obtained in the searches of Krar and Bruey's storage units, house, and vehicles. The FBI and Justice Department say the case is still under investigation.

The face of terror

Krar was no mere "yarn-spinner," as his defense attorney once portrayed him. The federal agents and prosecutors who interviewed Krar described him as highly intelligent, dedicated, well organized, extremely manipulative, and very dangerous. His radical right, anti-government commitment clearly grew out of the gun and paramilitary culture that spread rapidly following the Gun Control Act of 1968 and the white backlash to civil rights that arose the same year.

Krar carried copies of *Hunter* and *The Turner Diaries*, the fictional *ur*-texts for white American revolution and terrorism written by the late neo-Nazi William L. Pierce under the pseudonym Andrew McDonald. The books have been favorites of white nationalist and anti-government terrorists for more than two decades. McVeigh carried stacks of the Diaries with him during his army days and later sold them at gun shows, pressing copies into the hands of potential allies in the years running up to the Oklahoma City bombing. When federal agents searched McVeigh accomplice Terry Nichols's home in Kansas after the bombing, they found copies of the *Diaries* and *Hunter*.

Krar had a copy of *Hunter* with him when he was stopped in Nashville. *The Turner Diaries* was found in his Texas storage unit along with all four volumes of Henry Ford's classic anti-Semitic conspiracy text, *The International Jew*, and Holly Sklar's left-wing expose of the "new world order," *Trilateralism*. Like McVeigh, Krar drew his anti-government worldview from across the spectrum of right and left.

A terrorist with limited resources would probably consider Krar's chemical weapon an attractive tool. The equipment needed is simple, and the chemicals are readily available from chemical supply houses. Procedures are easily obtained in the open literature, including on the internet. Unlike the more stringent requirements for production of satin or other nerve agents, fabricating hydrogen cyanide devices demands no greater skills beyond those needed to construct an ammonium nitrate-anhydrous hydrazine truck bomb like that used in Oklahoma City.

There is no doubt that Krar was capable of producing such devices. He had the means and technical information to do so. He was well organized, disciplined, highly skilled, and comfortable in the production of improvised explosive devices.

Would he have used such a weapon? FBI agents and Justice officials who interviewed Krar don't think so. But their assessment is not reassuring.

"I don't believe Krar would've used this himself," said Brit Featherston, assistant U.S. attorney and Justice's anti-terrorism coordinator for the Eastern District of Texas. But, "If Krar came across a Tim McVeigh or an Eric Rudolph [now facing trial for fatal bombings at the Atlanta Olympics and an abortion clinic] it would be a disaster. I don't believe he'd have a problem with putting this into their hands and sending them on their way."[9]

FBI Special Agent Bart LaRocca, lead agent in the Krar investigation, agrees. "Krar was a facilitator and a provider," said LaRocca. "There was no indication that he was marketing his bombs or chemical devices. They were intended to be used against the government or in the event of 'martial law.' They were for those willing to use them or those he could manipulate into using them."[10]

An attack with such a weapon on an office building, an abortion clinic, a large auditorium, or a shopping mall could be managed by a single, disciplined individual. A terrorist cell armed with several devices could de liver a coordinated attack at different locations. Either scenario would have a tremendous psychological impact that would go far beyond immediate casualties. The bombings in Oklahoma City and during the Atlanta Olympics are stark examples of how homegrown terrorists are just as willing to indiscriminately kill men, women, and children as are their radical Islamic counterparts elsewhere in the world.

Ashcroft's Justice Department has shown almost no interest in what was, until the calamitous events of September 11, the primary domestic terrorism threat—the white nationalist, anti-government militia movement and its corollaries with theocratically driven terrorism, primarily abortion-related assassinations and bombings.

The upheavals in U.S. counter-terrorism and anti-terrorist intelligence agencies after the 2001 attacks on the World Trade Center and the Pentagon have not resulted in more nimble thinking about domestic terrorist threats. Apart from the FBI's longtime obsession with environmental and animal rights extremists, the Bureau's primary target for surveillance, investigation, and detention seems to be either immigrants of Arab descent or those who profess Islam as their religion. Although this focus is understandable, it is not commendable.

Had a similar sodium cyanide device been found in a storage unit rented by someone named Khalid or Omar, there is little doubt that Ashcroft and Mueller would have conducted a press conference and that it would have been the story of the week. For some reason, the Krar case was not deemed important—even though the facts of the case show that no other case has demonstrated a comparable and immediate threat. Certainly not the case of Jose Padilla, the small-time thug who merely *talked* in vague terms about a radiological bomb, or that of the young Muslims who thought it might be a good idea to travel to Pakistan for jihad training. Those incidents have been front page fodder, touted by the FBI as cases involving "significant" terrorism. Homegrown terrorists with functional cyanide gas devices are surely as serious a threat.

While Al Qaeda has no need to reach out to indigenous terrorist cells within the United States—or vice versa—a tactical confluence between them would not be surprising. Many anti-government extremists hold beliefs compatible with Islamic terrorist factions worldwide. They are violent]y against the "new world order," especially with regard to U.S. government and corporate policies. They are uniformly anti-Semitic or anti-Israel and are totally opposed to the war in Iraq.[11]

Apart from the one-off attack in September 2001 by 19 young foreigners, most of them Saudis, the country's most deeply entrenched and most persistent domestic terrorist threat has come from within its own borders and at the hands of its own citizens. It would be folly to believe that the American terrorist underground, after 15 years of sustained and bloody action, has somehow just given up and disappeared.

Perhaps Ashcroft and Mueller called no press conferences because the discovery of Krar's arsenal was a fluke. It was not the result of a proactive federal anti-terrorism intelligence effort targeting the American right-wing paramilitary movement.

Just like Ashcroft and the FBI, the press thinks of "angry white guys" like McVeigh, Nichols, and Rudolph as old news.

Well, maybe Bill Krar and his compatriots don't fit the politically marketable paradigm, the post-9/11 face and faith of terrorism—non-white and Muslim. But such thinking may prove unnecessarily fatal in times to come. Consider the Krar case fair warning.

Notes

1. Affidavit of FBI Special Agent Leslie Lahr, *United States of America v. Joseph Konopka*, Case No. 02CR, March 9, 2002.
2. Plea Agreement, *United States of America v. William J. Krar*, Case No. 6:03CR36, November 13, 2003; Plea Agreement, United States of America v. Judith L. Bruey, Case No. 6:03CR36 (02), November 7, 2003.
3. Interview and e-mail exchanges with Margaret E. Kosal, July 6, 2004.
4. *New Jersey Militia Newsletter*, January 1996.
5. Interview with former co-worker of William Krar, August 26, 2004.
6. Interview with John Siegenthaler, August 20, 2004.
7. Affidavit of PB1 Special Agent Bart LaRocca, *United States of America v. William J. Krar*, Case No. 6:03M12, April 3, 2003.
8. Interview with FBI Special Agent David McIntosh, Nashville, Tennessee, July 20, 2004.
9. Interview with Brir Featherston, assistant U.S. attorney, Tyler, Texas, August 4, 2004.
10. Interviews with FBI Special Agent Bart LaRocca, Tyler, Texas, July 7, July 22, 2004.
11. Michael Reynolds, "Virtual Reich," *Playboy*, February 2002.

Michael Reynolds writes on political and religious extremism and terrorism. He has contributed to Playboy, U.S. News & World Report, Rolling Stone, 60 Minutes, *and* Newsweek. *He was senior analyst at the Southern Poverty Law Center's Intelligence Project from 1994 to 2000.*

Speaking for the Animals, or the Terrorists?

"History will be written about them. They will be defamed now, but they will be taught to children later. They will write storybooks about these people, like Harriet Tubman."

—Steven Best, an associate professor of philosophy at the U. of Texas at El Paso, speaking about animal-rights activists

Scott Smallwood

THE EL PASO, TEX., suburbs stretch west, across the Rio Grande and into New Mexico. Just on the other side of the river lies this community of 2,500 people. In a gated housing development here, not far from a golf course and around the corner from a swimming pool, Steven Best lives alone—just him and his 10 cats.

He has turned one of his bedrooms into a home office. Tall bookshelves line three walls. Along another, near his computer, are posters of big cats, including a tiger and a mountain lion. Several cat beds sit on the desk.

None of this looks much like a press office for terrorists.

That is what Mr. Best has been accused of running. He is not a cat-loving professor of philosophy, some argue, but a mouthpiece for terrorists who attack university laboratories, factory farms, and pharmaceutical companies.

Mr. Best, an associate professor of philosophy at the University of Texas at El Paso, is one of the leading scholarly voices on animal rights. In the past year, though, he has taken on a role that, he believes, has gotten him into hot water in Washington and in his own department.

In December he co-founded the North American Animal Liberation Press Office, which answers questions and helps disseminate information about actions by the Animal Liberation Front. The animal-liberation group, along with another extremist group called the Earth Liberation Front, has been labeled by the Federal Bureau of Investigation as one of the most serious domestic-terrorism threats in the country.

While the groups have not killed anyone, since 1990 they have committed more than 1,200 attacks causing millions of dollars in damages, according to the FBI. Many of them have been attacks against university labs, including a raid on a Louisiana State University building in Baton Rouge in April.

The ALF has no real structure. A few people engage in a "direct action"—such as destroying computers, setting fire to a building, or "liberating" mink from fur farms—and then one or two days later they send a communiqué taking credit for the attack.

That's where Mr. Best comes in. The Animal Liberation Press Office has a fax machine in California ready to receive messages from ALF activists. After an attack, Mr. Best and three other press officers post the information on their Web site and answer reporters' questions. "We explain what the ALF is about," he says. "We interpret the nature of an action, and we explain why the action was taken."

Mr. Best balks at being labeled a "spokesman," however. That suggests a centralized organization and a hierarchy the ALF simply does not have, he says. He sees himself as a philosopher in action, a scholar who has the courage to put his theories into practice.

"I'm not in the ALF," he says, standing in his kitchen sipping coffee with soy milk. "If I were, I'd be wearing a mask, and you wouldn't know who I was. You're either above ground or you're underground, or you're a moron looking to get caught."

That may be true, says David Martosko, but it does not clear Mr. Best. Mr. Martosko, research director of the Center for Consumer Freedom, a Washington-based nonprofit organization that campaigns against the animal-rights movement, argues that groups like People for the Ethical Treatment of Animals and press officers like Mr. Best form an above-ground support system for the ALF. Mr. Martosko and other opponents maintain that Mr. Best is a spokesman for terrorists, who should not be able to use his faculty post to indoctrinate his students and offer violent extremists a dash of intellectual legitimacy.

"If a university professor were out there saying that abortion-clinic bombers had a good plan going," he says, "the university would sever the guy's tenure in a New York minute."

Friends and Enemies

In November members of the Animal Liberation Front broke into laboratories used by the University of Iowa's psychology department. They took 88 mice and 313 rats. They destroyed computers and poured acid on papers and equipment, causing about $450,000 worth of damage. In an anonymous message sent after the attacks, the perpetrators wrote: "Let this message be clear to all who victimize the innocent: We're watching. And by ax, drill, or crowbar—we're coming through your door."

The attacks sparked debate and outrage at Iowa, which was heightened two months later when a law-student group invited Mr. Best to speak at the university. Some psychology professors tried unsuccessfully to get the university's president to cancel the speech. Although the professors managed to retrieve much of their research data from the damaged computers, they were later hit with 400 unsolicited magazine subscriptions after their home addresses were posted by ALF activists.

At the Iowa speech, Mr. Best compared the Animal Liberation Front to 19th-century abolitionists, likening their "direct actions" to the Boston Tea Party. In a statement after the speech, he wrote: "Please, let's stop the hypocrisy and put our moral outrage in perspective. For every window or computer smashed in the name of animal liberation, a billion animals suffer horrendous torture and death at the hands of exploiters operating the fur farms, factory farms, slaughterhouses, rodeos, circuses, and laboratories."

In May Mr. Best's notoriety increased when he was invited to appear at a hearing of the U.S. Senate Committee on Environment and Public Works about animal- and eco-terrorism. Mr. Best declined, in part because he was scheduled to be out of the country.

But that did not keep his name from being tossed around on Capitol Hill. James M. Inhofe, a Republican from Oklahoma and chairman of the Senate committee, wondered why Mr. Best had been invited to speak at Iowa. "We cannot allow individuals and organizations to, in effect, aid and abet criminal behavior or provide comfort and support to them after the fact," he said at the hearing. "Just as we cannot allow individuals and organizations to surf in between the laws of permissible free speech and speech that incites violence when we know the goal is to inspire people to commit crimes of violence."

At the hearing, Mr. Martosko of the Center for Consumer Freedom urged Congress to investigate the ties between above-ground groups like PETA, and the ALF itself. The center, a nonprofit organization based in Washington, regularly battles animal-rights activists like PETA and gets its financial support mainly from food and restaurant companies, although it declines to identify them.

During his testimony, Mr. Martosko showed several photographs of Mr. Best. In one, the professor has his arm around Kevin Kjonaas, an activist who faces charges in New Jersey related to a campaign against a British drug-testing company. In another, Mr. Best stands next to Rodney Coronado and Gary Yourofsky, who both did time in prison for animal-liberation attacks.

"These are the guys, these are the ALF felons that Best hangs out with," says Mr. Martosko in a telephone interview. "This is his crowd."

The El Paso professor "may not have his fingerprints on a matchbook," Mr. Martosko continues, "but I think it's clear that he's trying to recruit young people." He points to an essay in which Mr. Best wrote that the movement needed to raise "an army of activists" and that the animal-liberation movement would benefit by "growing roots in academia."

Mr. Best fumes at those allegations. He calls Mr. Martosko a "vulgar McCarthyite" who is seeking to demonize the entire animal-rights movement. "I certainly do not recruit students into the ALF," he says. "I don't even know anyone in the ALF."

And he is proud to put his arm around men that some consider terrorists. "They are heroes of mine," he says. "History will be written about them. They will be defamed now, but they will be taught to children later. They will write storybooks about these people, like Harriet Tubman. And I respect them infinitely more than I respect a philosopher lost in abstraction."

No More Burgers

About 25 years ago, Mr. Best sat down in a White Castle restaurant in Chicago for his regular late-night fast-food fix. He had dropped out of high school a few years earlier and was working in factories and driving a truck while dreaming of a career as a jazz guitarist. At 22 he had scrapped the music fantasies and enrolled in a Chicago-area community college. After earning an associate degree in film and theater, he went south to the University of Illinois at Urbana-Champaign. On this night, at 2 a.m., he was a half-drunk undergraduate in philosophy, stopping for a double cheeseburger.

"I had an epiphany that what I was eating was the blood, the juices, the bones, the tendons, the muscles of an animal," he says. "It repulsed me, and I spit it out." Despite a few more attempts to eat fast-food burgers, he became a vegetarian and later a vegan, shunning all animal products.

At the time he was active in human-rights issues. He even sheltered illegal El Salvadoran refugees for a time. He went on to earn a master's degree in philosophy from the University of Chicago and a doctorate from the University of Texas at Austin in 1993. That year he started his first and only faculty job, at El Paso.

He has a mop of reddish-brown hair and dresses casually. At 49, he looks youthful despite the few spots of gray in his goatee. He describes himself as a philosophical bachelor who wants to preserve his autonomy by staying away from marriage. Maybe, he says, that's why he gets along with the cats: they all prefer a little isolation. All 10 of them were rescued by Mr. Best—some caught in alleys, some delivered as tiny kittens in a paper sack. That's how Shag and Willis arrived. They each fit into the palm of his hand and had to be bottle fed. Willis, with his "infinitely deep eyes," still sucks on Mr. Best's finger. Slim Shady, the newest addition, is young and aggressive. Chairman Meow, a long-haired Himalayan, is the loudest and the friendliest around strangers.

When talking about his cats, he does not look like the stubborn philosophy professor whom one colleague at El Paso describes as "very acerbic." Another calls him simply "angry." No one doubts his passion about animals, though.

For him, the campaign for animals' rights is the modern abolition movement. Expanding the notion of rights to include animals—just as it was expanded in centuries past to include women and blacks—is the next evolutionary step in man's moral progress, he says. And years from now, Mr. Best argues, people will look back on the way humans currently treat animals the way we now look at slavery.

Most of us focus on the differences between humans and animals, he says, but they are outweighed by the similarities. "We can say that we can build spaceships and they can't," he says. "We can write algebraic equations and they can't. Therefore, we must be better." But that doesn't make any sense to him.

"They know the difference between pleasure and pain," he says of animals. "They can experience a life of happiness and joy or a life of suffering and misery. If they can suffer like us, then they have a right to live a life free of suffering."

For him, and many in the animal-rights movement, the key is sentience. "That's why I'm not engaged in an act of cruelty right now," he says as he stands at his kitchen counter eating a pear. "That's why there is no Pear Liberation Front—because this does not have a brain, this does not have a nervous system. It cannot feel pain."

In El Paso Mr. Best has his own talk-radio show; he serves as vice president of a local vegetarian group; and he led a lobbying effort to get Sissy the elephant, who had been videotaped being beaten by an El Paso zookeeper, sent to a sanctuary in Tennessee. But Mr. Best staunchly maintains that he has never participated in any ALF action. He says he has never dressed up in black at night and broken into a mink farm or released animals from a university lab. That's not to say that he hasn't been arrested a few times.

In the 1990s, he participated in a PETA campaign against Wendy's restaurants. He walked through a waiting phalanx of police officers to the front of the restaurant, jumped on top of the counter, and loudly declared: "This store is closed for cruelty."

"The point," he says, "was not to have people right then and there stop eating their hamburgers. It was for them to think about the issues." He was taken down, handcuffed, and tossed into jail for the rest of the day. "For me this is philosophy in action. This is taking a stand for what you believe in."

Guilt By Association?

That's just what Rodney Coronado admires in Mr. Best. Mr. Coronado, one of the best-known ALF members, spent nearly five years in prison in the 1990s for an arson at a mink-research laboratory at Michigan State University. Now a regular on the animal-rights speaking circuit, Mr. Coronado says he is no longer a member of the animal-liberation group.

"Steve is true blue," he says. "A lot of professors are just chasing recognition and capitalizing on the latest social movement. He's not like that."

Mr. Best has always supported him and other animal-rights activists, Mr. Coronado says. "He kicks in $100 when we need it, and he always buys guys like me dinner because he knows we're broke."

But that should not make Mr. Best or other academics who support animal rights into de facto members of a terrorist organization, he says. "You can call me whatever you want, but these other people are professors," Mr. Coronado says. "That should be protected, sacred ground."

Nevertheless, having convicted arsonists as friends raises eyebrows. Brian O'Connor is a retired professor of anatomy and cell biology at the Indiana University School of Medicine. He worked with dogs in his research on the nervous system and always feared that he would be the victim of an ALF attack. That never came, but in his retirement he has become an expert on the movement and keeps tabs on developments through his blog, Animal Crackers. He says he agrees with the Center for Consumer Freedom that the best way to stop the ALF and people like Mr. Coronado and Mr. Best "is to attack them and reveal them for what they really are."

He is determined to expose what he believes is the incoherence of the animal-rights argument. Mr. Best is a "total ideologue," he says. "He's being very rational in living the AR philosophy." But that philosophy is flawed, Mr. O'Connor says. "If you honestly believe that each life is of equal value, then my dog's life has the same value as my wife's life."

That would mean, say opponents of Mr. Best like Mr. Martosko, that humans should stop using animals in medical research, a move that would have prevented the development of various lifesaving treatments, such as the polio vaccine.

"If you're talking about rights, then you have to be arbitrary about where you draw the line," says Mr. O'Connor. "Why sentience? Why not just being a human being?" Human beings are making that distinction, he says, and it can be pursued to strange extremes. "You can bestow rights to rocks. Or say that rivers have rights not to be desecrated. Why should we buy into their assumptions?"

Ousted

Since starting the animal-liberation press office and taking a higher profile role in defending the Animal Liberation Front, Mr. Best has lost his post as chairman of the philosophy department at El Paso. He maintains that the two things are inextricably linked. Others at the university, including Mr. Best's dean, say there is no connection.

This much is clear: The six other members of the philosophy department voted unanimously in March to recommend to their dean that Mr. Best be removed as chairman, a position he had held for four years.

Mr. Best calls that decision an ambush. He says that while he may not have been the perfect chairman, his colleagues never told him they had problems with him. After the vote, he says, students in the department told him that they heard other professors saying, "If we don't get rid of Best, we're going to have another Ward Churchill on our hands," referring to the controversial University of Colorado at Boulder professor.

But philosophy professors at El Paso say their decision had nothing to do with Mr. Best's activism. "We were extremely unsatisfied with his performance as chairman," says John Symons, an assistant professor. "From our perspective, this has nothing do with his politics. It was a matter of running the practical affairs of the department."

At the same time, Mr. Symons defends his colleague's activism. "Steve is a passionate defender of the Animal Liberation Front," he says. "But he is by no means a recruiter for the Animal Liberation Front."

Howard C. Daudistel, dean of liberal arts at El Paso, says the change in leadership in the philosophy department was solely about administrative issues. "Our position is that Dr. Best, just as any faculty member, has a right to express his views and engage in a discourse off campus or in any setting, as long as he is not representing himself as speaking for the institution," he says.

The dean also says he spoke to Mr. Best about the charge that he was recruiting students to the ALF. "There's no evidence that he has used the classroom to recruit students to take part in any actions," Mr. Daudistel says.

Adrian Paredes, a student of Mr. Best's, says the entire notion is absurd. "The ALF isn't some big group," he says. "It's not an organization you could join. How could you even recruit for that? That's just crazy."

Mr. Best displays a dark sense of humor amid the criticism. As he sits in his living room, defending his activism and talking about his heroes within the animal-rights struggle, yet another cat—a slender black-and-white one named Gadget—strolls into the room.

"He's incredibly sweet," the professor says. "But he like to start fights. He terrorizes the other cats." Gadget meows at the sliding-glass door, and Mr. Best lets him into the backyard. "That's why I call him Osama bin Gadget."

Women and Organized Racial Terrorism in the United States

KATHLEEN M. BLEE

Department of Sociology
University of Pittsburgh
Pittsburgh, Pennsylvania, USA

Racial terrorism—violence perpetrated by organized groups against racial minorities in pursuit of white and Aryan supremacist agendas—has played a significant role in U.S. society and politics. Women have been important actors in much of this violence. This article examines women's involvement in racial terrorism from the immediate post-Civil War period to the present. Although organized racial violence by women has increased over time, this trend may not continue. The strategic directions and tactical choices of Aryan and white supremacist groups are likely to alter the extent and nature of women's involvement in racial terrorism in the future.

In April 2003, 28-year-old Holly Dartez of Longville, Louisiana was sentenced to a year and a day in prison and fined $1,000 for her part in a Ku Klux Klan (KKK) cross burning the previous year. Ms. Dartez, whom the U.S. Attorney's Office characterized as secretary to the local Klan chapter, pled guilty to conspiracy for driving four other KKK members to the residence of three African-American men, recent migrants from Mississippi, where a cross was erected and set ablaze. Among the Klan members convicted in this episode was her husband Robert, described as a leader of the local Klan, who received a 21-month sentence and a $3,000 fine. Despite these arrests and convictions, the African-American men targeted in the attack clearly received the message intended by the Klan's action. All abandoned their desire to move their families to Longville and returned to Mississippi.[1]

That same year, 23-year-old Tristain Frye was arrested for her part in an attack and murder of a homeless man in Tacoma, Washington. The attack was carried out by Ms. Frye and three men, among them her boyfriend David Pillatos, with whose child she was pregnant, and Kurtis Monschke, the 19-year-old reputed leader of the local neo-Nazi Volksfront. The four, all known racist skinheads, had set out to assault a Black drug dealer, but instead attacked Randy Townsend, a 42-year-old man suffering from paranoid schizophrenia. Frye's involvement in the attack was apparently motivated by her desire to earn a pair of red shoelaces, a symbol of her participation in violence against a minority person. Although Frye reportedly made the initial contact with Townsend and admitted to kicking him in the head, hard, three or four times, her agreement to testify against Monschke and the prosecutors' conclusion that she had not been dedicated to White supremacy—despite the Nazi and racist tattoos on her back—were sufficient to get her charges reduced to 2nd degree murder.[2]

A year earlier, Christine Greenwood, 28, of Anaheim, California and her boyfriend John McCabe, already imprisoned for a separate offense, were charged with possessing bombmaking materials, including 50 gallons of gasoline and battery-operated clocks that could be used as timers. Greenwood was described as the co-founder of "Women for Aryan Unity," a group to integrate women into White supremacism, and a member of the militant racist skinhead gang "Blood and Honor." She pled guilty to this charge as well as an enhancement charge of promoting a criminal gang and received a short sentence and probation. She has not been visible in racist activities since her arrest, but both groups with which she was associated continue, with elaborate websites claiming chapters and affiliates across the globe.[3]

The women in these three vignettes were arrested for very different kinds of racist violence. Holly Dartez was involved with a Ku Klux Klan group in a cross-burning, an act whose violence was symbolic rather than physically injurious. Tristain Frye took part in the murder of a homeless man—an act of brutal physical violence—with a racist skinhead group, but the victim was White. Christine Greenwood—with her White supremacist group affiliations and bombmaking equipment—seemed intent on racial mayhem, although her target was unclear. As these cases suggest, women in the United States today participate in acts of racial-directed violence whose nature, targets, and social organization vary considerably.

This article explores women's involvement in racial violence associated with the major organized White supremacist groups in the United States: the Ku Klux Klan, White power skinheads, and neo-Nazis.[4] Such violence is best understood as racial terrorism. As commonly specified in the scholarly literature and by federal counterterrorist agencies, terrorism requires three components: acts or threats of violence, the communication of fear to an audience beyond the immediate victim, and political, economic, or religious aims by the perpetrator(s) (Cunningham 2003, 188; Hoffman 1998, 15; also Crenshaw 1988), each of which is characteristic of White supremacist racial violence. Racial terrorism, then, is considered here as *terrorism undertaken by members of an organized White supremacist or pro-Aryan group against racial minorities to advance racial agendas.*

Considering the violence of organized racist groups as a form of racial terrorism brings together scholarships on terrorism and organized racism that have largely developed in parallel tracks. With few exceptions (e.g., Blazak 2001; Cunningham 2003), research on terrorism has paid relatively little attention to the growing tendency of White supremacism in the United States to adopt the organizational structures, agendas, and tactics more commonly associated with terrorist groups in other places. Similarly, studies of U.S. organized racism have rarely portrayed racist groups as perpetrating racial terrorism, although at least some of their actions clearly fall under the U.S. State Department's definition of terrorism as "premeditated, politically motivated violence perpetrated against noncombatant targets by subnational groups or clandestine agents, usually intended to influence an audience."[5]

To analyze the nature and extent of women's involvement in U.S. racial terrorism, it is useful to consider two dimensions of terrorism. The first is the nature of the intended ultimate target; what organized racist groups consider their enemy. Some acts of racial terrorism are "intended to coerce or to intimidate"[6] governments; others are directed toward non-state actors such as members of minority groups. The second dimension is how violence is organized. Some acts of racial terrorism are strategic, focused on a clear target and directed by the group's agenda. Others are what the author terms "narrative," meant to build solidarity among racist activists and communicate a message of racial empowerment and racial vulnerability but instigated outside of a larger strategic plan (Blee 2005; Cooper 2001; Perry 2002). This article explores women's roles in racial terrorism from the immediate post–Civil War era to the present along these two dimensions. It concludes with a proposition about the relationships among women's participation, definitions of the enemy, and the organization of terroristic violence in the U.S. White supremacist movement.

Perceptions of the Enemy

Organized White supremacism has a long history in the United States, appearing episodically in response to perceptions of gains by racial, ethnic, or religious minorities or political or ideological opportunities (Chalmers 1981). White supremacism is always organized around a defined enemy. African Americans have been the most common enemy of organized racists over time, but other enemies have been invoked on occasion. The massive Ku Klux Klan of the 1920s, for example, targeted Catholics, Jews, labor radicals, Mormons, and others, in addition to African Americans. Today's small and politically marginal KKK, neo-Nazi, and White supremacist groups express little hostility toward Catholics, Mormons, or labor radicals, focusing their anger instead on Jews, Asian Americans, gay men and lesbians, and feminists, in addition to African Americans and other persons of color.[7]

Each wave of organized White supremacism has been accompanied by terrorist acts against its enemies, although the nature of such violence has varied considerably over time and across groups. The KKK of the 1920s, for example, amplified its periodic and vicious physical attacks on African Americans, Catholics, Jews, and others with frequent terrifying displays of its economic and political strength, including rallies and parades, boycotts of Jewish merchants, and electoral campaigns (Blee 1991; Chalmers 1981). Today, a few White supremacist groups, particularly some KKK chapters and Aryan-rights groups such as the National Association for the Advancement of White People (NAAWP), a former political outlet for racist media star David Duke, follow the lead of the 1920s Klan in seeking public legitimacy for agendas of White rights, but most openly advocate or engage in physical violence against enemy groups. The form of such racial terrorism ranges from street-level assaults against racial minority groups to efforts to promote a cataclysmic race war.

Women's involvement in racial terrorism is strongly associated with how organized White supremacists define the nature of their enemies. Although variation in the racist movement, even within a single historical period, makes it impossible to make broad generalizations that hold for every racist group, there have been changes since the Civil War in how racist groups define their enemies. Particularly important for understanding women's involvement is the changing focus on members of racial/ethnic groups versus institutions of the state as the primary enemy of organized racist groups. The following sections focus on definitions of the enemy in three major periods of racial terrorism: the immediate postbellum period, the first decades of the twentieth century, and the present.

Postbellum Racial Terrorism

Most White supremacist groups in the immediate postbellum period directed their violence at racial minority groups, but the ultimate target of their actions was the state apparatus imposed on the defeated southern states during the Reconstruction era. The quintessential White supremacist organization of this time—the Ku Klux Klan—emerged in the rural south in the aftermath of the Civil War, inflicting horrific violence on newly emancipated African Americans and their White, especially northern, allies. Organized as loose gangs of White marauders, the first Klan may have had a chaotic organizational structure, but its goals and efforts were focused and clear—to dismantle the Reconstructionist state and restore one based on White supremacism. Women played no direct role in this Klan. Indeed, its moblike exercise of racial terrorism on behalf of traditional southern prerogatives of White and masculine authority left no opening for the participation of White women except as symbols for White men of their now-lost privileges and lessened ability to protect "their" women against feared retaliation by former slaves (Blee 1991).

Racial Terrorism in the Early Twentieth Century

The first wave of the KKK collapsed in the late nineteenth century, but its legacy of mob-directed racialized violence continued into the first decades of the twentieth century through extra-legal lynchings and racially biased use of capital punishment to execute African Americans.[8] The re-emergence of the Klan in the late 1910s (a Klan that flourished through the 1920s) substituted political organization for mob rule, enlisting millions of White, native-born Protestants in a crusade of racism, xenophobia, anti-Catholicism, and anti-Semitism that included contestation of electoral office in some states. The violence of this second Klan also took a new form, mixing traditional forms of racial terrorism with efforts to instill fear through its size and political clout and create financial devastation among those it deemed its enemies (Blee 1991).

The targets of lynchings, racially biased capital punishment, and the 1920s Klan were mostly members of racial, ethnic, and religious minority groups; they also constituted its primary enemies. The racial terror of lynching and racially biased capital punishment both depended on state support, either overtly or covertly. Similarly, for the second Klan, located primarily in the north, east, and western regions rather than the south, the state was not an enemy; instead, it was a vehicle through which White supremacists could enact their agendas. Rather than attack the state, in this period, organized racism was explicitly xenophobic and nationalist, embracing the state through an agenda they characterized, in the Klan's term, as "100% American."

Women were active in all aspects of racial terrorism in the early twentieth century, including lynchings and the public celebrations that often accompanied, and added enormously to the terror of, these events. It is difficult to assess the precise role of women in such forms of violence because the historical record is mute about how often a woman tied the noose around a lynched person's neck or struck the match to burn an African-American corpse, or a living person. Yet, it is clear that women were integrally and fully involved in these events. Photographic records of lynchings, often the only means by which these were recorded, show large numbers of women, often with their children, gathered around lynched bodies, partaking in the spectacle with a fervor and brutality that shocks contemporary observers (Allen 2000). The inclusion of women and children helped make such racial murders possible, even respectable, in many areas of the country.

Women also were active in the second KKK, adding more than half a million members to its ranks in female-led chapters, the Women of the Ku Klux Klan. They participated actively and avidly in the terrorist actions of this Klan which, unlike Klans that preceded and followed it, practiced racial terror largely through mechanisms of exclusion and expulsion. Women Klansmembers were instrumental, even leaders, in the effort to rid communities of Jews, Catholic, African Americans, and immigrants through tactics such as financial boycotts of Jewish merchants, campaigns to get Catholic schoolteachers fired from their jobs, and attacks on the property and sometimes the bodies of African Americans and immigrants (Blee 1991).

Part of the explanation for women's increased involvement in racial politics and terror in the early decades of the twentieth century lies in changing gender roles and possibilities in this time. The granting to women of the right to vote in all elections in 1920 made women attractive recruits for the second Klan as it sought to increase its size, financial base, and electoral strength. At the same time, women's increasing involvement in other forms of public life, including prohibition politics, the paid labor force, and civic improvement societies, made women more likely to join racist groups. But women's participation was also the result of tactics of racial organization and violence that were more compatible with the lives of (White) women than had been the case in previous decades. Women could, and did, contribute to the Klan's strategy of creating economic devastation, for example, by spreading vicious rumors about Catholic schoolteachers or Jewish merchants without stepping far from their roles as mothers and consumers. Such factors also made women's participation in mob-directed racial terrorism like lynching more likely. The rigid patriarchal ideas that precluded White southern women's entrance into the first Klan had crumbled significantly by the 1920s, making more acceptable the notion that women could act in the public sphere. Moreover, racial lynchings and other forms of mob-directed racial terrorism often were enacted as largescale community events in which women could join without straying from their primary roles as mothers and wives, for example, by bringing their children to what Tolnay and Beck (1995) termed the "festival of violence" of lynching (also Allen 2000; Patterson 1998).

Racial Terrorism Today

In the later decades of the twentieth century, the nationalist allegiances of many White supremacist groups began to crumble. Much of this shift can be traced to the widespread adoption of new forms of anti-Semitic ideology, especially the idea that the federal government[9] had been compromised by its allegiance to the goals of global Jewish elites. This understanding, commonly summarized in the belief that the United States is a "Zionist Occupation Government (ZOG)," shifted the central axis of organized White supremacism. Additional pressures toward global pan-Aryanism diminished the allegiance of U.S. White supremacism to nationalist agendas and, increasingly, Jews became the focus of its vitriol, with African Americans and other persons of color regarded as the lackeys or puppets of Jewish masters. With this ideological shift—codified in the precepts of the widely embraced doctrines of "Christian Identity," a vicious racist theology that identifies Jews as the anti-Christ—the U.S. government itself became a target of White supremacist violence. The bombing of the Oklahoma City federal building, assaults on federal land management agencies in the West, and a series of aborted efforts to attack other government installations were the outcome of this shift toward the U.S. state as an enemy of White supremacism.

Identifying the state as a primary enemy has had complex effects on the participation of women in organized White supremacism and racial terrorism. Some racist groups have made considerable effort to recruit women in recent years (Blee 2002; Cunningham 2003), especially those, like some chapters of the KKK, that want to develop a durable and intergenerational racist movement. These groups see women as key because of their centrality in family life and their (perceived) lesser likelihood to become police informants. Some neo-Nazi and Christian Identity groups are also recruiting women heavily, but generally to create a more benign image for White supremacism (Blee 2002).

Following the influx of women into racist groups, there has been an apparent rise in the participation of women in racist terrorism, as suggested by the vignettes at the beginning of this article. However, the number of women involved appears to be relatively low, despite their increasing numbers in racist groups. Firm statistics on the gender composition of perpetrators of racially motivated violence are not available (see, e.g., FBI, 2000), but reports compiled by the Southern Poverty Law Center in Montgomery, Alabama (SPLC 2004),[10] the most highly regarded non-official source of such data, indicate that the clear majority of perpetrators are still male. In particular, the SPLC reports indicate that, relative to men, women have low levels of involvement in racial terrorism targeted at state institutions, with somewhat greater involvement in violence directed at racial minority groups.

What can be concluded from this brief history? Although any generalization needs to be treated with caution, given the heterogeneity of organized White supremacism, the historical data examined suggest that in the United States *women are more likely to be involved in organized racial terrorism that is directed at racial/ethnic minorities than racial terrorism directed against the state.*

The Organization of Racial Terror

White supremacism has taken a variety of organizational forms in the United States, each typically associated with a particular form of violence. Much organized White supremacism is highly structured and hierarchical, with clear (if often violated) lines of authority, like the second and subsequent Ku Klux Klans. However, some White supremacist groups are very loosely organized with highly transient memberships and little hierarchy, such as contemporary racist skinheads, which operate like gangs bound together by ideology rather than territory. The following sections consider how the form of racist organization is associated with the level and nature of women's involvement in racial terrorism, although particular racist groups may be involved in different forms of violence. What is proposed is an analytic abstraction meant to highlight specific aspects of racial terrorism rather than a firm typology of racial violence and racist groups.

Structured, Hierarchical Organization

White supremacism is an ideology that puts tremendous value on ideas of hierarchy. Indeed, the very premise of modern-day Western racism is the idea that human society is naturally divided into racial categories that can be ranked by their moral, political, cultural, and social worthiness (Frederickson 2003; Winant 2002). This ideology is mirrored in how racist groups are typically constituted, with strong demarcations between leaders and followers, a high valuation on acceptance of internal authority, and firm boundaries against participation by those of inferior categories, including not only those from enemy groups, but also, at many times, White Aryan women. This form of organization is characteristic of racist groups like the second and subsequent Ku Klux Klans, World War II–era Nazi groups, and some racial terrorist groups in the late twentieth century.

In recent years, a number of those involved in the racist movement have embraced a new structure known as "leaderless resistance," a concept developed in response to racist group's desires to shield themselves from authorities. The principle of leaderless resistance is simple: the activities of racist activists are coordinated by their allegiance to a set of common principles rather by than communication among racist groups. In practice, leadership resistance requires that racist activists develop very small cells in which plans are developed and enacted, with little or no communication between cells that would allow the police to trace a chain of racist groups.

Strategic racial terrorism is generally, although not always, associated with structured, hierarchical groups, including those that follow the model of leaderless resistance. This is violence that is planned, focused on precise targets, and calculated to have predictable consequences. Typically, such violence is developed in a small leadership group and disseminated to members for activation, or, in the case of leaderless resistance, created and executed by a small, tightly knit group. Strategic racial terrorism is exemplified by efforts to foment race war or to terrorize racial minority communities by burning crosses, scrawling swastikas on buildings, or assaulting racial minority persons. It also includes attacks on government agencies or efforts to precipitate cataclysmic economic collapse and social chaos, thereby hastening the demise of the Jewish-dominated government. One example in which a number of women were implicated was a paramilitary survivalist, Christian Identity–oriented group known as the Covenant, Sword, and Arm of the Lord (CSA). Insisting that Jews were training African Americans to take over the nation's cities, CSA members initiated a series of strategic terrorist activities, including firebombing a synagogue and a church and attempting to bomb the pipeline that supplied the city of Chicago with natural gas. When the FBI raided the CSA compound in 1985, they found supplies for further terrorism: weapons, bombs, an anti-tank rocket, and quantities of cyanide apparently intended for the water supply of an undisclosed city.

One woman from a highly structured racist group talked of her involvement in terms that succinctly summarize strategic racial terrorism. In an interview conducted for a study of women in contemporary racist activism (Blee 2002), she told me that she felt it was necessary to:

> prepare yourself for war constantly—don't speak if you can't defend yourself in every way. Prepare by knowing—first of all, then work on guns and ammo, food and water supply, first aid kits, medication, clothing, blankets, try to become self-sufficient and [move] away from the city, if possible. Don't get caught into the "debit" or "marc" cards, etc—[that is, in the] new world order.

This woman, as well as Christine Greenwood whose efforts on behalf of the Women for Aryan Unity included making bombs, are examples of women who participate in strategic racial terrorism. But men are far more likely than women to be arrested for direct involvement in such acts. The strict principles of social hierarchy embraced by most tightly organized racist groups tend to exclude women from leadership, even from inclusion, and thus from a role in executing violence (also see Neidhardt 1992; Neuburger and Valentini 1998; Talbot 2000). Women's involvement in strategic racial terrorism is generally indirect, like Holly Daretz's role as a driver for the Klansmen arrested for the Louisiana cross-burning. This indirect involvement in strategic racist terrorism takes three forms: serving as legitimation, promoting group cohesion, and providing abeyance support. Women are used to *legitimate* strategic racial terrorism by creating an air of normalcy that belies the violence of organized racism (Blee 2002; Dobie 1997), a tactic increasingly common among terrorist groups across the globe (Cunningham, 2003). In the United States, this legitimation role can be seen in efforts like those of the Women's Frontier/Sisterhood, female affiliates of the violent World Church of the Creator (WCOTC), whose Web publications stress benign topics like motherhood that serve to blunt the violent activities of its members, including Erica Chase and Leo Felton, arrested for attempting to detonate bombs to incite a "racial holy war" (Ferber 2004, 7; Rogers and Litt 2004; also Bakersfield *Californian* 24 July 2004). Women also function to *promote group cohesion* in organized racism—making possible its agendas of strategic terrorism—by working to create solidarity within existing racist groups and recruit new members (Blee 2002). An example of this cohesive function is the effort of Women for Aryan Unity's campaign "White Charities—by Whites for Whites"[11] to provide support to imprisoned White racists. This campaign, one of a number in which racist women are involved, target those they term "prisoners of war" through pen pal programs, prison visitation, and aid to the families of POWs as well as by reintegrating former prisoners into the racist movement. And, finally, women create *abeyance support* (Taylor 1989) by standing in for male racist leaders when they die or are in prison. One example is that of Katja Lane, whose husband David was arrested for murder and other crimes during his involvement in the underground Aryan supremacy group, Silent Brotherhood. During David's imprisonment, Katja has risen to prominence in the racist movement for her work in maintaining movement publications and a prison outreach program for White supremacist prisoners (Dobratz & Shanks-Meile 2004; Gardell 2003).

Loose Organization

White supremacist groups that operate with loose, ganglike forms of organization typically exhibit high levels of violence. Indeed, such groups often eschew tighter forms of organization in the effort to avoid detection and arrest for their violent actions.[12] Klansmen who terrorized African Americans and their allies in the immediate postbellum period operated in this way, as do racist skinheads whose thinly linked groups operated under names like "Confederate Hammerskins" or "Blood and Honor."

Loosely organized White supremacist groups often practice what can be termed *narrative* instead of, or in addition to, strategic racial terrorism. Narrative racial terrorism is at least somewhat spontaneous, in which victims are chosen impulsively and without clear purpose, and whose consequences are rarely calculated by the perpetrators in advance. Practices of narrative racial terrorism include street assaults on African Americans, gay men or lesbians, or Jews, like the description of the actions of one racist woman who would provoke her husband to go with her to "find a homosexual or someone and beat them up" (ABC News.com 2004a) or the acts of brutality inflicted on African Americans by the night riders of the first Klan. It also includes acts of violence that seem inexplicable, like the murder of the White homeless man by Tristain Frye and her fellow skinheads, or those that seem attributable to the immaturity or psychological pathologies of their perpetrators, such as violence and savagery against fellow White supremacists or self-inflicted violence (Blee 2002; Christensen 1994; Hamm 1994).

What distinguishes narrative from strategic racial terrorism is not the character of the acts of violence, but its incorporation into a larger set of plans and tactics. Strategic racial terrorism is intensely focused on disabling, undermining, or exterminating those considered to be the enemies of White supremacism. Narrative racial terrorism is less clearly focused on specific enemies; it targets enemies for violence, but that violence also has an internal purpose: to strengthen, sometime even to create, organized White supremacism, to attract new members, to instill a sense of collective identity among existing members and bind them closer to each other, and to instill the passion and commitment that will sustain their efforts into the future.

Women are directly involved in narrative racial terrorism, although in lesser number than are male racists (Christensen 1994; Dobie 1997; but see Blazak 2004). Yet, there is evidence that women's role in narrative racial terrorism may be increasing, as racists skinheads and similar groups attract larger number of women who see themselves as empowered through the enactment of physical violence (Blee 2002). A description of narrative racial terrorism was related by a racist activist, in response to the present author's question about whether she had been involved in physical fights:

> Yes. [With] about 20–25 women, six men. Some of who were nonwhites, i.e., gangbangers—people who don't like people like me so they start trouble with me—and others were White trash traitors who had either screwed me over, started trouble because they don't believe in my ways or caused trouble in the movement. Some were hurting, physically, friends of mine, so I involved myself in it.

It is unclear whether women's increasing participation in groups that practice narrative racial terrorism is due to pull or push factors. It is likely that both are operating. Women may be attracted to groups that practice narrative violence are less likely than those engaged in strategic violence to have the rigid ideological and organizational structures that have excluded women from power and decision making in the U.S. White supremacist movement since its inception. Indeed, there have been at least fledgling attempts to organize all-women racist skinhead groups under the joint banner of "White power/ women power" (Blee 2002), efforts that would be unimaginable in other parts of the White supremacist movement. But it is also the case that groups that practice narrative racial terrorism, like White-power skinheads, can be surprisingly receptive to the inclusion of women because their boundaries are loosely guarded, relatively permeable, and often fairly undefined. For example, it can be more difficult to ascertain who is a member of a group that is bound together by the practice of violence and often-fragile and superficial connections between people than a group that has a more clearly defined agenda, strategy, and sense of what constitutes membership. There are instances in which White-power skinheads have later become active in anti-racist skinhead groups that fight racist skinheads, often with a great deal of violence. Such ideological switching is an indication that commitment to violence may outweigh commitment to racist ideas, a phenomenon rarely found among those who practice strategic racial terrorism.

What can be concluded about the relationship between gender and the organization of racial terror? Again, the diversity within organized racism means that any generalization can only be provisional, but the evidence presented here suggests that *women participate in strategic racial terrorism to a lesser extent than they do in narrative racial terrorism, and women participate in strategic racial terrorism largely through indirect means, whereas women participate in narrative racial terrorism more directly.*

Conclusion

Thus far, the relationships of gender to definitions of the enemy and to the organization of racial terror have been considered separately. The brief case studies of White supremacist groups can also be used to think about the three-way relationships among gender, enemies, and violence, as presented in Table 1.

The case in which the state is perceived as the main enemy and violence is narrative in nature (cell A) is rare in the history of modern U.S. White supremacism. The first Ku Klux Klan is the paradigmatic example, and in this Klan women had no direct involvement either as members or as participants in Klan violence. For the first KKK, women's exclusion is explicable by the specific historical and sociopolitical situation of the Reconstruction-era South and by this Klan's intense emphasis on White men as the protector of vulnerable White women. Whether women would always be excluded from this type of racial terrorism is unclear because there are no major subsequent racist movements that have this set of characteristics. Indeed, this form of racial terrorism is unlikely to recur in the foreseeable future in the United States as it is associated with situations of profound political uncertainty and fluctuations in the organization of the state, as in the Reconstruction era. With the consolidation of federal state power, racial terrorism directed at the state is much more likely to be strategic in nature, both because the enemy is more clearly defined and because the state has the power to monitor and suppress its opponents.

The case in which racial minorities are the primary enemy group and violence is expressed in a narrative form (cell B) is exemplified today by racist skinheads. In these groups, women generally participate substantially less than do men, but women's role appears to be increasing in recent years. A similar situation exists when the state is the enemy, but violence is strategic in nature (cell C). This is the case with many racial terrorist groups today, especially those that target the state as a agent of Jewish domination. For these groups too, women tend to participate at considerably lower rates than men, but their participation has increased in recent years and is likely to continue to increase. Both require very public and assertive actions—the street-level violence of skinheads or bombing campaigns of ZOG-focused groups—that contradict traditional ideas about women's passivity and subservience. Further, participation in these forms of racist terrorism challenges the traditional male leadership and public image of such groups. Yet, it is likely that the barriers to women's participation in these forms of racist terrorism will decline over time. Gender ideologies are crumbling in racist groups as else-

Table 1
Gender, enemies, and violence

Type of violence	Definitions of the enemy: State	Definitions of the enemy: Racial minorities
Narrative	A (no women)	B (some women, increasing)
Strategic	C (some women, increasing)	D (many women, steady)

where in U.S. society (Blee 2002). Moreover, media attention to recent instances of women in gender-traditional societies who are involved in terrorism against the state, in such places as Chechnya, Israel, Germany, and Sri Lanka (ABC News 2004b; Cunningham 2003), as well as women's involvement in domestic terrorism against the U.S. government by groups such as the Weather Underground and Black Panther Party (Brown 1994; Zwerman 1994) have provided models for the incorporation of women into these forms of organized racial terror. These factors are likely to result in an increase in women's activity in narrative forms of terror against racial minorities and strategic forms of terror against the state.

The case in which racial minorities are the enemy and violence is expressed in a strategic form (cell D) is different. This is characteristic of groups like the 1920s Klan or some Klans and other White supremacist groups today. In these, women's participation is often high—although always lower than men's—as this organization of racial terror provides structural openings for women to participate without challenging existing ideas about gender hierarchies. Women in these groups often work to facilitate and promote violence behind the scenes or in less directly confrontational ways. They recruit and cultivate new racist group members and steer them toward ideas of strategic violence, spray-paint swastikas on houses and cars of new immigrants to convince them to move, and burn crosses in the yards of interracial couples. Each of these forms of racial terrorism can be undertaken from within the perimeters of the group's existing gender hierarchies, resulting in a level of women's participation that is higher than other forms of racial terrorism, although unlikely to increase further in the future.

This brief history of women's role in organized U.S. racial terrorism suggests that women are fully capable of participating in the most deadly kinds of terrorist activities on behalf of agendas of White or Aryan supremacy. But it also points to the variability of women's involvement in racial terrorism. Although women's participation in racist terrorism has increased over time in the United States, it is not the case that there is a simple temporal pattern to women's involvement in such violence. Rather, the conditions under which women are likely to become involved in racist terrorism reflect not only broader societal changes in the acceptability of women's involvement in politics and in violence, but also the strategic directions and tactical choices of organized White supremacist groups.

Notes

1. *State-Times Morning Advocate*, Baton Rouge, Louisiana, 19 April 2003; accessed 31 July 2004 at (ww.lexis-nexis. com/universe).
2. Heidi Beirich and Mark Potok, "Two Faces of Volksfront," available at (www.splcenter.org/intel/intelreport/article.jsp?aid=475@printable=1); "'To Do the Right Thing.' A Guilty Plea," *News Tribune* (Tacoma, Washington), 26 February 2004.
3. "Domestic Terrorism Ties?" NBC 4, 18 November 2002, available at (www.nbc4.tv/prnt/1793308/detail.html); "ADL Assists in OC White Supremacists Arrest," *The Jewish Journal of Greater Los Angeles*, available at (www.jewishjournal.com/home/print.php?id=9642); "Out of the Kitchen: Has the Women's Rights Movement Come to the Extreme Right?" ABC News, 12 December 2003, available at (http://abcnews.go.com/sections/us/DailyNews/extreme_women021212.html).
4. This excludes individual acts of racial violence, such as hate crimes.
5. Title 22 of the United States Code, Section 2656f(d), available at (http://www.state.gov/s/ct/rls/pgtrpt/2003/31880.htm/October 21, 2004).
6. From DoD definition of terrorism, cited in Cunningham (2003, 188, n. 4).
7. The idea that racist movements express sentiments of anger needs to be used with caution. For a discussion of the theoretical and political implications of understanding emotions such as anger as expressions of individual sentiment versus group-level emotions, see Blee (2003, 2004); also della Porta (1992).
8. The exact number of lynchings is difficult to determine, both because of the extralegal, secret nature of most lynchings and because of the overlap of lynchings with legal forms of execution of African Americans such as misapplications of the death penalty, what George C. Wright (1990) terms "legal lynchings" (also Tolnay and Beck 1995).
9. Some groups, especially those who regard local and county government as less likely to be under the control of ZOG, support devolving government power to these levels. Some of these complexities are explored by Levitas (2002).
10. Analysis not reported, but available from the author.
11. 4 September 2004 accessed at (http://www.faughaballagh.com/charity.htm).
12. In this sense, there is a continuum from the loose organization of groups like racist skinheads to the very ephemeral racist groups that operate with little or no lasting organization such as lynch mobs, but this article considers only groups with some level of organization.

References

ABC News. 2004a. "Out of the kitchen: Has the women's rights movement come to the extreme right?" Accessed from ABCNEWS.com, 5 August 2004.

ABC News. 2004b. "Black Widows: Hell hath no fury like Chechnya's ruthless widows of war," accessed from ABCNEWS.com, 4 September 2004.

Allen, James. 2000. *Without Sanctuary: Lynching Photography in America*. Santa Fe, NM: Twin Palms.

Bakersfield *Californian*. 2004. "Making fascist statements over frappuccinos," 24 July.

Blazak, Randy. 2001. "White boys to terrorist men: Target recruitment of Nazi skinheads," *American Behavioral Scientist*, 44(6) (February), pp. 982–1000.

Blazak, Randy. 2004. "'Getting it': The role of women in male desistence from hate groups," in *Home-Grown Hate: Gender and Organized Racism*, edited by Abby L. Ferber. New York: Routledge, pp. 161–179.

Blee, Kathleen. 1991. *Women of the Klan: Racism and Gender in the 1920s*. Berkeley: University of California Press.

Blee, Kathleen. 2002. *Inside Organized Racism: Women in the Hate Movement*. Berkeley: University of California Press.

Blee, Kathleen. 2003, 2004. "Positioning hate," *Journal of Hate Studies*, 3(1), pp. 95–106.

Blee, Kathleen. 2005. "Racial violence in the United States," *Ethnic and Racial Studies* 28(4) (July), pp. 599–619.

Brown, Elaine. 1994. *A Taste of Power: A Black Woman's Story*. New York: Anchor/Doubleday.

Chalmers, David M. 1981. *Hooded Americanism: The History of the Ku Klux Klan*. Durham, NC: Duke University Press.

Christensen, Loren. 1994. *Skinhead Street Gangs*. Boulder: Paladin Press.

Cooper, H. 2001. "Terrorism: The problem of definition revisited," *American Behavioral Scientist*, 45, pp. 881–893.

Crenshaw, Martha. 1988. "Theories of terrorism: Instrumental and organizational approaches," in *Inside Terrorist Organizations*, edited by David C. Rapoport. New York: Columbia University Press, pp. 13–31.

Cunningham, Karla J. 2003. "Cross-regional trends in female terrorism," *Studies in Conflict and Terrorism*, 26(3) (May–June), pp. 171–195.

Della Porta, Donatella. 1992. "Introduction: On individual motivations in underground political organizations," in *International Social Movement Research, Vol. 4, Social Movements and Violence: Participation in Underground Organizations*, edited by Donatella Della Porta. London: JAI Press, pp. 3–28.

Dobie, Kathy. 1997. "Skingirl Mothers: From Thelma and Louise to Ozzie and Harriet," in *The Politics of Motherhood: Activist Voices from Left to Right*, edited by Alexis Jetter, Annelise Orleck, and Diana Taylor. Hanover, NH: University Press of New England, pp. 257–267.

Dobratz, Betty A., and Stephanie L. Shanks-Meile. 2004. "The white separatist movement: Worldviews on gender, feminism, nature, and change," in *Home-Grown Hate: Gender and Organized Racism*, edited by Abby L. Ferber. New York: Routledge, pp. 113–142.

Federal Bureau of Investigation (FBI). 2002. Hate Crime Statistics. Available at (`http://www.fbi.gov/ucr/hatecrime2002.pdf`).

Ferber, Abby L. 2004. "Introduction." in *Home-Grown Hate: Gender and Organized Racism*, edited by Abby L. Ferber. New York: Routledge, pp. 1–18.

Fredrickson, George M. 2003. *Racism: A Short History*. Princeton, NJ: Princeton University Press.

Gardell, Mattias. 2003. *Gods of the Blood: The Pagan Revival and White Separtism*. Durham, NC: Duke University Press.

Hamm, Mark S. 1994. *American Skinheads: The Criminology and Control of Hate Crimes*. Westport, CT: Praeger.

Hoffman, Bruce. 1998. *Inside Terrorism*. New York: Columbia University Press.

Levitas, Daniel. 2002. *The Terrorist Next Door: The Militia Movement and the Radical Right*. New York: Thomas Dunne Books/St. Martin's Press.

Neidhardt, Friedhelm. 1992. "Left-wing and right-wing terrorist groups: A comparison for the German case," in *International Social Movement Research, Vol. 4: Social Movements and Violence: Participation in Underground Organizations*, edited by Donatella della Porta. London: JAI Press, pp. 215–235.

Neuburger, Luisella de Cataldo, and Tiziana Valentini. 1998. *Women and Terrorism*. New York: St. Martin's Press.

Patterson, Orlando. 1998. *Rituals of Blood: Consequences of Slavery in Two American Centuries*. Washington, DC: Calvados Counterpoints.

Perry, Barbara. 2002. "Defending the color line: Racially and ethnically motivated hate crime," *American Behavioral Scientist*, 46(1), 72–92.

Rogers, Joann, and Jacquelyn S. Litt. 2004. "Normalizing racism: A case study of motherhood in White supremacy," in *Home-Grown Hate: Gender and Organized Racism*, edited by Abby L. Ferber. New York: Routledge, pp. 97–112.

Southern Poverty Law Center (SPLC). 2004. On-line copies of the SPLC *Intelligence Report* and other publications, accessed 5 September 2004 from (`www.splcenter.org`).

Talbot, Rhiannon. 2000. "Myths in the representation of women terrorists," *Beire-Ireland: A Journal of Irish Studies*, 35(3), pp. 165–186.

Taylor, Verta. 1989. "Social movement continuity: The women's movement in abeyance," *American Sociological Review*, 54, pp. 761–75.

Tolnay, Stewart E., and E. M. Beck. 1995. *A Festival of Violence: An Analysis of Southern Lynchings, 1882–1930*. Urbana: University of Illinois Press.

Winant, Howard. 2002. *The World Is a Ghetto*. New York: Basic Books.

Wright, George C. 1990. *Racial Violence in Kentucky, 1865–1940: Lynchings, Mob Rule, and "Legal Lynchings."* Baton Rouge: Louisiana State University Press.

Zwerman, Gilda. 1994. "Mothering on the lam: Politics, gender fantasies and maternal thinking in women associated with armed, clandestine organizations in the United States," *Feminist Review*, 47, pp. 33–56.

José Padilla and the War on Rights

JENNY S. MARTINEZ

On June 9, 2002, an American citizen named José Padilla disappeared into a legal black hole. The government says he is a dangerous terrorist, but they have never charged him with a crime. For the nearly two years since he was arrested at Chicago O'Hare Airport, Mr. Padilla has been held without trial in solitary confinement in a military brig. Now the Supreme Court's decisions restricting the government's power to detain "enemy combatants" in the "war on terror" may finally have brought him back into the light. Mr. Padilla will get his day in court. But Mr. Padilla's saga remains a cautionary tale for all Americans concerned about preserving the liberty for which our ancestors fought and died, and which our troops overseas defend today.

I am one of Mr. Padilla's lawyers, but I have never met him. All I have seen of him is the same menacing picture that you have, the one that appears in every newspaper article about his case. I remember the first time I saw that picture, which was accompanied by a headline announcing that by arresting Mr. Padilla, the government had foiled a plot to set off a "dirty" radiological bomb, possibly in Washington, D.C. I was relieved to hear that this alleged plot had been thwarted. The Washington suburb of Arlington, Virginia, is my hometown, and I watched the smoking Pentagon from my office rooftop on the morning of September 11 and worried that my best friend's younger brother, a volunteer firefighter in Northern Virginia, might be in danger there. My mother was on an airplane to New York that morning, and until I got her cell phone call telling me she was safe, I was sick with fear. Even six months later, the threat of terrorism felt very personal to me.

I assumed that Mr. Padilla would be charged with a crime, that in the time-honored way he would be given his day in court, and that if the jury found him guilty, he would be locked up for a very, very long time. But that was not what happened.

Instead of charging Mr. Padilla with a crime, the president declared him an "enemy combatant." Some people think that Congress gave the president the power to imprison people as "enemy combatants" in the PATRIOT Act, but that law says nothing at all about such detentions. In fact, the term "enemy combatant" does not appear in any statute passed by Congress, nor in any regulation, nor in any international treaty. Searching for additional powers following September 11, the government plucked the term from *Ex parte Quirin*, a World War II–era Supreme Court decision upholding the government's right to put Nazi soldiers (whom the Court described as "enemy combatants") on trial in military commissions rather than in civilian courts. From this narrow decision, the government extrapolated that the president could hold anyone he decided was an "enemy combatant," without any real review by the courts, until the end of the "war on terror." This theory was novel, to say the least. While the government has captured and detained prisoners of war on the field of battle in many past wars, never before has it claimed the power to designate American citizens, arrested in civilian settings, as "enemies of the state" who can be held forever at the whim of the president.

It was only once the government took away all of Mr. Padilla's constitutional rights that I became involved in his case. For it seemed to me that if they could take away his rights, they could take away the rights of anyone. And that scared me more than the threat of another terrorist attack.

I am an international human rights lawyer. I do work in places like Bosnia and Rwanda, where tens or hundreds of thousands of people were killed because they were in the wrong ethnic group. I study countries like Chile, where thousands of people disappeared off the streets, never to be seen again, because they disagreed with the government. The reason I decided after law school to focus on international human rights rather than problems closer to home was that it seemed from where I stood that America was doing all right. We had some very ugly incidents in our past—slavery and segregation, the Japanese internment camps, and the displacement and slaughter of Native Americans for starters. Even today, not everyone in America is getting a fair shake. But we were founded on a set of ideals—liberty, equality and democracy—that ignited a fire of freedom worldwide. America was learning from its mistakes, and our current human rights problems seemed to pale in comparison to those in other countries.

I was shocked by the Padilla case. A system in which the government is allowed to lock up anyone it decides is an "enemy" without any trial or even the pretense of legal process seemed to me, frankly, like the kind of thing that happens in the messed-up third-world countries I spend most of my time thinking about, not what I expected out of America. And so, after following the case in the newspapers for a few months, I eventually volunteered to write an amicus, or "friend of the court," brief for the retired federal judges supporting Mr. Padilla's right to have his day in court. Amicus briefs are often filed by nonprofit activist groups like the ACLU, but it was highly unusual to have a distinguished group of retired judges (both Democrats and Republicans) weighing in with such a brief. As it turned out, they were not alone. By the time we got to the Supreme Court, there were dozens of briefs filed in Mr. Padilla's case and the cases of the other enemy combatants, by everyone from retired judges and law enforcement officials to former prisoners of war to Fred Korematsu, a Japanese American who was interned in World War II and whose name has become synonymous with the case he lost in the Supreme Court in 1944, one of the most disgraceful moments in the Supreme Court's history.

By the time I became involved in the case, Mr. Padilla had already been in jail for almost a year. He was initially arrested returning to the U.S. at Chicago O'Hare Airport on May 8, 2002, by civilian law enforcement agents. He was arrested pursuant to something called a material witness warrant, which allows an individual to be held so that he or she can give testimony in court or, in Mr. Padilla's case, before a grand jury. The warrant had been issued by a judge in the Southern District of New York, the federal court that sits in Lower Manhattan. Mr. Padilla was transported by the government from Chicago to New York. Upon his arrival, the court there appointed a lawyer to represent him in the material witness proceedings, Donna Newman. Ms. Newman is a criminal defense attorney in private practice, and it happened to be one of the few days a year she takes court-appointed indigent clients. Little did she know what she was getting into.

Ms. Newman met with Mr. Padilla several times at the Metropolitan Correctional Center, where he was being held, and filed papers with the court seeking his release. A hearing was scheduled for Tuesday, June 11, 2002. Two days before that hearing, Ms. Newman got a call from a young lawyer in the U.S. Attorney's office who was working on the case. He told her that the hearing was off. The president had declared her client an "enemy combatant" and the military had taken him away. At first, Ms. Newman thought the lawyer was joking with her. As the truth dawned on her, she was shocked.

At the time scheduled for the hearing, Ms. Newman appeared in court and filed a petition for a writ of habeas corpus, seeking Mr. Padilla's release. Ms. Newman sat alone at her table in the courtroom, without even a client next to her. She had heard on television that her client had been taken to a military brig in South Carolina. As she looked at the swarm of high-ranking government lawyers across the room, she realized she needed help. The judge quickly appointed as cocounsel Andrew Patel, another local defense attorney who had worked on some high-profile terrorism cases.

Ms. Newman and Mr. Patel quickly plunged into a world of arcane legal precedents. It seemed obvious to them from grade school civics class that in America, the government was not allowed to lock someone up forever without giving him a lawyer and a trial, but the government's case had a perverse, airtight logic to it: Mr. Padilla had no constitutional right to challenge his designation as a prisoner-of-war-like "enemy combatant" because enemy combatants have no constitutional rights. The government's argument was circular but maddeningly slippery. Newman and Patel found themselves reading cases about the writ of habeas corpus from England in the 1600s, cases from the 1800s involving swashbuckling seizures of ships as prizes of war on the high seas, and obscure treatises on the Geneva Conventions and other laws of war.

Throughout this time, the government refused to let Ms. Newman and Mr. Patel communicate with their client in any way. The government chillingly explained that allowing Mr. Padilla to learn that a court was hearing his case might give him hope that he would some day be released: "Only after such time as Padilla has perceived that help is not on the way can the United States reasonably expect to obtain all possible intelligence information from Padilla.... Providing him access to counsel now ... would break—probably irreparably—the sense of dependency and trust that the interrogators are attempting to create." In court, the government claimed that they had the power to imprison Mr. Padilla until the "war on terror" was over, and the court had no power to intervene other than to make sure that there was "some evidence" to support the government's decision. The "some evidence" the government pointed to was a written affidavit from a midlevel Pentagon official, who recounted information reportedly given to the government by unnamed confidential sources. The affidavit alleged that Padilla was part of a plot to build and detonate a dirty bomb in the United States but acknowledged that the plot was "still in the initial planning stages" and "there was no specific time set for the operation to occur." (Deputy Secretary of Defense Paul Wolfowitz later stated publicly that "I don't think there was actually a plot beyond some fairly loose talk and his coming in here obviously to plan further deeds.")[1] The government admitted that the information provided by its confidential sources "may be part of an effort to mislead or confuse U.S. officials" and that one of the sources "recanted some of the information that he had provided." (Later press reports indicated that one of the confidential sources had given up Padilla's name while being subjected to "water-boarding," a form of torture in which the suspect is held down in a tub and made to think he will drown.) The government argued that the district court had no power to question the information in the affidavit and no authority to allow Mr. Padilla to come into the court to tell his side of the story. In effect, the government argued, the court's power was limited to rubber-stamping the government's decision to detain Padilla.

Rejecting the government's Orwellian logic, in December 2002, the district court held that even under the lax "some evidence" standard, Padilla was entitled to present his side of the case in court, and ordered that Padilla be allowed to meet with his lawyers. The district court agreed, however, that the government had the power to hold an American citizen arrested on American soil as an "enemy combatant." Seeking to avoid even the minimal challenge to its authority posed by the district court's ruling, the government took an immediate appeal to the federal circuit court in New York.

It was at this point that I became involved in the case, filing my brief on behalf of the retired federal judges. Although I started out as an amicus, as the case progressed, I began working more and more closely with Ms. Newman and Mr. Patel. Not only had I read all the same obscure cases and treatises they had, I had actually read even more because of my background working on war crimes issues for the United Nations International Criminal Tribunal for the Former Yugoslavia in the Hague. I had the distinction of actually having owned a copy of the Geneva Conventions prior to September 11, 2001, a rare thing among U.S. lawyers. By the time the U.S. Court of Appeals for the Second Circuit heard the case in November 2003, Ms. Newman and Mr. Patel decided to let me share some of the argument time as an amicus.

A month after the argument, the court issued a ruling in our favor. Going even further than the district court, the court of appeals held that the government lacked authority to hold a U.S. citizen seized in the U.S. as an "enemy combatant." Congress had not given the president such extraordinary power, and the president had no inherent power to deprive citizens of liberty in this way, the court held. The court rejected the government's reading of *Ex parte Quirin*, the lynchpin of its case, noting that it had involved the congressionally authorized trial by military commission (with lawyers and full opportunity for the defense to be heard) of admitted soldiers in the German army. The case provided no support for the unilateral presidential detention without trial of an individual who denied that he was a soldier at all. Mr. Padilla had to be charged with a crime or released in thirty days, the court ordered.

Again, the government quickly appealed. Once the case reached the Supreme Court, Ms. Newman and Mr. Patel (casting a wary eye at the ever-growing crowd of government lawyers across the courtroom) decided to ask me to stop being a mere friend of the court and join the core legal team for Mr. Padilla. I accepted. Jonathan Freiman, an appellate lawyer from Connecticut and part-time instructor at Yale who had written an amicus brief in the court of appeals for a broad spectrum of groups (including the conservative CATO and Rutherford Institutes), also came on board, as did David DeBruin, a top partner at Jenner & Block, a leading Supreme Court litigation firm in Washington, D.C.

Shortly before the first round of briefs were due in the Supreme Court, the government finally decided, out of the goodness of its heart and without acknowledging that he had any right to counsel, that Mr. Padilla could finally speak to his lawyers. Ms. Newman and Mr. Patel had put in for the necessary security clearances back in December 2002, when the district judge had ordered access. They were finally allowed to go visit him in March 2004, but under the strict rules imposed by the military, they were not allowed to say or ask much—and they were not allowed to tell the court or the rest of the legal team, let alone the rest of the world, what Mr. Padilla had said. All they could really tell us was that after two years of incommunicado interrogation, Mr. Padilla was apparently very glad to see them.

The case was argued before the Supreme Court on April 28, 2004, on the same day as the case of Yaser Hamdi, an American citizen seized in Afghanistan and also held as an "enemy combatant," and one week after the case concerning the Guantánamo detainees. Although any member of the team could have done the honors, in the end, I ended up making the oral presentation for Mr. Padilla to the high court.

The night after the case was argued in the Supreme Court, CBS broadcast the first photos of Abu Ghraib prison. That very morning, the government's lawyer had responded to questions from the justices about torture by explaining that our government didn't do that sort of thing. In the days that followed, more photos leaked, followed by memos justifying the potential abuse of detainees that were full of legal reasoning as contorted as the bodies in the photos.

After several weeks of this bad news, the government finally won more favorable headlines when the Justice Department held a press conference at which they finally revealed the "evidence" against Mr. Padilla—evidence that they had claimed for months would endanger national security if shared with a federal judge. Mr. Padilla, the government now claimed, had not really planned to set off a dirty bomb, but rather to blow up apartment buildings with natural gas. The government had his confession to this scheme now after months of interrogation,[2] and it was a good thing he had not been given his constitutional rights or he might have gotten off. The government was not trying to influence the Supreme Court, the government lawyers explained, but rather the court of public opinion. Mr. Padilla, still locked away in solitary, had no chance to hear about or respond to the only trial the government had seen fit to give him so far, this trial in the court of public opinion. A good friend of mine who lives in a high-rise in New York City told me at a picnic that she had supported my work on the case until she learned my client might have plotted to blow up apartment buildings like hers; that had hit a little too close to home, and she was no longer sure he ought to have constitutional rights. She was only partly joking.

Two months later, on June 28, 2002, the Supreme Court issued decisions in the three cases—the Guantánamo case (*Rasul v. Bush*), *Hamdi v. Rumsfeld*, and *Rumsfeld v. Padilla*. The Court ruled in favor of the detainees in *Rasul* and *Hamdi* but bounced *Padilla* on a technicality. In *Rasul*, the Court held by a vote of six to three that the U.S. federal courts have jurisdiction to entertain habeas petitions from prisoners at Guantánamo, sending the case back to the lower courts to determine what precisely the rights of the prisoners were.

The Court's decision in *Hamdi* was equally a defeat for the government but more confusing in the details. Only one member of the Court, Justice Thomas, agreed with the government's position. Four justices thought the government had no authority at all to hold a U.S. citizen, even one seized on an overseas battlefield, as an "enemy combatant." Leading the charge for this group was Justice Scalia, the Court's most conservative justice, who explained that the government's actions ran contrary to several hundred years of Anglo-American legal tradition beginning with the Magna Carta. Unless Congress suspended the writ of habeas corpus (a grave action the Constitution allows to be taken only in cases of "Rebellion" or "Invasion"), the government's only constitutional option was to charge Mr. Hamdi with a crime or release him. Justice Stevens, the Court's most liberal member, joined Justice Scalia. Justices Souter and Ginsburg reached the same conclusion about the government's lack of authority but relied mainly on a statute (passed in the 1970s to prevent recurrence of the Japanese internment camps) that provided that "[n]o citizen shall be imprisoned or otherwise detained by the United States except pursuant to an Act of Congress." Since no act of Congress expressly allowed the detention of U.S. citizens as enemy combatants, these justices reasoned, Mr. Hamdi could not be imprisoned without criminal charges.

The other four justices, in an opinion written by Justice O'Connor, found that the government had authority to hold people who were "part of or supporting forces hostile to the United States or coalition partners" in Afghanistan and "who engaged in an armed conflict against the United States there." But they also held that an individual like Mr. Hamdi—who, although he was apprehended in Afghanistan, claimed that he was not engaged in armed conflict against the U.S.—was entitled to access to counsel and a meaningful hearing at which he could present his side of the story and challenge the government's evidence.

Although the *Hamdi* decision set the floor in terms of the rights of U.S. citizens to have access to counsel and a fair hearing on their status, it left open the question whether the government had any authority at all to detain as "enemy combatants" citizens who were not captured on the battlefields of Afghanistan. Did the government have the authority to detain U.S. citizens arrested in the U.S. as "enemy combatants"? Justice O'Connor's opinion in *Hamdi* was careful not to say, noting that the Court's finding of authority to detain fighters in Afghanistan was premised on a reading of the law based on traditional warfare. "If the practical circumstances of a given conflict are entirely unlike those of the conflicts that informed the development of the law of war, that understanding may unravel." Detainees nabbed in the broader "war on terror"—a conflict that takes place everywhere, all the time, in which anyone walking down the street may be a combatant and which may last forever (in short, a conflict whose practical circumstances are entirely unlike traditional warfare)—were implicitly left for another day.

The Supreme Court dismissed the case presenting that very question—Mr. Padilla's case—on a technical issue of court procedure. The lower courts had all found that the case had been properly filed in New York—which was hardly surprising, given that the government had initially brought Mr. Padilla to New York and then whisked him away in the middle of the night just before his court hearing there. But the Supreme Court disagreed. The Court held that the only proper defendant for the suit was the commander of the brig where Mr. Padilla was currently imprisoned in South Carolina, rather than Secretary of Defense Donald Rumsfeld, to whose custody the presidential order had entrusted Padilla. Thus, the Court found, the suit could be brought only in South Carolina, and not in New York. After two years, Mr. Padilla must wait a little longer for his day in court. Four justices dissented, arguing that the Court should reach the merits as soon as possible, for "[a]t stake in this case is nothing less than the essence of a free society."

Within days, Mr. Padilla's petition was refiled in South Carolina. At a minimum, he will receive the hearing that the Supreme Court's decision in *Hamdi* guarantees. It is still possible—indeed probable—that the Supreme Court will rule that there is no authority to detain persons arrested in the U.S. as "enemy combatants." But the litigation will take several months more to reach the Supreme Court again, probably not until after this fall's election.

As for me, I still haven't met my client. Perhaps this delay will allow me time to get a security clearance so I can finally see him. As I have worked on this case, I have often thought of Attorney General John Ashcroft's menacing warning to civil libertarians: "To those who scare peace-loving people with phantoms of lost liberty, my message is this: Your tactics only aid terrorists, for they erode our national unity and diminish our resolve. They give ammunition to America's enemies and pause to America's friends."[3] I take that more or less personally.

To those like Mr. Ashcroft who would scare liberty-loving people with phantoms of lost security, *my* message is this: Your tactics only aid terrorists, for they erode our national unity and diminish our resolve. They give ammunition to America's enemies and pause to America's friends. No one who watched the attacks on September 11 can deny that terrorism represents a grave threat to American security, but winning the war on terror also requires that we remain true to our ideals. Guantánamo, the photos of Abu Ghraib, the image of America as a nation above the law—none of these things has helped us in the fight against terror. Moreover, the government's argument—that we must sacrifice human rights for security—presents a false choice. There is a balance to be struck, but it is far more nuanced than the current government recognizes.

Take the issue of detention. Many other democratic nations confronted with terrorist threats have enacted special measures for some kind of administrative detention of terrorists.[4] These nations include the U.K.,[5] Israel,[6] and Spain.[7] The administrative detention practices of many of these countries have been criticized by human rights activists, and many of these criticisms are legitimate, but it is notable that they all provide greater protection for human rights than does current U.S. practice with respect to so-called "enemy combatants." First, these other nations have passed actual legislation authorizing detention of suspected terrorists. By contrast, the U.S. has relied on presidential fiat. Second, the laws of our democratic allies provide for access to counsel and judicial review of detention within a matter of hours or days—not months or years. In the U.K., for example, terrorism detainees are entitled to counsel and judicial review as soon as "reasonably practicable," and in any event no later than 48 hours.[8] Detainees in Israel are entitled to see a judge within 48 hours.[9] In Spain, they are entitled to counsel and to be brought before a judge within 120 hours.[10] And so on. Third, most of these laws provide for time limits on detention. The U.K.'s 2000 and 2001 antiterrorism laws allow the government to hold citizen detainees in administrative detention for only 48 hours, with extension to seven days possible only with a judge's approval.[11] Spanish detainees must be charged within 72 hours, which can be extended by another 48 hours only by a judge.[12] Even where detention is indefinite, regular judicial review is required. Israel's 2002 Incarceration of Unlawful Combatants Law, for example, requires that a district court judge review the status of each detainee every six months to determine if the captive is still a threat to state security or if there are other circumstances that justify release.[13]

Moreover, overseas courts have stepped in to guarantee detainees' rights above and beyond those provided by legislation. In *Marab v. IDF Commander in the West Bank*, for example, the Israeli Supreme Court invalidated a military order that allowed investigative detention of Palestinians in the West Bank for 12 days without a judicial hearing. Rejecting the government's claim that security necessitated the delay, the Court held that "this approach is in conflict with the fundamentals of both international and Israeli law," which view "judicial review of detention proceedings essential for the protection of individual liberty."[14] Instead, the Court held, the detainee must be brought before a judge as promptly as possible.[15] Similarly, in invalidating Turkey's detention of suspected terrorists for more than 14 days without access to counsel or court, the European Court of Human Rights explained that although "the investigation of terrorist offences undoubtedly presents the authorities with special problems, it cannot accept that it is necessary to hold a suspect for fourteen days without judicial intervention."[16]

These examples show that detention of individuals for more than two years without access to counsel or a hearing before a neutral judge is well beyond the bounds of what civilized countries allow nowadays, even when fighting terrorism. Moreover, it is fundamentally contrary to American values. As the U.S. Supreme Court wrote in a case involving First Amendment rights during the Cold War,

> Implicit in the term "national defense" is the notion of defending those values and ideals which set this Nation apart.... It would indeed be ironic if, in the name of national defense, we would sanction the subversion of one of those liberties—the freedom of association—which makes the defense of the Nation worthwhile.[17]

The Israeli Supreme Court expressed a similar view in its decision banning torture and other cruel, inhuman, or degrading treatment in interrogation:

> This is the destiny of democracy, as not all means are acceptable to it, and not all practices employed by its enemies are open before it. Although a democracy must often fight with one hand tied behind its back, it nonetheless has the upper hand. Preserving the Rule of Law and recognition of an individual's liberty constitutes an important component in its understanding of security. At the end of the day, they strengthen its spirit and its strength and allow it to overcome its difficulties.[18]

The U.S. Supreme Court has come to the rescue of liberty for now, upholding the rule of law in the first round of "war on terror" cases, but in the end it is the American people that must defend our Constitution by making our views known. As Judge Learned Hand said, "Liberty lies in the hearts of men and women; when it dies there, no constitution, no law, no court can save it."[19] That perhaps is what democracy is all about. We alone can ensure that the "war on terror" does not become a "war on rights."

NOTES

1. `Http://usinfo.state.gov/topical/pol/terror/ 02061103.htm.`
2. The government acknowledged in a footnote to its press release that Mr. Padilla continued to deny that he had actually planned to engage in any terrorist acts.
3. Testimony of Attorney General John Ashcroft before the Senate Judiciary Committee (Dec. 6, 2001).
4. Two excellent sources on comparative detention practices are Stephen J. Schulhofer, *Checks and Balances in Wartime*, 102 Mich. L. Rev. 1501 (forthcoming 2004), and *Brief Amicus Curiae of Comparative Law Scholars and Experts on the Laws of the United Kingdom and Israel in Support of Respondent, Rumsfeld v. Padilla*, No. 03-1027 (2004). This section draws particularly on the latter.
5. Terrorism Act, 2000, c.11, para. 41, sched. 8 (Eng.) and Anti-Terrorism, Crime and Security Act, 2001, c. 24, pt. 4 (Eng.) [hereinafter U.K. Act].
6. Emergency Powers (Detention) Law, 1979, 33 L.S. I. 89 (1978–79) (Isr.), and Incarceration of Unlawful Combatants Law, 2002 (Isr.), at `www.Justice.gov.il/NR/rdonlyres/8459847C-84FD-956D-0F2CB10C948A/0/IncarcerationLaw L.438/01` [hereinafter Israeli Detention Law and Israeli Unlawful Combatants Law].
7. Spanish Constitution art. 17(2); L.E. Crim. Art. 496.
8. U.K. Act ¶ 7.
9. Israeli Detention Law § 4(a). Separate measures apply in the occupied territories.
10. Spain art. 17.
11. U.K. Act ¶ 436. U.K. law allows for indefinite detention of some aliens, however.
12. Spanish Constitution art. 17(2); L.E. Crim. Art. 496.
13. Israeli Unlawful Combatants Law § 5.
14. Marab v. IDF Commander in the West Bank.
15. ¶ 36.
16. Askoy v. Turkey, 23 Eur. H.R. Rep. 553 ¶ 78 (1996). *See also* Advisory Opinion OC-8/87, Habeas Corpus in Emergency Situations (arts. 27(2) and 7(6) of the American Convention on Human Rights), Inter-Am Ct. H.R. (Ser. A) No. 8 ¶ 12 (Jan. 30, 1987) ("[E]ven in emergency situations, the writ of habeas corpus may not be suspended or rendered ineffective.... To hold the contrary view—that is, that the executive branch is under no obligation to give reasons for a detention and may prolong such a detention indefinitely during states of emergency, without bringing the detainee before a judge ... would ... be equivalent to attributing uniquely judicial functions to the executive branch, which would violate the principle of separation of powers, a basic characteristic of the rule of law and of democratic systems.").
17. United States v. Robel, 389 U.S. at 264.
18. Supreme Court of Israel: Judgment Concerning the Legality of the General Security Service's Interrogation Methods, 38 I.L.M. 1471, 1488 (1999).
19. Learned Hand, *The Spirit of Liberty* 190 (1960).

UNIT 6

Terrorism and the Media

Unit Selections

Key Points to Consider

- What was the impact of the news coverage of September 11? To what extent is the media complicit in allowing the terrorists to set the agenda?
- Can terrorism be solved by a global community? Why or why not?
- Should the Internet be censored as part of the war on terrorism?
- Does the media exaggerate threats to boost ratings? What responsibilities does the media have?

Student Website

www.mhcls.com/online

Internet References

Further information regarding these websites may be found in this book's preface or online.

Institute for Media, Peace and Security
http://www.mediapeace.org

Terrorism Files
http://www.terrorismfiles.org

The Middle East Media Research Institute
http://www.terrorismfiles.org

The media plays an important role in contemporary international terrorism. Terrorists use the media to transmit their message and to intimidate larger populations. Since the hijackings at Dawson's field in Jordan in 1970 and the massacre at the Munich Olympics in 1972, international terrorists have managed to exploit the media and have gained access to a global audience. The media provides terrorists with an inexpensive means of publicizing their cause and a forum to attract potential supporters. In the age of independent fund-raising, terrorists have become increasingly dependent on accessible media coverage.

As media coverage has become more sophisticated, terrorist organizations have become increasingly conscious in their interactions with the press. Managing public relations, drafting press releases, and arranging interviews have become important functions, often delegated to individuals or groups in the semi-legal periphery of the organization.

The impact of the increasingly symbiotic relationship between terrorists and media has been twofold. On the one hand the media has provided terrorists with real-time coverage and immediate 24-hour access to a global public. As long as the explosion is big enough and the devastation horrific enough and there are cameras close by, media coverage of the incident is guaranteed. Holding true to the old axiom "if it bleeds, it leads," the media seems all to willing to provide terrorists with free, unlimited, and at times indiscriminate coverage of their actions.

On the other hand, the media also provides terrorists with a means of ventilation, potentially reducing the number of violent incidents. It subtly influences terrorists to function within certain, albeit extended, boundaries of social norms, as grave violations of these norms may precipitate unintended or unwanted public backlash and a loss of support. In light of these apparently contradictory tendencies, the debate about media censorship or self-censorship continues.

The articles in this unit explore the relationship between terrorism and the media. Brigitte Nacos, one of the most well-known experts in the field, provides an analysis of the relationship between the media and terrorists who seek to exploit it in the wake of 9/11. John Gray shows how media portrayals of terrorist acts can shape reality. David Talbot explores the extent to which tighter security and regulation of Internet content could aid the war on terror. Finally, Lori Robertson asks whether or not the media causes the public to panic unnecessarily.

Terrorism as Breaking News: Attack on America

BRIGITTE L. NACOS

Terrorism's efficacy, like beauty, is in the eyes of the beholder in that those who commit political violence deliberately directed against civilians believe in the success of their deeds. In one respect, most if not all terrorists do achieve an important goal. Whenever they strike, especially if they stage so-called terrorist spectaculars, their deeds assure them massive news coverage and the attention of the general public and governmental decision makers in their particular target societies. Moreover, given the global nature of the communication system, the perpetrators of international and domestic terrorism also tap into the international media and thereby receive the attention of publics and governments beyond their immediate target countries. Former British Prime Minister Margaret Thatcher once said that publicity is the oxygen of terrorism. Without publicity, terrorism would be like the proverbial tree that falls in the forest and the press is not there to report—it would be as if the incident never happened. The term "mass-mediated terrorism" signifies the centrality of media considerations in the calculus of political violence that is committed by nonstate actors against civilians.[1] While publicity is not the ultimate objective, terrorists recognize that it is the means to advertise and further their larger political ends. This article examines whether and how the architects of the horrific events of September 11 succeeded in achieving their media goals.

Tuesday, September 11, 2001, began as a perfect day along the American East coast. The sun was golden bright. The sky was blue and cloudless. On a clear day like this, the World Trade Center's twin towers resembled two exclamation marks above Manhattan's skyline, and they could be seen from many miles away in the surrounding counties of three states—New York, New Jersey, and Connecticut. At 8:48 A.M., when the workday began for thousands of employees in the offices of the 110 stories of the Center's towers, a hijacked Boeing 767 crashed into the North Tower. Eighteen minutes later, at 9:06 A.M., another Boeing 767 crashed into the South Tower. Just before 10:00 A.M., the South Tower collapsed, and twenty-nine minutes later, its twin fell down. In between these events, at 9:40 A.M., a Boeing 757 dived into the Pentagon; at 10:10 another Boeing 757 crashed in Somerset County near Pittsburgh, Pennsylvania. September 11, 2001 was forever America's "Black Tuesday."

Within eighty-two minutes, the United States suffered a series of synchronized attacks that terminated in the most deadly, most damaging case of terrorism in history. More than three thousand persons were killed, and the damage to properties, to businesses, and to the economic conditions in the United States and abroad was incalculable. With the symbol of America's s economic and financial power toppled in New York, the symbol of U.S. military strength partially destroyed in Washington, and a symbol of political influence—most likely the White House or Capitol—spared by courageous citizens aboard another jetliner that crashed near Pittsburgh, Pennsylvania, the impact was cataclysmic. America after the terror attack was not the same as it was before. Although the World Trade Center bombing of 1993 demonstrated that the United States is not immune to international terrorism on its own soil, and the Y2K terrorism alert reinforced that recognition, Americans were stunned by the velocity and audacity of the 2001 strike.

Apart from the relatively small number of people who were alerted by relatives and friends via phone calls from the stricken WTC towers and those eyewitnesses who watched in horror, millions of Americans learned of the news from television, radio, or the Internet. In fact, minutes after the first kamikaze flight into Tower I, local radio and television stations as well as the networks reported first a possible explosion in the WTC, then a plane crash into one of the towers. Soon thereafter, the first pictures of the North Tower appeared on the screens, with a gaping hole in the upper floors enveloped in a huge cloud of dark smoke. As anchors, hosts of morning shows, and reporters struggled to find words to describe what was indescribable, a mighty fireball shot out of Tower II—presumably the result of a second powerful explosion. The towering inferno was eventually replaced by another horror scene: one section of the headquarters of the Department of Defense engulfed in a large plume of smoke. With the cameras again on the WTC, the South Tower collapsed in what seemed like slow motion. Switching again to the Pentagon, the camera revealed a collapsed section of the facade. Amid rumors that a fourth airliner had crashed in Pennsylvania, the cameras caught the collapse of the World Trade Center's North Tower.

For at least part of this unfolding horror, many millions of Americans watched television stations or their related Internet sites. And, ironically, most Americans were familiar with the shocking images: the inferno in a skyscraper, the terrorist attack on a towering high rise, the total destruction of a federal building in the nation's capital by terrorists, the nuclear winter landscape in American cities, Manhattan under siege after a massive terrorist attack. In search of box office hits, Hollywood produced a steady stream of disaster movies and thrillers, often based on best-selling novels about ever more gruesome images of destruction. The entertainment industry's cavalier exploitation of violence was shockingly obvious following the terror strikes, when it was revealed that the "planned cover for a hip-hop album due to be released in November [2001] depicted an exploding World Trade Center." [2]

In a popular culture inundated with images of violence, Americans could not comprehend what was happening before their eyes and what had happened already. The horror of the quadruple hijack and suicide coup was as real as in a movie, but it was surreal in life. As Michiko Kakutani observed, "there was an initial sense of dejà vu and disbelief on the part of these spectators—the impulse to see what was happening as one of those digital special effects from the big screen." [3] The following quotations reflect the reactions of people who escaped from the World Trade Center, witnessed the disaster, or watched it on television:

> I looked over my shoulder and saw the United Airlines plane coming. It came over the Statue of Liberty. It was just like a movie. It just directly was guided into the second tower.
>
> —Laksman Achuthan, managing director of the Economic Cycle Research Institute [4]

> I think I'm going to die of smoke inhalation, because you know, in fires most people don't die of burning, they die of smoke inhalation. This cop or somebody walks by with a flashlight. It's like a strange movie. I grab the guy by the collar and walk with him.
>
> —Howard W. Lutnick, chairman of Cantor Fitzgerald[5]

> I looked up and saw this hole in the World Trade Center building. And I—I couldn't believe it. I thought, you know, this can't be happening. This is a special effect; it's a movie.
>
> —Clifton Cloud, who filmed the disaster with his video camera [6]

> It's insane. It's just like a movie. It's, it's actually surreal to me to see it on TV and see major buildings collapse.
>
> —Unidentified man in Canada [7]

> This is very surreal. Well, it's out of a bad sci-fi film, but every morning we wake up and you're like it wasn't a dream, it wasn't a movie. It actually happened.
>
> —Unidentified woman in New York [8]

Witnessing the calamity from a tenth-floor apartment in Brooklyn, novelist John Updike felt that "the destruction of the World Trade Center twin towers had the false intimacy of television, on a day of perfect reception." [9] Many people who joined newscasts in progress thought that they were watching the promotion for one of several terrorism thrillers scheduled for release later in the month. Whether they realized it or not, and many did not, most people, even eye-witnesses at the disaster scenes, were far from sure whether movies had turned into life, or whether life was now a movie. Updike alluded to this sentiment, when he recalled the experience:

> As we watched the second tower burst into ballooning flame (an intervening building had hidden the approach of the second airplane), there persisted the notion that, as on television, this was not quite real; it could be fixed; the technocracy the towers symbolized would find a way to put out the fire and reverse the damage. [10]

In a seemingly inexplicable lapse of judgment, the German composer Karlheinz Stockhausen characterized the tenor attacks on the United States as "the greatest work of art." [11] His remarks caused outrage in his country and the abrupt cancellation of two of his concerts in Hamburg. Perhaps this was a case of total confusion between the real world and the "pictures in our heads" that Walter Lippmann described long before the advent of television. In particular, Lippmann suggested that "[f]or the most part we do not first see, and then define, we define first and then see." [12] While many people initially identified the horrors of "Black Tuesday" as familiar motion picture images, Stockhausen may have processed the real life horror first as a symphonic Armageddon in his head, when he said: "That characters can bring about in one act what we in music cannot dream of, that people practice madly for 10 years, completely fanatically, for a concert and then die. That is the greatest work of art for the whole cosmos ." Following the uproar over his statement, Stockhausen apologized for his remarks saying, "Not for one moment have I thought or felt the way my words are now being interpreted in the press." [13] One can only guess that the angry reactions to his statement brought him back from the pseudo-reality in his head to the real life tragedy and its consequences.

When emotions gave way to rationality, the truth began to sink in. The most outrageous production of the terrorist genre was beyond the imagination of the best special effects creators. This was not simply two hours worth of suspense. Real terrorists had transformed Hollywood's pseudo-reality into an unbearable reality, into real life. This time there was neither a happy ending to be enjoyed nor an unhappy ending that the audience could forget quickly.

Perhaps the temporary confusion was a blessing. Perhaps the fact that reality replaced media-reality in slow motion helped people cope with the unprecedented catastrophe within America's borders. Perhaps the delayed tape in people's heads prevented citizens in the stricken areas from panicking, helped citizens all over the country to keep their bearings.

The greatest irony is that the terrorists who loathed America's pop culture as decadent and poisonous to their own beliefs and ways of life turned Hollywood-like horror fantasies into real life hell. In that respect, they outperformed Hollywood, the very symbol of their hate for western entertainment. After visiting the World Trade Center disaster site for the first time, New York's Governor George Pataki said:

> It's incredible. It's just incomprehensible to see what it was like down there. You know, I remember seeing one of these Cold War movies and after the nuclear attacks with the Hollywood portrayal of a nuclear winter. It looked worse than that in downtown Manhattan, and it wasn't some grade "B" movie. It was life. It was real. [14]

The question of whether imaginative novelists and filmmakers anticipate terrorist scenarios or whether terrorists borrow from the most horrific images of Hollywood's disaster films was no longer academic. Shortly after the events of September 11, an ongoing cooperation between filmmakers and the U.S. Army intensified in order to predict the forms of future terrorist attacks. The idea was that the writers who created Hollywood terrorism might be best equipped to conceptualize terrorists' intentions. According to Michael Macedonia, the chief scientist of the Army's Simulation, Training, and Instrumentation Command, "You're talking about screenwriters and producers, that's one of the things that they're paid to do every day—speculate. These are very brilliant, creative people. They can come up with fascinating insights very quickly." [15] However, it was not farfetched to suspect that the perpetrators of the September 11 terror took special delight in borrowing from some of the most horrific Hollywood images in planning and executing their terrorist scheme.

THE PERFECT "BREAKING NEWS" PRODUCTION

From the terrorists' point of view, the attack on America was a perfectly choreographed production aimed at American and international audiences. In the past, terrorism has often been compared to theater. According to this explanation,

> Modern terrorism can be understood in terms of the production requirements of theatrical engagements. Terrorists pay attention to script preparation, cast selection, sets, props, role playing, and minute-by-minute stage management. [16]

While the theater metaphor remains instructive, it has given way to that of terrorism as television spectacular, as breaking news that is watched by record audiences and far transcends the boundaries of theatrical events. And unlike the most successful producers of theater, motion picture, or television hits, the perpetrators of the lethal attacks on America affected their audience in unprecedented and lasting ways. "I will never forget!" These or similar words were uttered over and over.

After President John F. Kennedy was assassinated in 1963, most Americans and many people abroad eventually saw the fatal shots and the ensuing events on television. But beyond the United States and other Western countries, far fewer people abroad owned television sets at the time. When the Palestinian "Black September" group attacked and killed members of the Israeli team during the 1972 Olympic games in Munich (using their surviving victims as human shields during their ill-fated escape), an estimated eight hundred million people around the globe watched the unfolding tragedy. At the time, satellite TV transmission facilities were in place to broadcast the competitions into most parts of the world. But nearly thirty years later, a truly global television network, CNN, existed along with competitors that televised their programs across national borders and covered large regions of the world. Thus, more people watched the made-for-television disaster production "Attack on America" live and in replays than any other terrorist incident before. It is likely that the terrorist assaults on New York and Washington and their aftermath were the most watched made-for-television production ever.

From the perspective of those who produced this unprecedented, terrorism-as-breaking-news horror show, the broadcast was as successful as it could get. Whether a relatively inconsequential arson by an amateurish environmental group or a mass destruction by a network of professional terrorists, the perpetrators' media-related goals are the same: terrorists strive for attention, for recognition, and for respectability and legitimacy in their various target publics. [17] It has been argued that contemporary religious terrorists, unlike secular terrorists (such as the Marxists of the Red Brigade/Red Army or the nationalists of the Palestinian Liberation Front during the last decades of the cold war), want nothing more than to lash out at the enemy and kill and damage indiscriminately, to express their rage. But while all of these sentiments may well figure into the complex motives of group leaders and their followers, there is no doubt that their deeds are planned and executed with the mass media and their effects on the masses and governmental decision makers in mind. Unlike the typical secular terrorists, religious terrorists want to inflict the greatest possible pain, but they want a whole country, and in the case of international terrorism, the whole world, to see their act, to understand the roots of their rage, to solidify their esteem in their constituencies, and, perhaps, to win new supporters.

To be sure, publicity via the mass media is not an end in itself. Most terrorists have very specific short-term and/or long-term goals. It is not hard to determine the short-term and long-term objectives of those that planned and executed the suicide missions against the United States. Even without the benefit of a credible claim of responsibility, the mass media, decision makers, and the general public in the United States and abroad discussed the most likely motives for the unprecedented deeds. In the short-term, the architects and perpetrators wanted to demonstrate the weaknesses of the world's only remaining superpower vis-a-vis determined terrorists, to frighten the American public, and to fuel perhaps a weakening of civil liberties and domestic unrest. No doubt, the long-term schemes targeted U.S. foreign policy, especially the American influence and presence in the Middle East and other regions with large Muslim populations. More important, as communications from Osama bin Laden and his organization revealed, those who decided on these particular terror attacks regarded the anticipated strikes by the United States as the beginning of a holy war between Muslims and infidels. Bin Laden, in a fax to Qatar-based al Jazeera television, called the Muslims of Pakistan "the first line of defense ... against the new Jewish crusader campaign [that] is led by the biggest crusader Bush under the banner of the cross." The bin Laden statement that was widely publicized in the United Sates left no doubt that he purposely characterized the confrontation as a battle between Islam and "the new Christian—Jewish crusade." [18]

Whatever else their immediate and ultimate goals were, those who planned the attacks were well aware, as are most perpetrators of political violence, that the mass media of communication is central to furthering their publicity goals and even their political and religious objectives. Without the frightening images and the shocking reportage, the impact on America and the rest of the world wouldn't have been as immediate and intense as it was.

WHEN TERRORISTS STRIKE HARD, THEY COMMAND ATTENTION

In the past, media critics have documented and questioned the mass media's insatiable appetite for violence; they have explored the effects of this kind of media content on people who are regularly exposed to violence in the news and in entertainment. [19] While violence-as-crime and violence-as-terrorism tend to be grossly overreported, the coverage of terrorist incidents that provide dramatic visuals is in a league of its own in terms of media attention. With few exceptions, ordinary criminals do not commit their deeds to attract cameras, microphones, and reporter's notebooks. But for terrorists, publicity is their lifeblood and their oxygen. No other medium has provided more oxygen to terrorism than television because of its ability to report the news instantly, nonstop, and in visuals and words from any place to all parts of the globe, a facility that has affected the reporting patterns of other media as well.

When commentators characterized the terrorist events of "Black Tuesday" as the Pearl Harbor of the twenty-first century or the second Pearl Harbor, they ignored one fundamental aspect that separated the surprise attack on December 7, 1941, from that on September 11, 2001: the vastly different communication technologies. Three hours passed from the time the first bombs fell on Pearl Harbor and the moment when people on the U.S. mainland first learned the news from radio broadcasts. More than a week lapsed before the New York Times carried the first pictures of the actual damages. Sixty years later, the terror attacks had a live global TV, radio, and Internet audience and many replays in the following hours, days, and weeks.

In September 1970, members of the Popular Front for the Liberation of Palestine (PFLP) simultaneously hijacked four New York-bound airliners carrying more than six hundred passengers. Eventually, three of the planes landed in a remote part of Jordan, where many passengers, most of them Americans and Europeans, were held for approximately three weeks. This was high drama. The media reported extensively, but the reporting paled in comparison to the great attention devoted to equally as dramatic or far less shocking incidents in later years. The communication technology at the time did not allow live transmissions from remote locations. Satellite transmissions were in their early stages and were very expensive. For the PFLP, the multiple hijacking episode ended in disappointment. While the tense situation resulted in media, public, and government attention, no news organization covered the events in ways that might have forced President Richard Nixon and European leaders to act under pressure. [20]

But as television technology advanced further and competition among TV, radio, and print organizations became fiercer, the media became more obsessed with exploiting violence-as-crime and violence-as-terrorism in search of higher ratings and circulation. As a result, the contemporary news media, especially television, have customarily devoted huge chunks of their broadcast time and news columns to major and minor acts of political violence, supporting the media critics' argument that the mass media, as unwitting as they are, facilitate the media-centered terrorist scheme.

There was no need to count broadcast minutes or measure column length to establish the proportion of the total news that dealt with "Black Tuesday" and its aftermath. For the first five days after the terror attack, television and radio networks covered the disaster around the clock without the otherwise obligatory commercial breaks. There simply was no other news. Most sports and entertainment channels switched to crisis news, many of them carrying the coverage of one of the networks and suspending their suddenly irrelevant broadcasts. For example, Fox cable's sports channel in New York simply showed the image of the U.S. flag. Newspapers and magazines devoted all or most of their news to the crisis. Given the warlike dimension of the attacks on America, this seemed the right decision early on. Eventually one wondered whether terrorism coverage needed to be curtailed so that other important news got the attention it deserved. *Newsweek* and *TIME*, for example, devoted all cover stories in the eight weeks following the events of September 11 to terrorism and terrorism-related themes.

If not the perpetrators themselves, the architects of the terror enterprise surely anticipated the immediate media impact: blanket coverage not only in the United States but in other parts of the world as well. How could the terrorists better achieve their objective than by obtaining the attention of their targeted audiences? Opinion polls revealed that literally all Americans followed the initial news of the terrorist attacks (99 percent or 100 percent according to surveys) by watching and listening to television and radio broadcasts. While most on-line adults identified television and radio as their primary sources for crisis information, nearly two thirds also mentioned the Internet as one of their information sources. [21] This initial universal interest in terrorism news did not weaken quickly. Probably affected by the news of anthrax attacks along the U.S. East coast, more than 90 percent of the public kept on watching the news about terrorism "very closely" or "closely" nearly six weeks after the event s of September 11. [22]

Political leaders as well followed the original terror news, replays, and subsequent crisis reporting. There is no doubt, then, that the terrorists behind the attack on America received the attention of all Americans, the general public, and world leaders alike. This level of media coverage was a perfect achievement as far as the "attention getting" goal in the United States was concerned. The architects of the September 11 terror were delighted. Referring to the Kamikaze pilots as "vanguards of Islam," bin Laden marveled,

> Those young men (...inaudible...) said in deeds, in New York and Washington, speeches that overshadowed other speeches made everywhere else in the world. The speeches are understood by both Arabs and non-Arabs—even Chinese. [23]

With these remarks, bin Laden revealed that he considers terrorism a vehicle to dispatch messages—speeches in his words. And since he and his circle had followed the news of September 11, they were sure that their message had been heard.

Not surprisingly, from one hour to the next, the perpetrators set America's public agenda and profoundly affected most Americans' private lives. As soon as television stations played and replayed the ghastly scenes of jetliners being deliberately flown into the World Trade Center and the Pentagon, business as usual was suspended in the public and private sector. All levels of government and vast parts of the business community concentrated on the immediate rescue contingencies and on preventing further attacks that were rumored in the media. Within days, all levels of government and the business community began to implement new anti- and counterterrorist measures.

Those who were responsible for the acts of terror spread anxiety and fear in a public traumatized by their terror. Fifty-three percent of the American public across the United States, not simply those in the attacked regions in the East, changed their plans and activities for the rest of "Black Tuesday"; four of every ten employed men and women did not go to work that day or quit their jobs. [24] In the days after the assault, nine in ten Americans worried about additional terrorist events in their country, and a majority worried that they themselves, or somebody close to them, could become victims the next time around. Compared with the public's reaction to previous acts of terrorism, these sentiments were stronger than ever before. For example, in the days after the 1995 Oklahoma City bombing, one in four Americans worried that they or a member of their family could become victims of terrorism. In the days after the 2001 attacks on New York and Washington, 51 percent to 58 percent of the public had such fears. [25] This is precisely what the architect of terror intended. In a videotaped message, Osama bin Laden said about the reactions of Americans to the terror of September 11, "There is America, full of fear from north to south, from west to east. Thank God for that." [26] These concerns did not evaporate as time passed. Instead, most Americans continued to be concerned about further terrorist attacks and that they or a member of their family could become a victim the next time around. [27] However, the number of those who felt depressed and/or had trouble sleeping in the days following the suicide attacks decreased quite dramatically during the following weeks. [28]

When President George W. Bush, New York's Mayor Rudy Giuliani, and other public officials urged Americans to return to quasi-normal lives, the media's crisis coverage did not reflect that public officials in Washington had returned to normalcy. There were pictures of Washington's Reagan National Airport remaining closed because of its proximity to the White House and other government buildings. There was an image of a fighter jet over Washington escorting the presidential helicopter on a flight to Camp David. There were reports explaining Vice President Richard Cheney's absence, when the president addressed a joint session of Congress and in the weeks thereafter, as a precaution, in case terrorists might strike again. And there were constant visuals of a tireless Mayor Giuliani as crisis-manager before the daunting background of ground zero in Manhattan's financial district, at the funeral of yet another police- or fireman, at a mass at St. Patrick's Cathedral, or at a prayer service in Yankee stadium.

Every public appearance by the president, New York City's mayor, New York State's Governor George Pataki, U.S. senators, U.S. representatives, members of the Bush administration; every hearing and floor debate in the two chambers of Congress; and every publicly announced decision was in reaction to the terrorist attack and was reported in the news. Even weeks after "Black Tuesday" and during the anthrax scare, CNN and other all-news channels interrupted their programs not only to report on President Bush's public appearances but on Mayor Giuliani's activities as well. And when the broadcast networks returned to their normal schedules, the around-the-clock news channels continued to report mostly on terrorism and counterterrorism in the form of military actions against targets in Afghanistan that began on 7 October.

When professional sports competition resumed after a moratorium of several days, watching sports broadcasts did not necessarily mean that fans could forget the horror for a while. Unwittingly, the media transmitted constant reminders: Baseball fans in Chicago displayed a "We Love New York" banner; American flags were placed on helmets and caps of competitors; a hockey game was interrupted so that players and fans could hear the presidential speech before Congress; players praised rescue workers at the terror sites and embraced members of a rival team in an expression of unity. The sports pages of newspapers captured the reactions of well-known sports stars; the American flag on the cover of Sports Illustrated signaled that the entire issue following September 11 was devoted to patriotism.

Entertainment as well was in the grasp of the horrendous acts. When David Letterman resumed his late-night show, he was unusually serious and made no attempts to be funny. Instead of offering hilarious punch lines, he found words of comfort for news anchor Dan Rather who was twice moved to tears when talking about the terror next door. *Saturday Night Live,* Comedy Central, and other entertainment shows were all less aggressive in provoking laughter at the expense of political leaders. Bill Maher of ABC's *Politically Incorrect* was the exception when he told his audience that the suicide bombers were not cowards but that the United States was cowardly by launching cruise missiles on targets thousands of miles away. Maher, who later apologized for his remarks, was criticized by White House spokesman Ari Fleischer and punished by some advertisers who withdrew their sponsorship; some local stations dropped the program. Even poking fun at bin Laden and the Taliban, as Jay Leno and the *Saturday Night Live* performers did, seemed not all that funny. Some publications suspended their weekly cartoon sections for a while. And when a star studded cast of entertainers performed in a two-hour telethon to raise funds for the victims of terror, the celebrities told touching stories of innocent victims and real-life heroes. But nothing reminded the American audience more succinctly of the extraordinary circumstances behind the benefit than the sight of superstars Jack Nicholson, Sylvester Stallone, Meg Ryan, Whoopi Goldberg, and other show business celebrities relegated to answer telephone calls of contributors, since the producers had not found slots for them to perform in the program. Finally, even the most outrageous TV and radio talk show hosts toned down their personalities as they embraced the terror crisis story, albeit only for a short time.

When the television series *The West Wing* postponed its scheduled season-opener and replaced it with a special episode in which the White House dealt with the aftermath of a terrorist nightmare, the blur of fact and fiction, life and entertainment, came full circle. After Hollywood's make-believe disasters became reality, when nineteen suicide terrorists stuck America, the real-life calamity inspired a television drama and ideas for more such episodes to come.

Whether tuned to the coverage of current affairs, sports news, or even entertainment, Americans had a hard time forgetting about terrorism for a while. Although most people agreed that TV programs about the attacks on the World Trade Center and the Pentagon, and later the news about anthrax bio-terror, were depressing and frightening, many people simply could not stop watching terrorism news. While this addiction was true for six of every ten Americans in the middle of September, only one in two Americans were still watching this sort of coverage by mid-October. At the same time, a vast majority felt that watching terrorism news was frightening. Not surprisingly, people who were hooked felt more fearful of future terrorism than those who were not addicted. [29]

Not only Americans but people abroad, too, knew quickly about the terrorist attacks on the United States and were affected by what they saw, heard, and read. The Gallup Organization found, for example, that more than 98 percent of the Hungarian public knew about the terror soon after it happened. This caused one commentator to conclude, "If there were any remaining doubts about the media's capacity to almost simultaneously disseminate global news, this poll's finding should serve to dispel it." [30] Reacting to the news, the majority of Hungarians (51.2 percent) were "very" or "somewhat" worried that they, or someone in their family, could become a victim of terrorism. [31] Equally informed, 83 percent of people in the United Kingdom and 76 percent of Russian public shared those fears. [32] Even a cursory examination of the media around the world affirmed what journalists reported from those regions: For weeks after the terror attack on the United States and before the first counterterrorist strikes against targets in Afghanistan, the event, its political, economic, and military implications, and the threat of a major military confrontation in the Middle East remained highest on the media, public, and elite agenda.

As media organizations, star anchors, and public officials became the targets of biological terrorism and postal workers became the most numerous victims of "collateral damage" in an unprecedented anthrax offensive by elusive terrorist(s), the news devoted to terrorism multiplied—especially in the United States but abroad as well. The aftermath of the September 11 terror—the anthrax cases, the debate of possibly more biological and chemical agents in the arsenal of terrorists, and the military actions against al Qaeda and Taliban targets in Afghanistan—crowded out most other events and developments in the news. Terrorists and terrorism had set the media agenda, the public agenda, and the government agenda. To the terrorists, the attention of the mass media, the public, and governmental decision makers was a total victory.

WHY DO THEY HATE US?

Sixteen days after the attacks on New York and Washington, the *Christian Science Monitor* published an in-depth article addressing a question that President Bush had posed in his speech before a joint session of Congress: "Why do they hate us?" Describing a strong resentment toward America in the Arab and Islamic world, Peter Ford summarized the grievances articulated by Osama bin Laden and like-minded extremists, but also held by many less radical people in the Middle East and other Muslim regions, when he wrote that

> the buttons that Mr. bin Laden pushed in statements and interviews—the injustice done to the Palestinians, the cruelty of continued sanctions against Iraq, the presence of US troops in Saudi-Arabia, the repressive and corrupt nature of the US backed Gulf governments—win a good deal of popular sympathy. [33]

This lengthy article was but one of many similar reports and analytical background pieces tracing the roots of anti-American attitudes among Arabs and Muslims and possible causes for a new anti-American terrorism of mass destruction. Lisa Beyer offered this summary of grievances in her story in *TIME* magazine:

> The proximate source of this brand of hatred toward America is U.S. foreign policy (read: meddling) in the Middle East. On top of its own controversial history in the region, the United States inherits the weight of centuries of Muslim bitterness over the Crusades and other military campaigns, plus decades of indignation over colonialism. [34]

A former U.S. ambassador at large for counterterrorism wrote,

> Certainly, the U.S. should reappraise its policies concerning the Israeli-Palestinian conflict and Iraq, which have bred deep anger against America in the Arab and Islamic world, where much terrorism originates and whose cooperation is now more critical than ever. [35]

While the print press examined the roots of the deeply seated opposition to U.S. foreign policy in the Arab and Islamic world extensively, television and radio dealt with these questions as well—in some instances at considerable length and depth. Thus, in the two-and-a-half weeks that followed the terrorist attacks, the major television networks and National Public Radio broadcast thirty-three stories that addressed the roots of anti-American terrorism of the sort committed on September 11, 2001, the motives of the perpetrators, and, specifically, the question that President Bush had asked. In the more than eight months before the attacks on New York and Washington, from 1 January 2001 to "Black Tuesday," none of the same TV or radio programs addressed the causes of anti-American sentiments in the Arab and Islamic world.[36] This turnaround demonstrated the ability of terrorists to force the media's hand, to set the media's agenda. Suddenly, in the wake of terrorist violence of unprecedented proportions, the news explored and explained the grievances of those who died for their causes and how widely these grievances were shared even by the vast majority of those Arabs and Muslims who condemned violence committed in the United States. With or without referring to a *fatwa* (religious verdict) that Osama bin Laden and four other extremist leaders had issued in 1998, or to the most recent communications from these circles, the news media now dealt with the charges contained in these statements as well as with additional issues raised by Muslims in the Middle East and elsewhere. The 1998 *fatwa*, posted on the Web site of the World Islamic Front, listed three points in particular:

> First, for over seven years the United States has been occupying the lands of Islam in the holiest places, the Arabian Peninsula.... If some people have in the past argued about the fact of the occupation, all people of the Peninsula have now acknowledged it. The best proof of this is the Americans' continuing aggression against the Iraqi people using the Peninsula as a staging post....
>
> Second, despite the great devastation inflicted on the Iraqi people by the crusader-Zionist alliance, and despite the huge number of those killed, which has exceeded 1 million..., despite all this, the Americans are once again trying to repeat the horrific massacres.
>
> Third, if the Americans' aims behind these wars are religious and economic, the aim is also to serve the Jews' petty state and divert attention from its occupation of Jerusalem and murder of Muslims there. [37]

These specific accusations were among a whole laundry list of grievances that the media explored in the wake of the terrorism in New York and Washington. And the existence of bin Laden's various declarations of hate was reported as news although the declarations were issued as far back as 1996 and 1998.

It has been argued that religious fanatics who resort to this sort of violence are not at all interested in explaining themselves to their enemies because their only conversation is with God. But it was hardly an accident that the leaders among the suicide attackers, who diligently planned every detail of their conspiracy, left behind several copies of their instructions in the hours before and during the attacks. By insuring that law enforcement agents would find the documents, the terrorists must have been confident that America and the world would learn of their cause. They were proven right when the FBI released copies of the four-page, handwritten document to the media for publication. Revealing the pseudoreligious belief that drove the hijackers to mass destruction and their own deaths, the instructional memorandum contained the following sentences:

> Remember that this is a battle for the sake of God....
>
> So remember God, as He said in His book: "Oh Lord, pour your patience upon us and make our feet steadfast and give us victory over the infidels."
>
> When the confrontation begins, strike like champions who do not want to go back to this world. Shout, "Allah Akbar," because this strikes fear in the hearts of the non-believers. God said: "Strike above the neck, and strike at all their extremities." Know that the gardens of paradise are waiting for you in all their beauty, and the woman of paradise are waiting, calling out, "Come hither, friend of God." [38]

The intent here is not to criticize the media for publicizing such documents, for trying to answer why terrorists hate Americans and why many nonviolent people in the Arab and Islamic world hold anti-American sentiments, but rather to point out that this coverage and the accompanying mass-mediated discourse were triggered by a deliberate act of mass-destruction terrorism. After September 11 there was a tremendous jump in the quantity of news reports about one of the other aspects of developments in the Muslim and Arab world and even more so about Islam. Television news especially paid little attention to these topics before the terror attacks in New York and Washington. But, as Table 1 shows, the switch from scarce or modest coverage before September 11 to far more news prominence thereafter occurred in radio and the print press as well. While many of these news segments and stories focused on anti-American terrorism committed by Muslim and Arab perpetrators and the role of fundamentalist Islamic teachings, there were also many stories reporting on and examining the grievances of the mass of nonviolent Muslims and Arabs as well as the teachings of mainstream Islam.

Before "Black Tuesday," the news from the Middle East and other Islamic regions was overwhelmingly episodic and focused on particular, typically violent, events. Following the Iran Hostage Crisis, one critic noted that "Muslims and Arabs are essentially covered, discussed, apprehended either as oil suppliers or as potential terrorists. Very little of the detail, the human density, the passion of Arab-Muslim life has entered the awareness of even those people whose profession it is to report the Islamic world." [39] Given the scarcity of foreign news in the post-cold war era in the American mass media, especially television,[40] there were even fewer contextual news or thematic stories[41] from this part of the world. But it would have been precisely the thematic approach that should have addressed all along the conditions that breed anti-American attitudes in the Arab and Muslim world. It took the terror of "Black Tuesday" for the media to offer a significant amount of contextual coverage along with episodic reporting. In the process, the perpetrators of violence achieved their recognition goal: By striking hard at America, the terrorists forced the mass media to explore their grievances in ways that transcended by far the quantity and narrow focus of their pre-crisis coverage.

TABLE 1

Muslim(s), Arab(s), Islam in the News before and after September 11, 2001

	Muslim Period I	*Muslim Period II*	*Arab Period I*	*Arab Period II*	*Islam Period I*	*Islam Period II*
	(N)	(N)	(N)	(N)	(N)	(N)
ABC News	31	163	11	99	1	31
CBS News	32	144	27	117	1	27
NBC News	9	98	5	90	--	18
CNN	23	203	43	200	1	31
Fox News	1	100	2	64	1	46
N.Y. Times	345	1,468	345	1,272	216	1,190
NPR	54	217	53	182	10	84

N = Number of news segments/articles mentioning the search words.
Source: Compiled from Lexis-Nexis and *New York Times* archives using the search words "Muslim," "Arab," and "Islam."
Note: Period I = Six months before the terrorist attacks of September 11, 2001; Period II = six months after the attacks of September 11.

MAKING THE NEWS AS VILLAIN AND HERO

What about the third goal that many terrorists hope to advance—to win or increase their respectability and legitimacy? Here, the perpetrators' number one audience is not the enemy or the terrorized public, in this case Americans, but rather the population in their homelands and their regions of operation. And in this respect, again, the terror of "Black Tuesday" was beneficial from the view of the architects and the perpetrators of violence. A charismatic figure among his supporters and sympathizers to begin with, Osama bin Laden was the biggest winner in this respect. Whether he was directly or indirectly involved in the planning of the terrorist strikes did not matter. The media covered him as "America's number one public enemy"[42] and thereby bolstered his popularity, respectability, and legitimacy among millions of Muslims. The American and foreign news publicized visuals and reports of the popular support for bin Laden following the terror attack against the United States. A lengthy bin Laden profile in the *New York Times*, for example, contained the following passage: "To millions in the Islamic world who hate America for what they regard as its decadent culture and imperial government, he [Osama bin Laden] is a spiritual and political ally." [43] A page-one article in the same edition of the New York Times reported from Karachi, Pakistan,

> In every direction in this city of 12 million people, the largest city in a nation that has become a crucial but brittle ally in the United States' war on terrorism, there are cries and signs for Osama bin Laden, for the Taliban, for holy war. [44]

The Associated Press reported that a book about the terrorist-in-chief was a bestseller in the Middle East. The volume contained the complete transcript of an interview with bin Laden that was broadcast in abbreviated form by al Jazeera television in 1998 and rebroadcast after the terror strike against the United States in September of 2001. Sold out in most bookstores of the region, readers were reportedly borrowing the book from friends and making photocopies. [45]

Bin Laden, his al Qaeda group, and the closely related web of terror spanning from the Middle East into other parts of Asia, Africa, Europe, North America, and possibly South America, were no match for the American superpower in terms of political, economic, and military power. But, as Table 2 shows, in the aftermath of the terrorist attacks on New York and Washington and up to the beginning of the bombing of Afghanistan on 7 October, the U.S. television networks mentioned Osama bin Laden more frequently and the leading newspapers and National Public Radio only somewhat less frequently than President George W. Bush. A terrible act of terror turned the world's most notorious terrorist into one of the world's leading newsmakers. The fact that the American news media paid more attention to bin Laden than to the U.S. president, or nearly as much, was noteworthy, if one considers that George W. Bush made fifty-four public statements during this time period (from major addresses to shorter statements to a few words during photo opportunities) compared to bin Laden, who did not appear in public at all, did not hold news conferences, or give face-to-face interviews.[46]

Although the American media did not portray bin Laden as a sympathetic figure, he did share center stage with President Bush in the mass-mediated global crisis. Since the 1998 bombings of the U.S. embassies in Kenya and Tanzania, the American media devoted considerable broadcast time and column inches to bin Laden. But the celebrity terrorist's ultimate ascent to the world stage was more dramatic and forceful than that of Yasir Arafat in the 1970s, the Ayatollah Khomeini and Muammar Gadhafi in the 1980s, Saddam Hussein in the early 1990s, and bin Laden himself in the years and months preceding "Black Tuesday." And through all of this, bin Laden was in hiding, did not hold news conferences or grant interviews. However, the Qatar-based Arab television network, al Jazeera, aired a videotape made available by al Qaeda immediately after President Bush told America and the world that military actions had begun in the multifaceted hunt for bin Laden and his terror organization. All U.S. networks broadcast the tape as they received the al Jazeera feed. Bin Laden's shrewdly crafted speech received the same air time as President Bush's speech. The same was true for a videotaped statement by bin Laden's lieutenant for media affairs who threatened that "Americans must know that the storm of airplanes will not stop. God willing, and there are thousands of young people who are as keen about death as Americans are about life." [47]

In the ten weeks following the attacks of September 11, *TIME* magazine depicted Osama bin Laden three times and President George W. Bush twice on its cover. During the same period, *Newsweek* carried bin Laden twice on its cover and President Bush not at all. Finally, the cover of *Newsweek's* eleventh issue after September 11 featured President George W. and First Lady Laura Bush.

TABLE 2

News Stories Mentioning President Bush and bin Laden Following the September 11, 2001 Terrorist Attacks

	Pres. G. W. Bush	*Osama bin Laden*
	(N)	(N)
ABC News	175	299
CBS News	210	270
NBC News	159	211
CNN	292	469
NPR	271	188
N.Y. Times	655	611
Wash. Post	684	490

N = Number of segments/ stories.

Source: Compiled from Lexis- Nexis data; TV and radio broadcasts for the period 11 September 2001-6 October 2001; newspaper articles for the period 12 September 2001-7 October 2001.

From the terrorists' point of view, it did not matter that bin Laden earned bad press in the United States and elsewhere. Singled-out, condemned, and warned by leaders, such as President Bush and British Prime Minister Tony Blair, Osama bin Laden was in the news as frequently as the world's legitimate leaders, or even more frequently. This in itself was a smashing success from the perspectives of bin Laden and his associates: The mass media reflected that bin Laden and his followers preoccupied not only America and the West but literally the entire world.

In sum, then, by attacking symbolic targets in America, killing thousands of Americans, and causing tremendous damage to the American and international economy, the architects and perpetrators of this horror achieved their media-centered objectives in all respects. This propaganda coup continued in the face of American and British counterterrorist military actions in Afghanistan.

HIGH MARKS FOR THE NEWS MEDIA

In the days following the attacks, when most Americans kept their televisions or radio sets tuned to the news during most of their waking hours, the public gave the media high grades for its reporting. Nearly nine in ten viewers rated the performance of the news media as either "excellent" (56 percent) or "good" (33 percent). The Pew Research Center for the People and the Press (that keeps track of the relationship between the public and the news media) called this high approval rating "unprecedented."[48] This record approval came on the heels of increasing public dissatisfaction with the mass media and a number of journalistic and scholarly works that identified the degree of and reasons for the increasing disconnect between the public and the news media.[49] The terrorism catastrophe brought Americans and the press closer together, closer than in recent times of normalcy and during previous crises, in particular, the Gulf War.[50] Five aspects in particular seemed to effect these attitudinal changes: First, the public appreciated the flow of information provided by television, radio, and print either directly or via media organizations' Internet sites. In the hours and days of the greatest distress, television and radio especially helped viewers and listeners feel as if they were involved in the unfolding news. Unconsciously, people took some comfort in seeing and hearing the familiar faces and voices of news anchors and reporters as signs of the old normalcy in the midst of an incomprehensible crisis. At a time when the overwhelming majority of Americans stopped their normal activities, watching television, listening to the radio, reading the newspaper, going online gave them the feeling of doing something, of being part of a national tragedy.

Second, people credited the news media, especially local television, radio, and newspapers in the immediately affected areas in and around New York, Washington, and the crash site in Pennsylvania, with assisting crisis managers in communicating important information to the public. For crisis managers, the mass media offered the only effective means to tell the public about the immediate consequences of the crisis—what to do (for example, donate blood of certain types, where to donate and when) and what not to do (initially, for example, not trying to drive into Manhattan because all access bridges and tunnels were closed). In this respect, the media served the public interest in the best tradition of disaster coverage.

Third, Americans experienced a media—from celebrity anchors, hosts, and other stars to the foot-soldiers of the fourth estate—that abandoned cynicism, negativism, and attack journalism in favor of reporting, if not participating in, an outburst of civic spirit, unity, and patriotism. From one minute to another, media critics and pollsters recognized a reconnection between the press and public after years of growing division. As even the most seasoned news personnel couldn't help but show their emotions while struggling to inform the public during the initial hours and days of the crisis, audiences also forgot about their dissatisfaction with the media in a rare we-are-all-in-this-together sentiment.

This explains the sudden high approval ratings for the fourth estate mentioned earlier.

To be sure, there were some bones of contention. As Americans everywhere displayed the star spangled banner, images of the American flag appeared on television screens as well. Many anchors and reporters wore flag pins or red, white, and blue ribbons on their lapels. Others rejected this display of patriotism. Barbara Walters of ABC-TV, for example, declared on the air that she would not wear any version of Old Glory.[51] When the news director of a cable station on Long Island, New York, issued a memo directing his staff not to wear any form of flag reproductions, there was a firestorm of opposition from viewers and advertisers. But even this incident seemed to fade after the news director issued an apology and it became obvious that the flag was not banned from the station's coverage.[52]

Fourth, the news provided public spaces where audience members had the opportunity to converse with experts in various fields and with each other, or witness question-and-answer exchanges between others. Whether through quickly arranged electronic town hall meetings or phone-in programs, television, radio, and online audiences wanted to get involved in public discourse. Many news organizations facilitated the sudden thirst for dialogue. While television and radio were natural venues for these exchanges, newspapers and newsmagazines published exclusively, or mostly, letters to the editors on this topic and reflected a wide range of serious and well-articulated opinions. Seldom, however, was the value of thoughtful moderators and professional gatekeepers more obvious than in the days and weeks after the terror nightmare. The least useful, often bigoted comments were posted on Internet sites and message boards.

Fifth, news consumers were spared the exasperation of watching reporters and camera crews chasing survivors and relatives of victims, camping on front lawns, shoving microphones in front of people who wanted to be left alone. In the 1980s, when terrorists struck against Americans abroad, the media often pushed their thirst for tears, grief, tragedy, and drama to and even beyond the limits of professional journalism's ethics in their hunt for pictures and sound bites. But this time around, neither the public nor media critics had reason to complain about the fourth estate's insistence on invading people's privacy and exploiting grief-stricken relatives of victims and survivors. This time, many husbands and wives, mothers and fathers, daughters and sons of disaster victims spoke voluntarily to reporters, appeared voluntarily, and in many instances repeatedly, on local and national television to talk about their traumatic losses. Many survivors described their ordeals and their feelings in touching detail. Most of these people were born and grew up in the era of television and seemed comfortable, in some cases even eager, to share their sorrow and their tears, their memories and their courage with anchors, hosts, correspondents—and millions of fellow Americans.

Again, this was not the result of a changed and more restrained media but of a cultural change. Expressing one's innermost feelings, showing one's despair, controlled crying or sobbing before cameras and microphones seemed natural in the communication culture of our time and in the age of so-called reality TV and talk shows with a human touch such as Oprah Winfrey or Larry King. Thus, unlike past TV audiences who were exposed to ruthless exploits of grief during and after terrorist incidents, following the terror attacks of September 11, 2001, viewers participated in mass-mediated wakes, full of collective sadness and shared encouragement.

When the broadcast media played and replayed the recorded exchanges between victims in the World Trade Center and emergency police dispatchers, they exploited the unimaginable suffering of those who were trapped and soon died in the struck towers. Criticizing this practice as "primetime pornography," one commentator wrote,

> Can there be anybody on the planet who failed to immediately grasp the full horror of what went on Sept. 11 that they need to hear, over and over, the emotional mayhem of ordinary people trying to cope amidst overwhelming disbelief, fear and terror—not to mention grief? But in our show-and-tell culture there is nothing so private and sensitive that it can't be exposed and sensationalized-especially where ratings are involved. [53]

One can perfectly agree with this insistence on journalistic ethics, as I do, and still wonder whether the "show-and-tell culture" has not only desensitized broadcasters but also confused the public's distinction between private and public sphere.

CRISIS NEWS: WEAKNESSES, QUESTIONS, AND ISSUES

Twelve days after the kamikaze attacks on the World Trade Center and Pentagon, media critic Marvin Kitman, commenting on the perhaps longest continuous breaking news events in the history of television, wrote:

> They [the TV people] kept on showing those same pictures of the planes hitting, the buildings crumbling. I'm sure if I turned the TV on right now, the buildings would still be crumbling. It never got any better. One picture is worth a thousand words, except in "live" television, where people felt compelled to constantly talk even when they knew very little about what they were talking about.[54]

While the initial emergency coverage, especially in television and radio, deserved high marks, some of the infotainment habits that had increasingly made their way into television news crept rather soon into the presentations of what screen banner called the "Attack on America" or "America Attacked." Recalling the rather trivial headlines and cover stories before September 11, one expert in the field suggested early on that "suddenly, dramatically, unalterably the world has changed. And that means journalism will also change, indeed is changing before our eyes."[55] As it turned out, this was wishful thinking. There was no longer the feeding frenzy on Congressman Gary Condit's private life, Mayor Rudy Giuliani's nasty divorce, or the meaning of Al Gore's beard for his political future, but that did not mean an end to the overkill and hype that characterized past reporting excesses, whether in the context of the O. J. Simpson murder case and trial or the accidental deaths of Princess Diana and John F. Kennedy, Jr. Immediately after September 11, when a series of unspeakable events were reported as they unfolded, and a day or two thereafter, when the enormity of the attacks and their consequences began to sink in, there was simply not enough genuine news to fill twenty-four hours per day. As a result, television networks and stations replayed the scenes of horror again and again, revisiting the suffering of people over and over, searching for emotions beyond the boundaries of good taste. In their search for family members or friends who were among the thousands of missing in New York, many people pursued reporters and camera crews with photographs of their loved ones in the hope of some good information, some good news. While highlighting these photographs could be seen as servicing a grieving community, dwelling on picture galleries of the victims was certainly not. One shocked observer recalled that "one of the yokels on Ch. 2 showed pictures she had found in the street after the explosion and cheerfully pointed out 'that these little children may now be without parents.'" [56]

The shock over the events of September 11 wore off rather quickly in the newsrooms, giving way to everyday routine. Some television anchors welcomed their audiences rather cheerfully to the "Attack on America" or "America's New War" and led into commercial breaks with the promise that they would be right back with "America's War on Terrorism" or whatever the slogan happened to be that day or week.

There were signs of bias that were especially upsetting to Arab and Muslim Americans who felt, for example, that the scenes of Palestinians rejoicing over the news of the attacks in New York and Washington were over-reported and too often replayed. In contrast to Palestinian celebrations, anti-American outbursts in Europe received little or no attention. For example, when fans of a Greek soccer team at the European Cup game in Athens jeered America during a minute of silence for the terrorism victims of September 11 and tried to burn an American flag, no television news programs and only a handful of American newspapers (publishing only a few lines in reports about the soccer game on the sports pages) mentioned the incident. [57] But the coverage raised far more serious questions about the proper role of a free press in a crisis that began with the suicide terror in New York, Washington, and near Pittsburgh and intensified when anthrax letters were delivered in states along the U.S. east coast. Three areas, in particular, proved problematic.

The first of these issues concerned the videotapes with propaganda appeals by bin Laden and his lieutenants that al Qaeda made available to the Arab language TV network al Jazeera. On 7 October, shortly after President Bush informed the nation of the first air raids against targets in Afghanistan, five U.S. television networks (ABC, CBS, NBC, CNN, and Fox News) broadcast an unedited feed from al Jazeera that gave bin Laden and his associates access to the American public. Two days later, three cable channels (CNN, Fox News, and MSNBC) aired in full a statement by bin Laden's spokesman Suleiman Abu Gheith. Both tapes contained threats against Americans at home and abroad. Bin Laden said, "I swear to God that America will not live in peace before peace reigns in Palestine and before all the army of infidels depart the land of Muhammad, peace be upon him."[58] His spokesman warned that "the storms will not calm down, especially the storm of airplanes, until you see defeat in Afghanistan." He called on Muslims in the United States and Great Britain "not to travel by airplanes and not to live in high buildings or skyscrapers."[59] The Bush administration cautioned that these statements could contain coded messages that might cue bin Laden followers in the United States and elsewhere in the West to unleash more terror. But intelligence experts were unable to identify suspect parts in the spoken text or visual images at a time when Attorney General John Ashcroft, the FBI, and other government officials in Washington warned of more terrorist strikes to come. While the administrations's argument that these tapes were vehicles for hidden messages was not credible, it was certainly true that these videos and their transcripts contained terrorist propaganda, which newspapers printed in full. The most damaging effect was that these broadcasts further frightened an already traumatized American public. Students of propaganda have argued that propaganda of fear or the "fear effect" is most effective "when it scares the hell out of people."[60] But this is not what the administration argued. Prodded by National Security Adviser Condoleezza Rice who warned that the hateful threats from the bin Laden camp could incite more violence against Americans abroad, all American television networks agreed to edit future tapes and eliminate "passages containing flowery rhetoric urging violence against Americans."[61] This argument was just as weak as the suggestion of hidden signs contained in the tapes. After all, al Jazeera and other television channels aired the material in the Middle East and other regions with Muslim populations. It would have been far more credible to argue that threats from bin Laden and his associates increased the public's anxiety.

The networks' joint decision was not universally applauded because it raised the question whether the networks had given in to pressure from the administration when they agreed to exercise this form of self-censorship. While the argument that the press in a democracy needs to fully inform citizens, especially in times of crisis and great danger, has most weight here, it is also true that the news media make all the decisions on whom and what to include and exclude, or whom and what to feature more or less prominently in their broadcasts. In the case of the al Qaeda tapes, after the first ones were aired excessively by some cable networks, subsequent tapes were undercovered. All of these videotapes should have been broadcast fully, and their transcripts should have been printed entirely by the press. The public should have learned of bin Laden's propaganda without being exposed to endless replays. The mistake was made initially when passages of the first bin Laden video were broadcast so many times with full screen or split screen exposure, when bin Laden and al Qaeda loomed too large in the overall news presentations compared to other news sources and developments. The second mistake was the suppression or partial suppression of the content of later videotapes by broadcast and print media.

A similar controversy arose over CNN's decision to join al Jazeera in submitting questions to Osama bin Laden following an invitation by al Qaeda and the promise that bin Laden would respond to them. While a face-to-face interview with the man who openly praised the terrorism of September 11 could have yielded valuable information—especially for U.S. decision makers, the exchange of written questions and answers was a far more questionable journalistic exercise under the circumstances (if only because the media organizations could not be sure whether bin Laden or someone else would answer their questions). Under these circumstances, it was just as well that the answers were never provided.

A second issue concerned the media's sudden obsession with endlessly reporting and debating the potential

TABLE 3

Biological and Chemical Terrorism and Anthrax in the News

	B/C Terror I	*B/C Terror II*	*Anthrax I*	*Anthrax II*
	(N)	(N)	(N)	(N)
ABC News	20	30	2	383
CBS News	12	30	0	267
NBC News	8	57	3	250
CNN	17	99	1	567
Fox News	23	37	0	103
NPR	11	37	8	176
N.Y. Times	76	194	27	729
Wash. Post	55	147	25	465

N = Number of segments/stories.
Source: Compiled by retrieving transcripts and newspaper content from the Lexis-Nexis archives using the search words "biological" and "chemical" and "terrorist" for biological and chemical terror and "anthrax."
Note: B/C Terror I and Anthrax I segments/stories aired/published from 11 September 2001 to 3 October 2001 for television and radio, and 12 September 2001 to 4 October 2001 for newspapers.
B/C Terror II and Anthrax II segments/stories aired/published from 4 October 2001 to 31 October 2001 for television and radio, and 5 October 2001 to 1 November 2001 for newspapers.

for biological, chemical, and nuclear warfare in the wake of September 11. As real and would-be experts filled the air waves, some hosts and anchors were unable to hide their preference for guests who painted doomsday scenarios. As Table 3 shows, this was common in broadcasts even before the first anthrax case in Florida made the news on 4 October 2001. It was as if anchors and news experts expected the other shoe to drop as they went out of their way to report to the public that the public health system and other agencies were ill prepared to deal with bioterrorism and other mass destruction terrorism.

When the news of a Florida man dying of anthrax and subsequent cases validated these predictions, anthrax terrorism and other forms of bioterrorism moved higher up on the agenda of TV, radio, and print news. In less than a month, from the discovery of the first case on 4 October to 31 October, the television networks covered or mentioned the anthrax terror in hundreds of segments. The leading newspapers published even more stories on anthrax and other possible threats from biological and chemical agents. (See Table 3.)

The most serious bioterrorism attacks in the United States deserved headlines and serious, regular, in-depth coverage, but the attacks did not merit an army of talking heads who beat the topic to death many times over. In the process, public officials who tried to mask their own confusion, experts who scared the public, and media stars who overplayed the anthrax card contributed to a general sentiment of uncertainty. By shrewdly targeting major news organizations and two of the most prominent television news anchors, Tom Brokaw of NBC and Dan Rather of CBS (and perhaps Peter Jennings of ABC, considering that the baby son of one ABC news producer was diagnosed with exposure to anthrax bacteria following his visit at ABC News headquarters in New York), the perpetrator(s) were assured massive attention even before the first anthrax letter hit Washington. Along the way, mass-mediated advice about whether to buy gas masks, take antibiotics, avoid public places, and speculations over the next form of bioterrorism (smallpox?) or chemical terror warfare (sarin gas attacks, such as Aum Shinrykio's attack in Tokyo's subway system?) fueled the nightmares of those citizens who could not switch off their television sets since September 11.

This concern was not lost on a few people inside the media. One political columnist identified the greatest danger of journalism—"our new obsession with terrorism will make us its unwitting accomplices. We will become (and have already partly become) merchants of fear. Case in point: the anthrax fright. Until now, anthrax has been a trivial threat to public health and safety." [62] The same columnist also warned,

> The perverse result is that we may become the terrorists' silent allies. Terrorism is not just about death and destruction. It's also about creating fear, sowing suspicion, undermining confidence in public leadership, provoking people—and governments—into doing things that they might not otherwise do. It is an assault as much on our psychology as on our bodies. [63]

Not many in the media listened. At the height of the anthrax scare, the media kept publicizing far more scary scenarios for terrorism of mass destruction. *Newsweek's* 5 November 2001 edition was a case in point. The issue's extensive cover story, "Protecting America: What must be done," described the most vulnerable targets for terrorist attacks as "airports, chemical plants, dams, food supplies, the Internet, malls, mass transit, nuclear power plants, post offices, seaports, skyscrapers, stadiums, water supplies." Collapsed into ten priorities "to protect ourselves" in the actual cover story, the described vulnerabilities read more like a target description for terrorist planners than useful information for a nation in crisis.

Finally, in taking a softer stand vis-à-vis the president, administration officials, members of Congress, and officials at lower level governments, the news media made the right choice when encountering a crisis that presented the country with problems it had never faced before. But suspending the adversarial stance of normal times is one thing, to join the ranks of cheerleaders is another. While comparing the hands-on and very effective crisis-managing mayor of New York City with Winston Churchill during World War II was understandable under the circumstances, likening President George W. Bush (on the basis of his speech before a joint session of the U.S. Congress) to Abraham Lincoln during the Civil War and Winston Churchill during WWII, as some media commentators and many cited sources did, was quite a stretch. But nothing demonstrated more clearly that some reporters and editors had lost their footing than an article about Laura Bush as "a very different" first lady after the terrorist crisis began. When Mrs. Bush visited New York in her "new role of national consoler," a reporter concluded, "As the need for a national hand-holder has made itself evident, Mrs. Bush's role as a kind of Florence Nightingale at least comes as a natural one." [64] Even more farfetched was a comparison by presidential scholar Michael Beschloss, one of the most frequent guests on political talk TV programs, who, according to the *New York Times*, compared "the first lady's sang-froid to that of Queen Elizabeth the Queen Mother during World War II. (The queen mother refused to leave London, against the wishes of her advisers.)" [65] Given this kind of hyperbole, even in the most respected media, it was hardly surprising that the news media's most important role in the democratic arrangement—that of acting as a governmental watchdog—took a back seat in the weeks after the September 11 terror attacks and in the first weeks following the anthrax scare.

When the Republican-controlled House of Representatives stopped its work after anthrax spores were found in Senator Tom Daschle's office (but were not yet found in the lower chamber of Congress), the *New York Post*, a conservative, pro-Republican daily, called members "Wimps" in a huge front-page headline and chided representatives because they had "chicken[ed] out" and "headed for the hills yesterday at the first sign of anthrax in the Capitol." [66] Even for a tabloid, this choice of words was perhaps not the best; however, the substance proved on the mark in the following days, when government offices from Capitol Hill to the Supreme Court were closed while thousands of fearful postal workers in Washington, New York, and New Jersey were told to continue working because anthrax spores in their buildings and on their mail sorting machines did not pose any danger to their health. At the time, two postal workers in Washington had already died of anthrax inhalation and several others had been diagnosed with less lethal cases. Yet, by and large, the news media did not question what looked like a double standard. In the face of an ongoing terrorism crisis at home and a counterterrorism campaign abroad, the mainstream watchdog press refrained from barking in the direction of public officials.

In late October and early November, when public opinion polls signaled that the American public was far less satisfied with the Bush administration's handling of homeland defense in the face of anthrax bioterrorism than with its military campaign against bin Laden, al Qaeda, and the Taliban in Afghanistan, columnists, journalists, and editorial writers asked the questions that needed to be answered and voiced criticism that needed to be expressed. In an in-depth piece in the *New York Times*, for example, John Schwartz wrote, "[If] there's one lesson to be learned from the Bush administration's response to the anthrax threat, it's this: People in the grip of fear want information that holds up, not spin control." [67] More specifically, he wrote:

> Critics of the administration say that the reasons for the lackluster response include lack of communication between agencies, a lack of preparedness on the part of the Health and Human Services Secretary, Tommy G. Thompson, a former governor of Wisconsin with little background in medicine or science, and officials' tendency to respond in the same way they would respond to a mere political problem.[68]

This piece, similar news stories, and commentary signaled that some voices in the news media began to reclaim their watchdog role with respect to September 11, the anthrax scare, and even the politics of anti- and counter-terrorist politics. But the events of September 11, 2001 changed the mindset of most Americans—including many in the media who remained reluctant to return to their more critical coverage patterns of pre-September 11 times.

Terrorists and the media are not bedfellows, they are more like partners in a marriage of convenience in that terrorists need all the news coverage they can get and the media need dramatic, shocking, sensational, tragic events to sustain and bolster their ratings or circulation. There is no doubt that the most fundamental responsibility of a free press is to inform the public of events and issues in order for citizens to make informed judgments and decisions. Therefore, it would be absurd to suggest that there should be no reporting on mass-mediated terrorism. Government censorship is not an acceptable option either. At issue is how to report on this kind of political violence in ways that are less accommodating to terrorists' mass-mediated goals and less likely to curtail the watchdog role of the press that is essential to a healthy democracy.

References

1. States, too, can and do perpetrate this sort of political violence, but they are generally not eager to have their deeds publicized and typically try to prevent news coverage. For this reason, the term "mass-mediated terrorism" means in

this context political violence by groups or individuals who deliberately target civilians with the immediate goal to get the attention of the mass media, the public, and government in their target societies.

2. Quoted here from Amy Harmon, "The Search for Intelligent Life on the Internet," *New York Times*, 23 September 2001.
3. Michiko Kakutani, "Critic's Notebook: Struggling to Find Words for a Horror beyond Words," *New York Times*, 13 September 2001.
4. "A Day of Terror: The Voices," *New York Times*, 12 September 2001.
5. "After the Attacks: One Man's Account," *New York Times*, 15 September 2001.
6. Cloud described his reaction as a guest on NBC's *Today* program, 12 September 2001.
7. The unidentified Canadian made the remark on the Canadian Broadcasting Corporation's program *The National*, 11 September 2001.
8. From *CNN Money Morning*, 14 September 2001.
9. John Updike, "Talk of the Town," *The New Yorker*, 24 September 2001, 28.
10. Ibid., 28.
11. Stockhausen's remarks and the reactions they caused in Germany were reported in "Attacks Called Great Art," *New York Times*, 19 September 2001, according to http://www.nytimes.com/2001/09/19/arts/music/19KARL.html [accessed 1 April 2002].
12. Walter Lippmann, *Public Opinion* (New York: Free Press, 1949), chaps. 1, 6.
13. Stockhausen is quoted here from Bill Carter and Felicity Barringer, "In Patriotic Times, Dissent Is Muted," *New York Times*, 28 September 2001. According to the article, the Eastman School of Music's Ossia Ensemble canceled a planned performance of Stockhausen's work "Stimmung" scheduled for early November at New York's Cooper Union.
14. George Pataki, ABC News, *Nightline*, 14 September 2001.
15. Quoted here from Associated Press reporter Robert Jablon, "Hollywood Think Tank Helping Army," http//dailynews.yahoo.com/h/ap/20011009/us/attacks_hollywood_1.html [accessed 1 April 2002].
16. Gabriel Weiman and Conrad Winn, *The Theater of Terror: Mass Media and International Terrorism* (New York: Longman, 1994).
17. Brigitte L. Nacos, *Terrorism and the Media: From the Iran Hostage Crisis to the World Trade Center Bombing* (New York: Columbia University Press, 1994); John O'Sullivan, "Media Publicity Causes Terrorism" in Bonnie Szumski, ed., *Terrorism: Opposing Viewpoints* (St. Paul: Greenhaven, 1986).
18. The statement was written in Arabic, but an English translation was carried by the wire services and widely publicized in the media. See, for example, http://www.msnbc.com/news/633244.asp [accessed 26 September 2001].
19. George Gerbner and Larry Gross, "Living with Television: The Violence Profile," *Journal of Communication* 26 (1976): 173–199; Nacos, "Terrorism and the Media"; Sissela Bok, *Mayhem: Violence as Public Entertainment* (Reading, MA: Perseus Books, 1998); Gadi Wolfsfeld, "The News Media and the Second Intifada," *The Harvard International Journal of Press/Politics* 6 (2001): 113–118.
20. Eventually, the hijackers released most of their hostages. Afterwards, some European governments did free a few terrorists from prison as demanded by the PFLP and thereby resolved the standoff.
21. According to a *Los Angeles Times* telephone poll on 13–14 September 2001, 83 percent of the respondents said they watched the news "very closely," 15 percent "closely," and 2 percent "not too closely." Nobody chose the response option "not closely at all." In a survey conducted 14–15 September 2001, the Gallup Organization found that 77 percent of the public followed the news "very closely," 20 percent "somewhat closely," 2 percent "not too closely," and 1 percent "not at all." An ABC/*Washington Post* poll on 11 September 2001, found that 99 percent of the public followed the news on television and radio. Polling on-line adults on 11–12 September 2001, Harris Interactive found that 93 percent identified television and radio as their primary news source, 64 percent mentioned the Internet as one of their primary sources.
22. According to a survey conducted by the Pew Research Center for the People and the Press on 17-21 October 2001, 78 percent of the respondents said that they watched terrorism news "very closely," 22 percent watched "closely," 5 percent "not closely," 1 percent gave no answer. This result was nearly the same level of interest as in mid-September (13–17) when 74 percent of survey respondents revealed that they watched terrorism news "very closely," 22 percent "closely." In fact, more Americans watched this kind of news very closely in the second half of October than in mid-September.
23. The quote is taken from the translated transcript of a videotape, presumably recorded in mid-November 2001, and retrieved from http://www.washingtonpost.com/wp-srv/nation/specials/attacked/transcripts/binladentext_121301.html [accessed 13 December 2001].
24. These figures were reported in an ABC/*Washington Post* survey on 11 September 2001.
25. These statistics are the results of surveys conducted by the Gallup Organization, 21–23 April 1995, and 14–15 September 2001. A Gallup survey conducted on 11 September 2001, reported that 58 percent of Americans expressed fears of more terrorism.
26. See "Text: Bin Laden's Statement," http://www.guardian.co.uk/waronterror/story/0,1361,566069,00html [accessed 7 April 2002].
27. According to surveys conducted by the Pew Center for the People and the Press in mid-October 2001, 52 percent of the respondents were "very" or "somewhat" worried that they or a member of their family would become a victim of terrorism. In mid-September, 53 percent had those concerns. However, the number of people who were "very concerned" actually increased from September to October. See http://people-press.org/midoct01rpt.htm [accessed 24 October 2001].
28. Surveys conducted by the Pew Research Center for the People and the Press found that one in seven Americans felt depressed following the September 11 terrorist attacks, but in mid-October, only 29 percent of respondents revealed that they felt depressed. During that same time period, those having trouble sleeping decreased from 33 percent to 12 percent.
29. A survey conducted on 13–17 September 2001, by the Pew Center for the People and the Press revealed that 63 percent of the respondents could not stop watching terrorism reports, on 17–21 October, 49 percent said that they were hooked on terrorism news. In the September poll, 77 percent of the respondents said that watching terrorism news was frightening, in the October pool, 69 percent said the same.
30. Richard Burkholder, Jr., "Initial Reaction to the Attacks on America: Polls from Hungary, the United Kingdom, Australia, New Zealand, France and Russia," The Gallup Organization, http://www.gallup.com [accessed 22 September 2001].
31. This statistic is from surveys conducted by the Gallup Organization on 12 September 2001.

32. These statistics are from surveys conducted by MORI (Roy Morgan International) on 14 September 2001, and by the ROMIR (Russian Public Opinion and Market Research) on 12 September 2001.
33. For the full text, see Peter Ford, "Why Do They Hate US?" *Christian Science Monitor*, `http://www.csmonitor.com/2001/0927/p1s1-wogi.html` [accessed 1 April 2002].
34. Lisa Beyer, "Roots of Rage," *TIME*, 1 October 2001, 44–47.
35. Philip C. Wilcox, Jr., "The Terror," *New York Review of Books*, 18 October 2001, 4.
36. To retrieve relevant transcripts from the Lexis-Nexis database, the following search words were used: "why they hate us," "roots" and "terrorism," and "motivations" and "terrorism." Each transcript was examined as to the relevancy of its content. While all transcripts retrieved for the post-attack period (11–29 September 2001) addressed the reasons for anti-American sentiments in the Arab and Muslim world, those retrieved for the pre-attack period (1 January 2001-10 September 2001) did not include a single record that dealt with this problem.
37. For the full text of the document, visit `http://www.fas.org/irp/world/para/docs/980223-fatwa.html` [accessed 1 April 2002].
38. The document was written in Arabic. This translation was taken from "Full Text of Notes Found after Hijackings," *New York Times*, `http://www.nytimes.com/2001/09/29/national/29S/FULL-TEXT`. html [accessed 1 April 2002].
39. Edward W. Said, *Covering Islam: How the Media and the Experts Determine How We See the Rest of the World* (New York: Pantheon, 1981), 26.
40. Neil Hickey, "Money Lust: How Pressure for Profit is Perverting Journalism," *Columbia Journalism Review* (July/August 1998).
41. Shanto Iyengar, *Is Anyone Responsible? How Television Frames Political Issues* (Chicago: University of Chicago Press, 1991).
42. Quoted here from *People* magazine on CNN, 29 September 2001.
43. D. McFadden, "Bin Laden's Journey from Rich, Pious Boy to the Mask of Evil," *New York Times*, 30 September 2001.
44. "Rick Bragg, "Streets of Huge Pakistan City Seethe with Hatred for U.S.," *New York Times*, 30 September 2001.
45. The sudden bestseller was Jamal Abdul Latif Ismail, *Bin Laden, Al-Jazeera—and I*. For more on this book, see Donna Abu-Nasr, "Bin Laden's past words revisited," http://dailynews.yahoo.com/htx/ ap/20010928/wl/ bin_laden_s_words_2.html [accessed 1 April 2002].
46. George W. Bush's fifty-four public statements during the period were retrieved from the LexisNexis database in the political transcript category.
47. This statement by al Qaeda's spokesman Sulaiman Abu Ghaith was aired by al Jazeera TV and U.S. networks. The quote was taken from the Associated Press's version as publicized on `http://dailynews.yahoo` and retrieved on 10 October 2001 [accessed 1 April 2002].
48. See the Pew Research Center for the People and the Press, `http://www.people-press.org/terrorist01rpt.htm` [accessed 1 April 2002], which states: "Overwhelming support for Bush, military response, but…
49. James M. Fallows, *Breaking the News: How the Media Undermined American Democracy* (New York: Pantheon, 1996); Thomas E. Patterson, *Out of Order: How the Decline of the Political Parties and the Growing Power of the News Media Undermine the American Way of Electing Presidents* (New York: Knopf, 1993).
50. See `http://www.people-press.org/terrist01rpt.htm`.
51. Walters revealed her "no flag" decision on *The View*, a talk show she cohosts. See Rita Ciolli, "Flags Raise among Media," *Newsday*, 23 September 2001.
52. The station was Long Island Cablevision. See Warren Strugatch, "Patriotism vs. Journalistic Ethics," *New York Times*, 7 October 2001.
53. "Comment: Broadcast News," *Wall Street Journal*, 8 October 2001.
54. Marvin Kitman, "The Nation's Painful Video Vigil," *Newsday*, 23 September 2001.
55. Howard Kurtz, "Media Hype May No Longer Be Necessary," *Washington Post*, 16 September 2001.
56. Kitman, "Nation's Painful Video Vigil."
57. The *New York Times*, for example, mentioned the Palestinian celebrations in nine articles following the September 11 terrorism, the anti-American incident in Athens received twenty lines of an Associated Press dispatch on its sports pages. See "Fans in Athens Try to Burn U.S. Flag," *New York Times*, 23 September 2001, sports section.
58. Quoted here from John F. Burns, "A Nation Challenged: The Wanted Man," *New York Times*, 8 October 2001.
59. Quoted in Susan Sachs and Bill Carter, "A Nation Challenged: Al Qaeda; Bin Laden Spokesman Threatens Westerners at Home and in the Gulf," *New York Times*, 14 October 2001.
60. Anthony Pratkanis and Elliott Aronson, *Age of Propaganda* (New York: W. H. Freeman, 1992), 165.
61. Bill Carter and Felicity Barringer, "A Nation Challenged: The Coverage," *New York Times*, 11 October 2001.
62. Robert Samuelson, "Unwitting Accomplices?" *Washington Post*, 7 November 2001.
63. Ibid.
64. Alex Kuczynski, "A Very Different Laura Bush," *New York Times*, 30 September 2001.
65. Ibid.
66. Deborah Orin and Brian Blomquist, "Anthrax Plays to Empty House," *New York Post*, 18 October 2001.
67. John Schwartz, "Efforts to Calm the Nation's Fears Spin Out of Control," *New York Times*, 28 October 2001.

BRIGITTE L. NACOS is an adjunct professor of political science at Columbia University and is a long-time U.S. correspondent for publications in Germany. This article is adapted from her new book, *Mass-Mediated Terrorism: The Central Role of the Media in Terrorism and Counterterrorism.*

A violent episode in the virtual world

TERROR AND THE UK—The media's globalisation of terror makes us feel part of a worldwide community facing a common problem, but this is a dangerous illusion.

John Gray

For those directly affected by them, the London bombings will always be an unalterable reality—an event, barely comprehensible in its pain and horror, with which they will struggle to come to terms for the remainder of their lives. For all the rest of us—the hundreds of millions or billions of people who watched the same images of bloodied commuters and cordoned-off Tube stations—the bombings are an episode in the virtual world that is being continuously manufactured by the media. In this simulated environment we can feel part of a global community facing a common problem. We are able to imagine that terror could be banished from our lives, as all the world's peoples and their leaders act in solidarity against a universal evil.

These sentiments are humane and generous, but they can easily turn into a sort of moral narcissism that willingly colludes in the deceptions of our leaders. The politicians who gathered at Gleneagles spoke as if the world could be reshaped by their good intentions. The truth is that they were deeply divided in what they wanted, and the world is not so simple or so malleable. Like almost every gathering of global leaders at the present time, Gleneagles was a media event before it was anything else.

The trouble with the omnipresence of the media in politics is that it tends to blur the distinction between reality and appearance. The causes of human action are obscure, and the course of events at times indecipherable; a central task of the media is to contrive a coherent narrative from this chaos. In doing so, they can end up shaping reality—but not in a manner that anyone intended or predicted. For example, there may no longer be anything resembling a globally organised terrorist network, but by instantaneously disseminating the same images of carnage and panic throughout the world, the media have globalised our perception of terror. Governments behave as if this media apparition were an actual entity, with the result that the policies that are adopted in order to resist terrorism are ineffective and sometimes disastrously counter-productive.

Western military intervention in Afghanistan practically destroyed al-Qaeda as an effective force. With its training camps in ruins and its leadership in hiding, the structure of the network fragmented and its capacity for action was correspondingly diminished. The effect of the war in Iraq has been to revivify al-Qaeda, but in a new and possibly more dangerous form. It has become an idea or a cause that can be taken up by anyone, and if the fluid and shifting groups of which it is at present composed appear to act in a concerted fashion, it may be by responding to media reports of each other's activities rather than by any kind of direct, systematic co-ordination. Their goal is to shift the public mood, and they attempt to do this by acts of spectacular violence that are transmitted worldwide via television. The type of terrorism that London suffered on 7 July may well have evolved as a by-product of the global media.

The development of terrorism illustrates a complex feedback Between the virtual world constructed by the media and the actual course of history. Al-Qaeda is now very largely an artefact of the communication industry—but it is also real, with a demonstrated capacity for mass murder. This is a development that exemplifies both the power of the media and the fragility of that power. The war in Iraq was launched on the basis of deceptive claims about Saddam Hussein's links with the attacks of 11 September 2001, and the self-deluding belief that the US would be accepted as a liberator of the Iraqi people. These fantasies have been demolished by events, and no amount of news management has been able to mask the scale and ferocity of the insurgency against the occupying forces. There are well-founded reports that US forces have been in talks with rebel commanders, and leaks from British sources suggesting that troop withdrawal is now on the agenda. Reality has smashed through the media constructions. At the same time, Tony Blair and George W Bush continue to try to use media jamborees such as the G8 meeting to demonstrate a solidarity in the face of terror that masks profound disagreement between the US administration and the governments of nearly every other country about how best to respond to it.

There is a tendency among some media analysts to talk as if the global communications industry actually moulds the pattern of events. For them the world is what appears in the media, and there is no difference between perception and reality. Certainly many politicians have come to subscribe to a version of this postmodern philosophy—Blair foremost among them. Yet the world is not in the end a human construction, and this is nowhere clearer than in regard to the issue on which the Gleneagles meeting failed most miserably. Climate change is a physical process that goes on entirely independently of human consciousness. Whatever politicians, opinion-formers or humanity at large may think or feel, a shift in the planetary environment is taking place that will alter irreversibly the way everyone lives in future. The basis for this belief is scientific observation of measurable changes in the material world: human emotions and perceptions are irrelevant. The mix of cynical news management and moral narcissism that is the core of contemporary politics serves only to postpone a brutal encounter with reality.

However, terrorism and climate change have a common feature that helps to explain the way they are treated in the media and by politicians. Both are not wholly soluble problems. Terrorism has been greatly boosted by the Iraq war; it is as true today as it was before London was bombed that the prudent and honourable course of action for Britain is to sever its connection with the Bush administration's folly and withdraw its troops as quickly and as completely as possible. Yet while withdrawal may diminish the terrorist threat to Britain, it will not remove it—there is too much hatred loose in the world, and terrorists are not always motivated by clear strategic goals. We will always be at risk, whatever we do.

The situation is even starker with regard to climate change. The scientific consensus is that there is a great deal of global warming in the pipeline, which even the wholesale abandonment of fossil fuels—if that were possible—would not much reduce. We no longer have the option of forestalling climate change; we can only adjust to it. Adjustment may prove extremely difficult, however, and will necessarily involve alterations in our current way of life. Sensing this, politicians and the public prefer to continue the ritual of announcing targets that will not be reached and which, even if they were met, would not make much difference.

Thinkers of the left often berate the media for skirting round the truth, and some write as if there were a conspiracy to deny the facts of power and oppression. It would be more accurate to say that the media insulate the public from realities it cannot tolerate. We seem to have lost the art of living in an intractable world, so we contrive an alternate reality in which insoluble problems can be conjured away by displays of goodwill. But the problems never really go away, and we would be better off trying to think about them clearly than seeking false security in a collective dream.

***John Gray** is the author of* Al-Qaeda and What It Means To Be Modern *(Faber & Faber).*

Terror's Server

Fraud, gruesome propaganda, terror planning: the Net enables it all. The online industry can help fix it.

David Talbot

Two hundred two people died in the Bali, Indonesia, disco bombing of October 12, 2002, when a suicide bomber blew himself up on a tourist-bar dance floor, and then, moments later, a second bomber detonated an explosives-filled Mitsubishi van parked outside. Now, the mastermind of the attacks—Imam Samudra, a 35-year-old Islamist militant with links to al-Qaeda—has written a jailhouse memoir that offers a primer on the more sophisticated crime of online credit card fraud, which it promotes as a way for Muslim radicals to fund their activities.

Law enforcement authorities say evidence collected from Samudra's laptop computer shows he tried to finance the Bali bombing by committing acts of fraud over the Internet. And his new writings suggest that online fraud—which in 2003 cost credit card companies and banks $1.2 billion in the United States alone—might become a key weapon in terrorist arsenals, if it's not already. "We know that terrorist groups throughout the world have financed themselves through crime," says Richard Clarke, the former U.S. counterterrorism czar for President Bush and President Clinton. "There is beginning to be a reason to conclude that one of the ways they are financing themselves is through cyber-crime."

Online fraud would thereby join the other major ways in which terrorist groups exploit the Internet. The September 11 plotters are known to have used the Internet for international communications and information gathering. Hundreds of jihadist websites are used for propaganda and fund-raising purposes and are as easily accessible as the mainstream websites of major news organizations. And in 2004, the Web was awash with raw video of hostage beheadings perpetrated by followers of Abu Musab al-Zarqawi, the Jordanian-born terror leader operating in Iraq. This was no fringe phenomenon. Tens of millions of people downloaded the video files, a kind of vast medieval spectacle enabled by numberless Web hosting companies and Internet service providers, or ISPs. "I don't know where the line is. But certainly, we have passed it in the abuse of the Internet," says Gabriel Weimann, a professor of communications at the University of Haifa, who tracks use of the Internet by terrorist groups.

Meeting these myriad challenges will require new technology and, some say, stronger self-regulation by the online industry, if only to ward off the more onerous changes or restrictions that might someday be mandated by legal authorities or by the security demands of business interests. According to Vinton Cerf, a founding father of the Internet who codesigned its protocols, extreme violent content on the Net is "a terribly difficult conundrum to try and resolve in a way that is constructive." But, he adds, "it does not mean we shouldn't do anything. The industry has a fair amount of potential input, if it is to try to figure out how on earth to discipline itself. The question is, which parts of the industry can do it?" The roadblocks are myriad, he notes: information can literally come from anywhere, and even if major industry players agree to restrictions, Internet users themselves could obviously go on sharing content. "As always, the difficult question will be, Who decides what is acceptable content and on what basis?"

Some work is already going on in the broader battle against terrorist use of the Internet. Research labs are developing new algorithms aimed at making it easier for investigators to comb through e-mails and chat-room dialogue to uncover criminal plots. Meanwhile, the industry's anti-spam efforts are providing new tools for authenticating e-mail senders using cryptography and other methods, which will also help to thwart fraud; clearly, terrorist exploitation of the Internet adds a national-security dimension to these efforts. The question going forward is whether the terrorist use of the medium, and the emerging responses, will help usher in an era in which the distribution of online content is more tightly controlled and tracked, for better or worse.

The Rise of Internet Terror

Today, most experts agree that the Internet is not just a tool of terrorist organizations, but is central to their operations*. Some say that al-Qaeda's online presence has become more potent and pertinent than its actual physical presence since the September 11 attacks. "When we say al-Qaeda is a global ideology, this is where it existson the Internet," says Michael Doran, a Near East scholar and terrorism expert at Princeton University. "That, in itself, I find absolutely amazing. Just a few years ago, an organization like this would have been more cultlike in nature. It wouldn't be able to spread around the world the way it does with the Internet."

The universe of terror-related websites extends far beyond al-Qaeda, of course. According to Weimann, the number of such websites has leapt from only 12 in 1997 to around 4,300 today. (This includes sites operated by groups like Hamas and Hezbollah, and others in South America and other parts of the world.) "In seven years it has exploded, and I am quite sure the number will grow next week and the week after," says Weimann, who described the trend in his report "How Modern Terrorism Uses the Internet," published by the United States Institute of Peace, and who is now at work on a book, *Terrorism and the Internet,* due out later this year.

These sites serve as a means to recruit members, solicit funds, and promote and spread ideology. "While the [common] perception is that [terrorists] are not well educated or very sophisticated about telecommunications or the Internet, we know that that isn't true," says Ronald Dick, a former FBI deputy assistant director who headed the FBI's National Infrastructure Protection Center. "The individuals that the FBI and other law enforcement agencies have arrested have engineering and telecommunications backgrounds; they have been trained in academic institutes as to what these capabilities are." (Militant Islam, despite its roots in puritanical Wahhabism, taps the well of Western liberal education: Khalid Sheikh Mohammed, the principal September 11 mastermind, was educated in the U.S. in mechanical engineering; Osama bin Laden's deputy Ayman al-Zawahiri was trained in Egypt as a surgeon.)

The Web gives jihad a public face. But on a less visible level, the Internet provides the means for extremist groups to surreptitiously organize attacks and gather information. The September 11 hijackers used conventional tools like chat rooms and e-mail to communicate and used the Web to gather basic information on targets, says Philip Zelikow, a historian at the University of Virginia and the former executive director of the 9/11 Commission. "The conspirators used the Internet, usually with coded messages, as an important medium for international communication," he says. (Some aspects of the terrorists' Internet use remain classified; for example, when asked whether the Internet played a role in recruitment of the hijackers, Zelikow said he could not comment.)

Finally, terrorists are learning that they can distribute images of atrocities with the help of the Web. In 2002, the Web facilitated wide dissemination of videos showing the beheading of *Wall Street Journal* reporter Daniel Pearl, despite FBI requests that websites not post them. Then, in 2004, Zarqawi made the gruesome tactic a cornerstone of his terror strategy, starting with the murder of the American civilian contractor Nicholas Berg—which law enforcement agents believe was carried out by Zarqawi himself. From Zarqawi's perspective, the campaign was a rousing success. Images of orange-clad hostages became a headline-news staple around the world—and the full, raw videos of their murders spread rapidly around the Web. "The Internet allows a small group to publicize such horrific and gruesome acts in seconds, for very little or no cost, worldwide, to huge audiences, in the most powerful way," says Weimann.

And there's a large market for such material. According to Dan Klinker, webmaster of a leading online gore site, Ogrish.com, consumption of such material is brisk. Klinker, who says he operates from offices in Western and Eastern Europe and New York City, says his aim is to "open people's eyes and make them aware of reality." It's clear that many eyes have taken in these images thanks to sites like his. Each beheading video has been downloaded from Klinker's site several million times, he says, and the Berg video tops the list at 15 million. "During certain events (beheadings, etc.) the servers can barely handle the insane bandwidths—sometimes 50,000 to 60,000 visitors an hour," Klinker says.

Avoiding the Slippery Slope

To be sure, Internet users who want to block objectionable content can purchase a variety of filtering-software products that attempt to block sexual or violent content. But they are far from perfect. And though a hodgepodge of Web page rating schemes are in various stages of implementation, no universal rating system is in effect—and none is mandated—that would make filters chosen by consumers more effective.

But passing laws aimed at allowing tighter filtering—to say nothing of actually mandating filtering—is problematical. Laws aimed at blocking minors access to pornography, like the Communications Decency Act and Childrens Online Protection Act, have been struck down in the courts on First Amendment grounds, and the same fate has befallen some state laws, often for good reason: the filtering tools sometimes throw out the good with the bad. "For better or worse, the courts are more concerned about protecting the First Amendment rights of adults than protecting children from harmful material," says Ian Ballon, an expert on cyberspace law and a partner at Manatt, Phelps, and Phillips in Palo Alto, CA. Pornography access, he says, "is something the courts have been more comfortable regulating in the physical world than on the Internet." The same challenges pertain to images of extreme violence, he adds.

The Federal Communications Commission enforces "decency" on the nation's airwaves as part of its decades-old mission of licensing and regulating television and radio stations. Internet content, by contrast, is essentially unregulated. And so, in 2004, as millions of people watched video of beheadings on their computers, the FCC fined CBS $550,000 for broadcasting the exposure of singer Janet Jacksons breast during the Super Bowl halftime show on television.

"While not flatly impossible, [Internet content] regulation is hampered by the variety of places around the world at which it can be hosted," says Jonathan Zittrain, codirector of the Berkman Center for Internet and Society at Harvard Law School—and thats to say nothing of First Amendment concerns. As Zittrain sees it, "its a gift that the sites are up there, because it gives us an opportunity for counterintelligence."

Industry adoption of tighter editorial controls would be a matter of good taste and of supporting the war on terror, says Richard Clarke.

As a deterrent, criminal prosecution has also had limited success. Even when those suspected of providing Internet-based assistance to terror cells are in the United States, obtaining convictions can be difficult. Early last year, under provisions of the Patriot Act, the U.S. Department of Justice charged Sami Omar al-Hussayen, a student at the University of Idaho, with using the Internet to aid terrorists. The government alleged that al-Hussayen maintained websites that promoted jihadist-related activities, including funding terrorists. But his defense argued that he was simply using his skills to promote Islam and wasn't responsible for the sites radical content. The judge reminded the jury that, in any case, the Constitution protects most speech. The jury cleared al-Hussayen on the terrorism charges but deadlocked on visa-related charges; al-Hussayen agreed to return home to his native Saudi Arabia rather than face a retrial on the visa counts.

Technology and ISPs

But the government and private-sector strategy for combatting terrorist use of the Internet has several facets. Certainly, agencies like the FBI and the National Security Agency—and a variety of watchdog groups, such as the Site Institute, a nonprofit organization based in an East Coast location that it asked not be publicized—closely monitor jihadist and other terrorist sites to keep abreast of their public statements and internal communications, to the extent possible.

It's a massive, needle-in-a-haystack job, but it can yield a steady stream of intelligence tidbits and warnings. For example, the Site Institute recently discovered, on a forum called the Jihadi Message Board, an Arabic translation of a U.S. Air Force Web page that mentioned an American airman of Lebanese descent. According to Rita Katz, executive director of the Site Institute, the jihadist page added, in Arabic, "This hypocrite will be going to Iraq in September of this year [2004]—I pray to Allah that his cunning leads to his slaughter. I hope that he will be slaughtered the Zarqawi's way, and then [go from there] to the lowest point in Hell." The Site Institute alerted the military. Today, on one if its office walls hangs a plaque offering the thanks of the Air Force Office of Special Investigations.

New technology may also give intelligence agencies the tools to sift through online communications and discover terrorist plots. For example, research suggests that people with nefarious intent tend to exhibit distinct patterns in their use of e-mails or online forums like chat rooms. Whereas most people establish a wide variety of contacts over time, those engaged in plotting a crime tend to keep in touch only with a very tight circle of people, says William Wallace, an operations researcher at Rensselaer Polytechnic Institute.

This phenomenon is quite predictable. "Very few groups of people communicate repeatedly only among themselves," says Wallace. "It's very rare; they don't trust people outside the group to communicate. When 80 percent of communications is within a regular group, this is where we think we will find the groups who are planning activities that are malicious." Of course, not all such groups will prove to be malicious; the odd high-school reunion will crop up. But Wallaces group is developing an algorithm that will narrow down the field of so-called social networks to those that warrant the scrutiny of intelligence officials. The algorithm is scheduled for completion and delivery to intelligence agencies this summer.

And of course, the wider fight against spam and online fraud continues apace. One of the greatest challenges facing anti-fraud forces is the ease with which con artists can doctor their e-mails so that they appear to come from known and trusted sources, such as colleagues or banks. In a scam known as "phishing," this tactic can trick recipients into revealing bank account numbers and passwords. Preventing such scams, according to Clarke, "is relevant to counterterrorism because it would prevent a lot of cyber-crime, which may be how [terrorists] are funding themselves. It may also make it difficult to assume identities for one-time-use communications."

A Window on Online Fraud

In 2003, 124,509 complaints of Internet fraud and crime were made to the U.S. Internet Crime Complaint Center, an offshoot of the FBI that takes complaints largely from the United States. The perpetrators' reported home countries broke down as follows:

Rank	Country	Reports
1	United States	76.4%
2	Canada	3.3%
3	Nigeria	2.9%
4	Italy	2.5%
5	Spain	2.4%
6	Romania	1.5%
7	Germany	1.3%
8	United Kingdom	1.3%
9	South Africa	1.1%
10	Netherlands	0.9%

Technology Review, February 2005

New e-mail authentication methods may offer a line of defense. Last fall, AOL endorsed a Microsoft-designed system called Sender ID that closes certain security loopholes and matches the IP (Internet Protocol) address of the server sending an inbound e-mail against a list of servers authorized to send mail from the message's purported source. Yahoo, the world's largest e-mail provider with some 40 million accounts, is now rolling out its own system, called Domain Keys, which tags each outgoing e-mail message with an encrypted signature that can be used by the recipient to verify that the message came from the purported domain. Google is using the technology with its Gmail accounts, and other big ISPs, including Earthlink, are following suit.

Finally, the bigger ISPs are stepping in with their own reactive efforts. Their "terms of service" are usually broad enough to allow them the latitude to pull down objectionable sites when asked to do so. "When you are talking about an online community, the power comes from the individual," says Mary Osako, Yahoo's director of communications. "We encourage our users to send [any concerns about questionable] content to us—and we take action on every report."

Too Little, or Too Much

But most legal, policy, and security experts agree that these efforts, taken together, still don't amount to a real solution. The new anti-spam initiatives represent only the latest phase of an ongoing battle. "The first step is, the industry has to realize there is a problem that is bigger than they want to admit," says Peter Neumann, a computer scientist at SRI International, a nonprofit research institute in Menlo Park, CA. "There's a huge culture change that's needed here to create trustworthy systems. At the moment we dont have anything I would call a trustworthy system." Even efforts to use cryptography to confirm the authenticity of e-mail senders, he says, are a mere palliative. There are still lots of problems with online security, says Neumann. "Look at it as a very large iceberg. This shaves off one-fourth of a percent, maybe 2 percent—but its a little bit off the top."

But if it's true that existing responses are insufficient to address the problem, it may also be true that we're at risk of an overreaction. If concrete links between online fraud and terrorist attacks begin emerging, governments could decide that the Internet needs more oversight and create new regulatory structures. "The ISPs could solve most of the spam and phishing problems if made to do so by the FCC," notes Clarke. Even if the Bali bombers writings don't create such a reaction, something else might. If no discovery of a strong connection between online fraud and terrorism is made, another trigger could be an actual act of "cyberterrorism"—the long-feared use of the Internet to wage digital attacks against targets like city power grids and air traffic control or communications systems. It could be some online display of homicide so appalling that it spawns a new drive for online decency, one countenanced by a newly conservative Supreme Court. Terrorism aside, the trigger could be a pure business decision, one aimed at making the Internet more transparent and more secure.

Zittrain concurs with Neumann but also predicts an impending overreaction. Terrorism or no terrorism, he sees a convergence of security, legal, and business trends that will force the Internet to change, and not necessarily for the better. "Collectively speaking, there are going to be technological changes to how the Internet functions—driven either by the law or by collective action. If you look at what they are doing about spam, it has this shape to it," Zittrain says. And while technological change might improve online security, he says, "it will make the Internet less flexible. If its no longer possible for two guys in a garage to write and distribute killer-app code without clearing it first with entrenched interests, we stand to lose the very processes that gave us the Web browser, instant messaging, Linux, and e-mail."

The first needed step: a culture change in the industry, to acknowledge a problem bigger than they want to admit, says Peter Neumann.

A concerted push toward tighter controls is not yet evident. But if extremely violent content or terrorist use of the Internet might someday spur such a push, a chance for preemptive action may lie with ISPs and Web hosting companies. Their efforts need not be limited to fighting spam and fraud. With respect to the content they publish, Web hosting companies could act more like their older cousins, the television broadcasters and newspaper and magazine editors, and exercise a little editorial judgment, simply by enforcing existing terms of service.

Is Web content already subject to any such editorial judgment? Generally not, but sometimes, the hopeful eye can discern what appear to be its consequences. Consider the mysterious inconsistency among the results returned when you enter the word "beheading" into the major search engines. On Google and MSN, the top returns are a mixed bag of links to responsible news accounts, historical information, and ghoulish sites that offer raw video with teasers like "World of Death, Iraq beheading videos, death photos, suicides and crime scenes." Clearly, such results are the product of algorithms geared to finding the most popular, relevant, and well-linked sites.

But enter the same search term at Yahoo, and the top returns are profiles of the U.S. and British victims of beheading in Iraq. The first 10 results include links to biographies of Eugene Armstrong, Jack Hensley, Kenneth Bigley, Nicholas Berg, Paul Johnson, and Daniel Pearl, as well as to memorial websites. You have to load the second page of search results to find a link to Ogrish.com. Is this oddly tactful ordering the aberrant result of an algorithm as pitiless as the ones that churn up gore links elsewhere? Or is Yahoo, perhaps in a nod to the victims' memories and their families' feelings, making an exception of the words "behead" and "beheading," treating them differently than it does thematically comparable words like "killing" and "stabbing?"

Yahoo's Osako did not reply to questions about this search-return oddity; certainly, a technological explanation cannot be excluded. But it's clear that such questions are very sensitive for an industry that has, to date, enjoyed little intervention or regulation. In its response to complaints, says Richard Clarke, "the industry is very willing to cooperate and be good citizens in order to stave off regulation." Whether it goes further and adopts a stricter editorial posture, he adds, "is a decision for the ISP [and Web hosting company] to make as a matter of good taste and as a matter of supporting the U.S. in the global war on terror." If such decisions evolve into the industrywide assumption of a more journalistic role, they could, in the end, be the surest route to a more responsible medium—one that is less easy to exploit and not so vulnerable to a clampdown.

David Talbot *is* Technology Review's *chief correspondent.*

HIGH ANXIETY

Americans made a panic-fueled run on duct tape and plastic sheeting in the wake of government terrorism warnings. Sure, the media were merely the messengers, but many news organizations could have reported this story with more context and less hype.

BY LORI ROBERTSON

In one of the surreal moments amid coverage of the February orange terror alert—and in an age of color-coded levels of threat, there were a number of surreal moments—the MSNBC show "Buchanan and Press" sought to explore whether the media were hyping the alert stories. It was February 13, and MSNBC had that little "Terror Alert: High" label affixed to the bottom corner of its screen. But the network wasn't shy in discussing that target of much lambasting and ridicule. Bill Press introduced Washington Post Metro columnist Marc Fisher, asking, "Do you really think that the media would, for the sake of ratings, fan the flames about an orange terror alert?"

Fisher's response: "Sure."

"Would we stoop that low?" asked Press.

"Yes, absolutely."

You can't say that the media weren't concerned about how they were covering this confusing, scary and, most of the time, shadowy story—from the raising of the terror alert on February 7; to the water, batteries, duct tape and plastic sheeting advice on February 10; to all those reaction stories on fear and indifference, to tape or not to tape. Most news executives say they continually have conversations about how to cover alarming stories, how to inform people without inciting fear. They don't take these things lightly.

But many say that this time around, they did a poor job of it. The critics came out in force.

"We seem to have only one volume these days: very loud," says Washington Post media writer Howard Kurtz. "I think it's entirely possible to write about potential terror attacks and suggested precautions without in effect yelling from the rooftops, and yet that's something that modem media or today's media seem to have great difficulty with."

Likewise, Martin Kaplan, associate dean of the University of Southern California Annenberg School for Communication, seems almost resigned to the belief that the media don't always take seriously their responsibility to think about how stories will affect people. "That's such a quaint notion," says Kaplan. Especially on television, with its dramatic "Showdown Iraq" and "Countdown Iraq" branding, he says, "everything gets turned into a soap opera.... Yes, they have a responsibility not to scare the wits out of us, and no, they don't live up to that responsibility. But that long ago has gone away.... The notion that the press should be responsible probably existed in our gauzy memory.

" Living in this massive-amount-of-media culture, one does have the feeling that the are-you-afraid? coverage was too flamboyant. But hysteria was hardly universal: Some news organizations either held back on or underplayed their stories, depending on your viewpoint. The Boston Globe, for instance, ran the February 11 story on how to prepare for a terrorist attack on A5; the Chicago Tribune put it on A12; and the New York Times shoved it back to A16. The Los Angeles Times didn't run it at all. CBS' and NBC's evening newscasts talked about "duct tape and plastic sheeting" in larger pieces, sure, but they weren't alarmist. A day or two later, though, the networks, like everyone else, were chasing after the fear-factor phenomenon.

Endless discussions about duct tape dominated cable news shows. Reports surfaced that a man in Connecticut had wrapped his house in plastic. A whole new genre of jokes was born. "Anxiety" made the cover of Newsweek and Time, the latter taking a comic look at the frenzy by picturing an eyeball peeking out from crisscross swaths of silver-gray tape.

The government's recommendations for preparing for a terrorist attack didn't seem like big news until people started panicking, says the *Los Angeles Times*' Scott Kraft.

"The truth is really... we didn't think it was much of a story until it created this panic," says Los Angeles Times National Editor Scott Kraft, who adds that people on the West Coast were not nearly as concerned about being the target of an attack.

It seemed no matter how softly some news organizations had initially whispered "duct tape and plastic sheeting," the words provoked quite a reaction. And the disparate play this story received suggests the media were in just as much disagreement as the public about what to do with this information.

Some news organizations played both the terror alert and the government advice stories prominently. The Washington Post and USA Today, Cleveland's Plain Dealer and the Miami Herald, Pennsylvania's Lancaster New Era and Allentown Morning Call ran stories on government advice on page one. Editors had no doubt that this was big news, and they don't buy charges that they were scaring the public.

"Washington is increasingly being singled out as a potential target," says Washington Post Executive Editor Leonard Downie Jr. "And so our readers are really nervous even if we didn't put anything in the paper."

Some people, he continues, feel the media shouldn't be talking about these things—that it makes people afraid. "I don't feel that's right," he says. "We now know information before the September 11 attacks didn't get disseminated." The possibility that valuable information about an attack wouldn't be released is "a great fear of mine."

What the coverage lacked, say critics, was context, perspective and, initially, critical examination of government advice. The public was made aware of different types of possible attacks—chemical, biological and radiological—with not much explanation of what is more likely to occur (and what is extremely unlikely to occur) or what the scope of the damage could be. Reading the early media coverage gave one the impression that a nuclear holocaust could occur tomorrow. Despite the volume of news stories, say critics, the media raised more questions than they answered. It really wasn't clear how all of that duct tape and plastic sheeting was to be deployed, and under what circumstances it would help.

Stories on the alerts and related precautions "have to be couched and put into proper perspective," says Calvin Sims, a former visiting fellow at the Council on Foreign Relations who recently taught a course on the media and terrorism at Princeton. Perhaps the public reaction would have been muted, he says, had someone explained that you have to be skilled to effectively seal a room with duct tape, or that it's unlikely there would be an attack on such a large scale that you would need to use the tape.

"I think the coverage so far falls into a couple of categories," says Karen Brown Dunlap, dean/president-designate of the Poynter Institute. "Media have done a good job at just providing the information, and that's basically saying what the government is saying.... They're not quite doing as good a job in analyzing what the government is saying.... You sense a certain reserve in challenging the government too much."

The vagueness of the coverage wasn't a comfort to anyone. "The sort of questions I wanted to have answered weren't answered," says Sims, now a New York Times reporter. "In the event of an attack, what kind of a response can we expect?... Why do we need to stock water for three days?"

When the government raised the terror alert level to orange, or "high," on February 7, most news organizations led with that story. (The New York Times played it on A8, though an Al story about France and China opposing war on Iraq mentioned the high alert.) There were signs early on that some journalists were doing their best to ferret out why and if the orange designation was warranted.

The Washington Post, for example, included this interesting paragraph in its story: "However, others with access to the intelligence upon which the alert was based said it was largely an effort to make sure government officials could not be blamed for not warning Americans, as they were after the Sept. 11, 2001, attacks. 'That's what this whole process is about,' said one well-placed intelligence source."

While public anxiety probably increased somewhat, it was nothing compared with what would soon take place. Everything changed after the duct tape.

When the government raised the alert, Tom Ridge, secretary of the Department of Homeland Security, urged people to talk to their families and be prepared for a potential terrorist attack. Some news organizations referred their viewers or readers to Web sites for more information on what they could do.

But on February 10, Department of Homeland Security officials held what one reporter calls "a poorly organized press conference to put out an extremely complicated message."

Nine days later, officials would launch the "Ready" campaign that was much clearer and more detailed in terms of explaining which items were more important to have in the event of an attack. But once the alert was raised to orange February 7, officials decided to hold a briefing, led by U.S. Fire Administrator David Paulison and department spokesman Gordon Johndroe, to get some advice out to the public in a hurry.

The reporter quoted earlier feels some journalists didn't take the briefing as seriously as they should have, commenting as they left that this information had been on Web sites months ago. That's true, says this reporter, but "what percentage of people had actually heard of this?... It was not news because anybody in government had ever said it before in some remote Web site. It was news because... they were worried it was going to happen. It was connected to a real fear."

USA Today's Mimi Hall says journalists have to take government officials at their word on terror more than on other subjects because so much information is off-limits.

That fear quickly went public. The next day, after reading and seeing news reports, people ran out to buy duct tape and plastic sheeting. The reporter at the briefing "totally" expected the public reaction. "Were they overreacting to media hype? Hell no.... The government people who really get this aren't saying that it's an overreaction for people to say they need to have water and food in their home... and even have plastic sheeting and duct tape.... This represents the most sophisticated thinking in civil preparedness in this country."

Mimi Hall, homeland security reporter for USA Today, says officials took some steps to curb potential panic: Cameras were not allowed at the briefing, which was conducted by Johndroe, not Tom Ridge. It would have been much more dramatic, Hall says, to have the homeland security chief on camera telling Americans to assemble their survival kits. Despite that, Hall says she thought most reporters at the briefing thought it was "a pretty big deal."

"It was significant that you had representatives of the federal government telling all Americans to take steps like this." She says editors and reporters at USA Today debated how the story should be played and how it should be written, and they felt it was "significant enough to put on page one. We certainly felt we were being really responsible with it."

The media reports were chilling. "Terror Attack Steps Urged" was the off-lead on the Washington Post's front page February 11. The Post noted that "officials suggested privately that they do not want the gravity of the threat overlooked."

Downie says he didn't think that putting the story on the front page would unduly frighten readers. "People thinking the media are scaring people have a rather low opinion of the American public," he says, adding that those in the Washington area were already concerned about a possible attack. "Whether we put a story on the front page does not raise or lower that concern."

Downie may be underestimating the influence of an off-lead in a paper of the Post's stature. Even the Post's Marc Fisher remarked on MSNBC: "It was my own newspaper that got people to head out to Home Depot and start buying duct tape and plastic sheeting by putting a story on the front page."

Looking at the coverage that day, some news outlets certainly deemed the story to be more urgent than others.

The New York Times' front page featured a photo of increased security in Times Square, with the following refer: "The Bush administration issued guidelines on how to prepare for a terrorist attack. Page A16." That story twice stressed that officials were not issuing advice because of an impending strike. The fourth paragraph: "'There is no specific, credible intelligence that says an attack using chemical or biological weapons is imminent,' said Gordon Johndroe."

A Times spokeswoman said that editors at the paper would not discuss their handling of this story, saying they don't want to be put in a position of defending their coverage or comparing it with what other news organizations did. The reporter who wrote the February 11 story, Philip Shenon, did not return phone calls. But apparently the Times wanted to be doubly sure officials didn't know more than they were revealing. The story later included this: "Mr. Johndroe said in telephone interview that the administration had long been planning to organize a public education campaign about disaster preparedness, and that today's news conference was not meant to signal an imminent threat."

The Wall Street Journal's story, which ran on D1, included a similar line: "[T]he government restated that it had no specific information that a particular attack was imminent." USA Today included this quote from Johndroe: "We don't have any specific intelligence that says everyone should rush out to the grocery store." (But in a mixed message all too typical of the episode, later in the same story law enforcement officials said "an attack against Jewish-owned businesses or other high-profile targets is 'imminent.'")

The nature of the government's advice had to suggest to many readers that officials must know something more. Should such qualifiers (that there was no evidence of an impending attack) have been included in all stories? And did their absence—from articles in the Post and other news outlets—make stories sound too alarming? Says Downie: "I'm just not as concerned about that tone.... Our responsibility is to tell [readers] as much as we can about... realistic possibilities of the threat."

The Washington Times did not share the Post's zeal. The paper's A3 story, "Higher alert level spurs tighter aviation security," focused on airspace restrictions and relegated scant information on the need for public precautions to the last three paragraphs. Homeland security reporter Audrey Hudson says her paper played the story "absolutely" where it should have. Hudson monitored the February 10 briefing from the office and called the Federal Emergency Management Agency to confirm her sense that this was old news. She says a spokeswoman told her, "Yeah, we did that last year, and we mailed it to the media and nobody paid attention to it." For Hudson, "It was more newsy that there were restrictions on air that had gone into effect," she says.

Other papers had a more difficult time: They had to judge how important this was from wire stories. Joycelyn Winnecke, associate managing editor for national news at the Chicago Tribune, expresses frustration that her paper wasn't included in that February 10 government briefing. "The reason we were so upset to be excluded... was that we didn't feel like we had the information necessary to judge what [the Department of Homeland Security was] trying to do.... We felt like we were in the dark, and we were hearing from people we were interviewing that they were in the dark, too."

The frenzied run on duct tape and plastic sheeting led to cover stories on national anxiety in *Time* and *Newsweek*.

The briefing, says Winnecke, was geared to East Coast newspapers (officials said they were particularly concerned about threats against New York and Washington). But that left others without the information they needed to evaluate the recommendations. For that reason, and the fact that the story came over the wires late in the day, the Tribune ran it on page 12.

Winnecke says Tribune editors have tried to avoid panicking readers with their terrorism coverage. "Our managing editor [James O'Shea] specifically raised the point that we need to be careful with our stories," she says. For instance, in articles about people rushing out to buy duct tape, reporters need to "capture what's really going on" and make sure they don't "fuel something that shouldn't be fueled."

The first Al story the Tribune ran on the subject was on February 13, headlined, "Critics unglued by government's advice to buy duct tape." Many said such precautions wouldn't offer much protection and compared the advice to the "duck and cover" guidelines during the Cold War—which now seem quite silly.

The Post's Downie acknowledges he would have liked to have included such duct-tape questioning on day one. The paper, he says, initially did not consult enough "outside-the-government experts" on whether this advice was solid. The Post ran such a piece two days later.

Critics, and some journalists, agree the questioning should have come sooner.

Marcy McGinnis, CBS' senior vice president for news coverage, says perhaps the media were late to go beneath the surface. When there's a press conference, the initial reaction "is to take the information and spit it back out again.... One important thing is saying, 'Wait a second, what exactly are you telling us to do?'"But there's not always two-way communication. "There's not always an opportunity to say, 'What's this all about?'"

The Department of Homeland Security launched an advertising campaign to make people aware of steps they could take to protect themselves in the event of a terrorist attack.

CBS News, not unlike ABC, NBC, cable news and a number of papers far removed from Washington and New York, stepped up its coverage when it became apparent some people were freaking out. "It picked up steam as the days went on, I guess as people said, 'Oh my God, I better get survival kits,'" says McGinnis. "The more that started to happen, the more [the media were] writing about it."

Downie likewise emphasizes that the Post "is reflecting the concern of our community instead of driving the concern."

But did the reaction to the concern simply fuel greater concern?

Many newspapers ran boxes that explained what chemical, biological and radiological attacks were and how you could best protect yourself in the event of one. They ran Q&A columns that tried to provide simple answers to public concerns. And they detailed what constitutes a survival kit with bulleted items.

But critics say such lists weren't helping. "Newspapers and television love lists, and that's part of our sort of self-help and news-you-can-use mentality," says Bob Giles, curator of the Nieman Foundation for Journalism at Harvard University. "But I think under these circumstances, there needs to be a very careful vetting of the suggestions by the government as soon as possible so people can read these lists in as accurate a context as possible."

Giles says the overreaction of the public wasn't unusual; many people horde supplies when a hurricane is approaching. "The government and the press both need to recognize that this is a natural impulse when these kinds of alarms are raised," he says. "Instead of simply reporting... the lists of what would go into a safety locker, there should be some interrogation of public officials as to why would you put this in there, when would you use it.... Maybe there could have been some stronger reporting at the very beginning that might have prevented this."

Journalists say it's tough to raise questions immediately, and with this story there was an additional hurdle: Homeland security is brand-new territory. "We are inventing coverage of an area where very few if any reporters have spent time working on these issues before," says Doyle McManus, the Los Angeles Times Washington bureau chief. "I think you are beginning to see those stories sifting though all the different kinds of emergencies and various possible responses. But it's taking longer than I think most readers wanted, and I think they were right to want the information right away. I went searching through the papers myself."

Before January, Audrey Hudson was covering Congress for the Washington Times. Then she got the new homeland security beat. The halls of the Capitol, this is not. "It is very challenging," she says. "You're sort of isolated.... Your sources are limited and the information is limited."

USA Today has tried to answer readers' questions. The paper ran a Q&A on February 11, and a more comprehensive one the next day that addressed different types of threats. The paper also referred people to the FEMA Web site. Says Hall: "We tried to do what we could to provide the information, but you're right, it does raise questions, and these things are extremely complicated." For example, with chemical attacks, she says, there are different steps people should take depending on the various kinds of chemicals that could be employed.

The charge that people just weren't given enough—or the right—information is prevalent. Some go so far as to say that the media have been timid in criticizing the government for fear that an attack might occur. "I think there is abject fear about that. I absolutely do," says Hub Brown, a former local TV news reporter and documentary producer who teaches broadcast journalism at the S.I. Newhouse School of Public Communications at Syracuse University. The media are "afraid that if we question it too much and something does happen... [it] makes us sound like we're bashing the system and makes us sound unpatriotic."

The Post's Kurtz says, "There's a certain reluctance, except perhaps on the part of late-night comics, to criticize government warnings, just in case something big does happen. But that doesn't mean that journalists have to act as if doomsday is just around the corner."

Certainly the press, like the government, wouldn't want to be in a position of failing to warn the public of an attack. But some journalists don't buy the "timid" charge. CNN's Keith McAllister, executive vice president for national newsgathering, says, "I think my view of patriotism is if I'm doing my job well.... My job is to report the news and to ask tough questions" and to find the truth.

Reporters would love to have more information from officials, says Hall, but they're only going to get so much. "We all feel somewhat helpless in the face of terrorist threats both as reporters and as citizens, I think," she says. "You want to convey as much information as possible to an anxious public, but there's only so much you can do.... We have to take government officials sort of at their word on this more than we might otherwise" because so much of the information is classified or exempted from the Freedom of Information Act.

The Washington Post has been the most aggressive paper on this story, at least in terms of the sheer amount of coverage. It's a local story for the paper, says Downie, and his philosophy is the more information, the better.

On February 12, the Post ran a photo of a woman shopping for duct tape and plastic sheeting on the front page, with a refer to the survival-frenzy story on the Metro front. Many critics say these reaction stories are important because they give the press and government officials an idea of how people are responding. The scare in Washington lasted more than one day, however, and the Post carried two more stories on whether people were afraid or not.

Kurtz says the reaction stories—not just in his paper; everyone carried something on the duct-tape panic—were too much. It's a legitimate story, he says. "I just think they should have been scaled way back.... [T]o pound away at this on the front page day after day was the newsprint equivalent of saturation cable coverage. It's a very easy story to do and it tells you almost nothing other than that some people out there are starting to get nervous."

The news media did a good job reporting what the government was saying about the threat, but seemed leery of challenging officials, says Karen Brown Dunlap, president-designate of the Poynter Institute.

Slowly, perspective trickled into some media reports. On March 16, the Post published a special section on emergency preparedness that had a decidedly different tone: calm. The lead piece reminded readers of the incredibly low probability of dying from terrorism (less than the 1-in-4.5 million chance of death by lightning) and talked about the great number of people who escape attacks unharmed. The Post spelled out the unknowns—saying it couldn't predict how likely an attack is or what kind might occur. A story on respirators laid out the pros and cons of each type and cautioned that incorrectly using a gas mask could result in death. A massive evacuation is very unlikely, the Post reported, and smallpox is considered a "low-probability, high-impact risk."

You get a sense from talking with both media critics and people in Washington that they needed this type of information to cling to—something that gives the public an idea of how fearful they should really be, and of what, something that puts the risks in context. (Some experts and columnists did make comments about the small statistical chance of being the victim of a terrorist attack, but these were lost in the din.) And by discussing multiple terrorism possibilities and trotting out experts to talk about gas masks and hazmat suits, say critics, the media provided lots of information, sure, but left the public wondering what to make of it all.

CBS' Marcy McGinnis says coverage of the terrorism scare picked up steam when people started saying, "'Oh my God, I better get survival kits.'"

Syracuse's Hub Brown says more sober discussions were needed. "It's up to the news organizations to put this into perspective." When you combine the government precautions story with news of another Osama bin Laden audiotape, released on February 11, and more talk of a potential war with Iraq, how can someone who's watching this not be afraid? he asks.

But it is difficult to tell people what to do in a non-frightening way against a background of intelligence "chatter" that doesn't provide the time, location or method of an attack.

Some critics are willing to cut the media, and the government, some slack. Everybody's new at this, says Philip Meyer, a journalism professor at the University of North Carolina at Chapel Hill. "You can't blame us on being confused on how to... establish warning systems in the first place and how to react to them in the second place," Meyer says. "You can't expect the government or the media to get it right" the first time out.

From the Onion's February 26 issue: "Orange Alert Sirens To Blow 24 Hours A Day In Major Cities." It was the perfect comic commentary on the government's mixed messages. The satirical story included this fake quote from Tom Ridge: "These 130-decibel sirens, which, beginning Friday, will scream all day and night in the nation's 50 largest metro areas, will serve as a helpful reminder to citizens to stay on the lookout for suspicious activity.... Please note, though, that this is merely a precautionary measure, so go about your lives as normal."

Indeed, some say officials deserve the blame for spreading fear and confusion, not the press. "I don't see the problem with the media," says Meyer. "I see it with the administration not making up its mind on how it wants people to behave."

Says Barbara Cochran, president of the Radio-Television News Directors Association: "It's so important for the government to provide as much information as possible.... Rather than causing panic, information can help dispel panic." She says the problems that have arisen are more because of the government's message than the media's relaying of that information. On February 11, she says, we "get a warning that everyone should go and buy duct tape and plastic sheeting.... Three or four days later, the message is... well, we didn't really mean everybody."

Two days after the Department of Homeland Security included duct tape and plastic sheeting in its list of recommendations, US. officials emphasized that constructing a safe room wasn't the priority. Stockpiling food and water was more important. Two days later Ridge and President Bush were trying to institute some calm. "I want to make something very very clear at this point," Ridge said. "We do not want individuals or families to start sealing their doors or windows."

And there were other signs that public relations wasn't the new department's strong point. According to Time magazine, Ridge told senators at a private meeting that there was a "50 percent or greater" chance of an attack against the U.S. in the following weeks. But the secretary's spokesman denied he said that. On February 7 on ABC's "Nightline," Ridge put the threat of a major attack happening in the next few weeks at an eight on a scale of 10. He later told PBS' Jim Lebrer: Ted Koppel is "a very good journalist and he got me to do something that initially I started to say I'm not going to do.... But I gave him a number and rue the day, obviously."

The media are afraid that if they raise questions about the terror warning and then an attack occurs, it will seem as if they are unpatriotic, says Syracuse University's Hub Brown.

Officials seemed to be struggling with what they should say and how they should say it. For journalists, this reinforced the frustrating nature of the story.

Scott Kraft, national editor at the Los Angeles Times, says part of the discussion editors have about where to play terror alerts includes an evaluation of the strength of the government's information. Kraft says the paper needs to be careful about overplaying such news if officials aren't forthcoming about what their actions are based upon. The L.A. Times did run the news of the terror alert going up and then down on the front page.

But, in the end, there still wasn't a clear idea of why the alert had changed, he says. "I feel like the reporting [by all medial on the reasons for the alert level having gone up, on what it was based upon, has been kind of all over the map;" Kraft says. "A few people have suggested that maybe the White House was hoping to gain some support for the potential war in Iraq." On the other hand, others say the raising of the alert was legitimate, he says. "I don't as a reader sit here and feel I know the answer.... We understand why the government is stingy in sharing some of this.... There is a sense from Washington that the American people should just trust us on this."

Neither the government nor the news media are exactly sure how to handle the terrorist threat, says CNN's Keith McAllister. "[T]hey're learning as they go."

The paper isn't giving up, mind you. "We're always trying to find out what the extent of the threat is," he says. "That's a reporting track that we have several people on."

On March 17, when President Bush issued an ultimatum to Saddam Hussein to go into exile or face a military attack, the alert jumped back up to orange. This time, the reasoning was clear: A war against Iraq could spark terrorist strikes.

The duct tape story didn't get any easier as the doubting experts came forward; it just got more confusing.

The February 11 edition of ABC's "World News Tonight with Peter Jennings" quickly raised questions about the effectiveness of duct tape in the face of a terror attack. "The recommendation to use duct tape and plastic is going to cause more fear and will have virtually no effect on any protection," said Dr. Peter Katona of UCLA's medical school.

But the next day, the network's "Good Morning America" aired a segment on how to properly pick and seal a safe room, with the show's home improvement editor helping a family in Connecticut tape up their laundry room. "USE DUCT TAPE AND PLASTIC SHEETING 46MM THICK," read a graphic.

A case of media confusion?

"I don't think there was confusion on how to report the story," says Jeffrey W. Schneider, ABC News vice president. "I think the whole country was trying to understand... what the advice was that they were giving." It was clear, he says, "that different people had different opinions about how to interpret that information.... On the one hand, you could be a little skeptical about duct tape and plastic sheeting, and on the other hand, there appeared to be some efficacy of using those materials."

A variation on the above example could be given for any number of news organizations. One day on CBS' "Early Show," Harvey Kushner, owner of the personal security store Safer America, showed a somewhat skeptical Hannah Storm what kind of gas masks and hazmat suits people could buy That night, CBS News questioned whether duct tape could offer any real protection. If people are concerned about their security and health, said one expert, what they should really do is "stop smoking, wear their seatbelt and not drink and drive."

CBS' McGinnis says the network received a lot of e-mails from people asking what types of protection are available. "It's not us saying what you should do," she says of segments like the one with Kushner. "It's telling people what there is." The network gives people information and allows them to make their own decisions, she says.

There was also a range of messages coming from experts and health officials. So viewers could see terrorism

TONE IT DOWN

These may not slow down the run on duct tape, but they're easy fixes. AIR offers three suggestions for improved terror alert coverage.

- Lose the terror alert labels. It's a rare media critic who isn't willing to give his two-cents' worth on why the cable news networks went too far when they added "Terror Alert: High" labels to their screens shortly after February 7. CNN discontinued use of the graphic relatively early, on February 12. Fox News Channel kept it up until the 21st; MSNBC didn't drop the label until early the next week. And the cable news networks aren't offering any apologies. Sharri Berg, vice president of news operations for Fox News Channel, emphasizes that the terror alert doesn't change with the wind. It's "so infrequently upgraded," she says, that "it's important to keep it up on the screen to get [the information] out there." MSNBC's Mark Effron says it's "ascribing too much power to cable television... by saying that a little bug on the bottom of the screen is going to scare people." AJR isn't ready to charge that this was a ratings ploy, but we will say it was a bad idea. Critics who said a label was like an alarm going off, or that it was meaningless without context or explanation, were right. Put it in the crawl, if you must, but lose the label.
- Look up "imminent" in the dictionary. Politicians and the press could use a reminder of what "imminent" means. According to Merriam-Webster's: "ready to take place; esp: hanging threateningly over one's head"—as in, in imminent danger of being run over. Code red is supposed to signify that an attack is "imminent" or under way. ABC News, and others, reported that the government insisted "the threat of an attack is real and imminent." That must not have been what officials meant. The same day, the New York Times was cautioning readers that officials weren't signaling a threat is imminent. Throughout the coverage, other journalists and politicians said an attack "could be imminent." Most things could. When you're talking to a jittery nation about a terrorist attack, word choice is important.
- Don't remind us. The words "duct tape" or "code orange" popped up in stories about books and fashion and theater. It's a common journalistic technique: "If you're tired of worrying about duct tape..." CNN aired a segment on escaping from anxiety by going to the movies, which included this helpful advice: "Well, escapists, be 'careful' which movie you choose. 'The Hours,' also nominated for best picture, is about depression and suicide. 'The Pianist,' another nominee, is about being a Jew under Hitler. Get too caught up in one of those and you may use the duct tape to hang yourself." Does every story have to be linked somehow to terrorism? Answer: No.

—Lori Robertson

expert Brian Jenkins offering this bit of reassurance on CNBC: "We can't overreact to this. Even the heightened probability of a terrorist attack does not automatically translate into great danger to the individual citizen." And the same day, CNN's Mike Brooks, cautioning, "We can't also forget some of the small towns. Some people say, 'I live in Smalltown, USA, nothing's going to happen here.' I think we were proved wrong when we looked at Oklahoma City and the bombing of the Murrah federal building there.... So no matter where you are, from a small town to a large town, here in the United States, you have to remain vigilant."

More mixed messages to add to the stew.

It's also possible that stories about terror alerts and intelligence and what to do about it all will never be clear. After September 11, news reports revealed information that suggested the attacks could have been prevented. John Miller and Michael Stone pinpointed the warning signs in their book "The Cell: Inside the 9/11 Plot, And Why the FBI and CIA Failed to Stop It." Yet, in the March 10 New Yorker, Malcolm Gladwell makes a convincing argument that it would not have been that easy. Gladwell uses a number of examples to show that evaluating intelligence before—instead of after—an attack is complicated, mired by much false information and bogus tips.

The public is going to depend on the press to serve as a guide through whatever confusing messages we're sure to receive in the future. But public attention fades faster than news coverage. And skepticism, it seems, is destined to rise.

On February 28, the day after the country's threat level went back down to code yellow, the Los Angeles Times carried the story on the front page. But National Editor Kraft wondered how many times the alert could fluctuate before it was no longer news. "If it had happened five or six times, [the story] probably wouldn't be as strong," unless more powerful evidence was produced, he says.

The Boston Globe didn't put the lowering of the alert on page one. Editor Martin Baron says the news came "well after the period of greatest apparent concern had passed." It had been two weeks, he points out, since the end of the hajj, the annual Muslim pilgrimage to Mecca during which officials had said an attack might come. The day before, National Editor Kenneth Cooper said he mentioned the alert change in the morning meeting, and "there were a few chuckles." He doubted the news would make the front page. It ran on A13.

Even journalists' attention can fade before a change in the terror alert.

Unfortunately, the media are at the beginning of the learning curve with this story. "To some degree, the learning experience of the U.S. government is similar to that of journalists in this country," says CNN's Keith McAllister. "They're not exactly sure how to handle it either, and they're learning as they go."

There are no rule books to consult, says Mark Effron, MSNBC's vice president, live news programming. The media aren't quite making it up as they go along, Effron says, but they're constantly thinking about what's appropriate. "It's as much instinct as anything else."

Lori Robertson is AJR's managing editor. Editorial assistants Michael Duck and Sofia Kosmetatos contributed to this story.

UNIT 7

Terrorism and Religion

Unit Selections

23. **Holy Orders: Religious Opposition to Modern States**, Mark Juergensmeyer
24. **The Madrassa Scapegoat**, Peter Bergen and Swati Pandey

Key Points to Consider

- Why do terrorists use religion to justify their opposition to the state?
- Why are college graduates more likely to commit terrorism than students who have attended Madrassas?

Student Website

www.mhcls.com/online

Internet References

Further information regarding these websites may be found in this book's preface or online.

FACSNET: "Understanding Faith and Terrorism"
http://www.facsnet.org/issues/faith/terrorism.php3#

Islam Denounces Terrorism
http://www.islamdenouncesterrorism.com

Religious Tolerance Organization
http://www.religioustolerance.org/curr_war.htm

SITE Institute
http://www.siteinstitute.org/

Over the past decade, the topic of religion has played an increasingly prominent role in discussions of international terrorism. Fears of what some have called the resurgence of fundamentalist Islam have spawned visions of inevitable clashes of civilizations. Even before the events of September 11th, the term *religious terrorism* had become a staple in the vocabulary of many U.S. policymakers.

While there is currently no commonly accepted definition of religious terrorism, one should note that in the popular press the term *religious terrorism* is often used as a euphemism for political violence committed by Muslims. It is naïve to presume that all political violence committed by members of a particular religious group is necessarily religious violence. The relationship between religion and political violence is much more complex.

Experts have noted that many of today's religious terrorists were nationalists yesterday and Marxists the day before. Unlike their historical predecessors like the Thugs of India who killed to sacrifice the blood of their victims to the Goddess Kali, today's religious terrorists see violence as a means of achieving political, economic, and social objectives. Religion is often seen as a means, rather than an end in itself. In many cases religious ideologies have taken over where other ideologies have failed.

Ideologies are systems of belief that justify behavior. They serve three primary functions: (1) They polarize and mobilize populations toward common objectives; (2) They create a sense of security by providing a system of norms and values; and (3) They provide the basis for the justification and rationalization of human behavior. Ideologies do not necessarily cause violence. They do, however, provide an effective means polarizing populations and organizing political dissent.

While the emergence of religious ideologies signals an important shift in international terrorism, the role of religion in international terrorism is often exaggerated or misunderstood. Religion is not the cause of contemporary political violence. It does, however, provide an effective means for organizing political dissent. In some parts of the world, political extremists have infiltrated the mosques, temples, and churches and have managed to hijack and pervert religious doctrine, superimposing their own views of the world and encouraging the use of violence.

The articles in this unit provide an overview of the relationship between religion and terrorism. The first selection, written by Mark Juergensmeyer, argues that religion is a tool of the powerless fighting against the perceived moral corruption of Western secular society. The final article examines the role of Islamic schools in terrorist training. It contends that highly educated individuals are more likely to be involved in major attacks than those who attended religious schools.

Holy Orders

Religious Opposition to Modern States

Mark Juergensmeyer

No one who watched in horror as the towers of the World Trade Center crumbled into dust on September 11, 2001, could doubt that the real target of the terrorist assault was US global power. Those involved in similar attacks and in similar groups have said as much. Mahmood Abouhalima, one of the Al Qaeda-linked activists convicted for his role in the 1993 attack on the World Trade Center, told me in a prison interview that buildings such as these were chosen to dramatically demonstrate that "the government is the enemy."

While the US government and its allies have been frequent targets of recent terrorist acts, religious leaders and groups are seldom targeted. An anomaly in this regard was the assault on the Shi'a shrine in the Iraqi city of Najaf on August 29, 2003, which killed more than 80 people including the venerable Ayatollah Mohammad Baqir al Hakim. The Al Qaeda activists who allegedly perpetrated this act were likely more incensed over the Ayatollah's implicit support for the US-backed Iraqi Governing Council than they were jealous of his popularity with Shi'a Muslims. Since the United Nations has also indirectly supported the US occupation of Iraq and Afghanistan, it too has been subject to Osama bin Laden's rage. This may well be the reason why the UN office in Baghdad was the target of the devastating assault on August 19, 2003, which killed the distinguished UN envoy Sergio Vieira de Mello. Despite the seeming diversity of the targets, the object of most recent acts of religious terror is an old foe of religion: the secular state.

Secular governments have been the objects of terrorism in virtually every religious tradition—not just Islam. A Christian terrorist, Timothy McVeigh, bombed the Oklahoma City Federal Building on April 19, 1995. A Jewish activist, Yigal Amir, assassinated Israel's Prime Minister Yitzhak Rabin. A Buddhist follower, Shoko Asahara, orchestrated the nerve gas attacks in the Tokyo subways near the Japanese parliament buildings. Hindu and Sikh militants have targeted government offices and political leaders in India. In addition to government offices and leaders, symbols of decadent secular life have also been targets of religious terror. In August 2003, the Marriott Hotel in Jakarta, frequented by Westerners and Westernized Indonesians, was struck by a car bomb. The event resembled the December 2002 attacks on Bali nightclubs, whose main patrons were college-age Australians. In the United States, abortion clinics and gay bars have been targeted. The 2003 bombings in Morocco were aimed at clubs popular with tourists from Spain, Belgium, and Israel. Two questions arise regarding this spate of vicious religious assaults on secular government and secular life around the world. Why is religion the basis for opposition to the state? And why is this happening now?

Why Religion?

Religious activists are puzzling anomalies in the secular world. Most religious people and their organizations either firmly support the secular state or quiescently tolerate it. Bin Laden's Al Qaeda, like most of the new religious activist groups, is a small group at the extreme end of a hostile subculture that is itself a small minority within the larger Muslim world. Bin Laden is no more representative of Islam than McVeigh is of Christianity or Asahara of Buddhism.

Still, it is undeniable that the ideals of activists like bin Laden are authentically and thoroughly religious. Moreover, even though their network consists of only a few thousand members, they have enjoyed an increase in popularity in the Muslim world after September 11, 2001, especially after the US-led occupations of Afghanistan and Iraq. The authority of religion has given bin Laden's cadres the moral legitimacy to employ violence in assaulting symbols of global economic and political power. Religion has also provided them the metaphor of cosmic war, an image of spiritual struggle that every religion contains within its repository of symbols, seen as the fight between good and bad, truth and evil. In this sense, attacks such as those on the World Trade Center and UN headquarters in Baghdad were very religious. They were meant to be catastrophic acts of biblical proportions.

From Worldly Struggles to Sacred Battles

Although recent acts of religious terrorism such as the attacks on the World Trade Center and United Nations had no obvious military goal, they were intended to make an impact on the public consciousness. They are a kind of perverse performance of power meant to ennoble the perpetrators' views of the world while drawing viewers into their notions of cosmic war. In my 2003 study of the global rise of religious violence, *Terror in the Mind of God*, I found a strikingly familiar pattern. In almost every recent case of religious violence, concepts of cosmic war have been accompanied by claims of moral justification. It is not so much that religion has become politicized but that politics has become religionized. Through enduring absolutism, worldly struggles have been lifted into the high proscenium of sacred battle.

This is what makes religious warfare so difficult to address. Enemies become satanized, and thus compromise and negotiation become difficult. The rewards for those who fight for the cause are trans-temporal, and the timelines of their struggles are vast. Most social and political struggles look for conclusions within the lifetimes of their participants, but religious struggles can take generations to succeed.

I once had the opportunity to point out the futility—in secular military terms—of the radical Islamic struggle in Palestine to Dr. Abdul Aziz Rantisi, the head of the political wing of the Hamas movement. It seemed to me that Israel's military force was strong enough that a Palestinian military effort could never succeed. Dr. Rantisi assured me that "Palestine was occupied before, for two hundred years." He explained that he and his Palestinian comrades "can wait again—at least that long." In his calculation, the struggles of God can endure for eons before their ultimate victory.

Insofar as the US public and its leaders embraced the image of war following the September 11 attacks, the US view of the war was also prone to religionization. "God Bless America" became the country's unofficial national anthem. US President George Bush spoke of defending America's "righteous cause" and of the "absolute evil" of its enemies. However, the US military engagement in the months following September 11 was primarily a secular commitment to a definable goal and largely restricted to objectives in which civil liberties and moral rules of engagement still applied.

In purely religious battles waged in divine time and with heavenly rewards, there is no need to compromise goals. There is also no need to contend with society's laws and limitations when one is obeying a higher authority. In spiritualizing violence, religion gives the act of violence remarkable power.

Ironically, the reverse is also true: terrorism can empower religion. Although sporadic acts of terrorism do not lead to the establishment of new religious states, they make the political potency of religious ideology impossible to ignore. The first wave of religious activism, from the Islamic revolution in Iran in 1978 to the emergence of Hamas during the Palestinian *intifada* in the early 1990s, focused on religious nationalism and the vision of individual religious states. Now religious activism has an increasingly global vision. The Christian militia, the Japanese Aum Shinrikyo, and the Al Qaeda network all target what they regard as a repressive and secular form of global culture and control.

Part of the attraction of religious ideologies is that they are so personal. They impart a sense of redemption and dignity to those who uphold them, often men who feel marginalized from public life. One can view their efforts to demonize their enemies and embrace ideas of cosmic war as attempts at ennoblement and empowerment. Such efforts would be poignant if they were not so horribly destructive.

Yet they are not just personal acts. These violent efforts of symbolic empowerment have an effect beyond whatever personal satisfaction and feelings of potency they impart to those who support and conduct them. The very act of killing on behalf of a moral code is a political statement. Such acts break the state's monopoly on morally sanctioned killing. By putting the right to take life in their own hands, the perpetrators of religious violence make a daring claim of power on behalf of the powerless—a basis of legitimacy for public order other than that on which the secular state relies.

Coincidence of Globalization and Modernization

These recent acts of religious violence are occurring in a way different from the various forms of holy warfare that have occurred throughout history. They are responses to a contemporary theme in the world's political and social life: globalization. The World Trade Center symbolized bin Laden's hatred of two aspects of secular government—a certain kind of modernization and a certain kind of globalization— even though the Al Qaeda network was itself both modern and transnational. Its members were often highly sophisticated and technically skilled professionals, and its organization was composed of followers of various nationalities who moved effortlessly from place to place with no obvious nationalist agenda or allegiance. In a sense, they were not opposed to modernity and globalization, so long as it fit their own design. But they loathed the Western-style modernity that they perceived secular globalization was forcing upon them.

Some 23 years earlier, during the Islamic revolution in Iran, Ayatollah Khomeini rallied the masses with the similar notion that the United States was forcing its economic exploitation, political institutions, and secular culture on an unknowing Islamic society. The Ayatollah accused urban Iranians of having succumbed to "Westoxification"—an inebriation with Western culture and ideas. The many strident movements of religious nationalism that have erupted around the world in the more than two decades following the Iranian revolution have echoed this cry. This anti-Westernism

RE-EVALUATING RELIGION

In an age of globalization, pre-modernists, modernists, and post-modernists offer contrasting perspectives on the role of religion.

Pre-Modernist Perspective

- Views religious organizations as essential to effective opposition of communist and authoritarian regimes
- Relies on past historical experience
- Thinks that the spread of secularization by means of globalization will cause only negative effects by destroying the power of religious organizations to check the power of government

Modernist Perspective

- Believes globalization is a drive force in the secularization of society and slow disappearance of religious groups throughout the world
- Argues that religions that abide in the modern age exist as marginal communities that sometimes initiate conflict against secularization
- Holds that religious organizations can play the positive role of correcting accidental distortions or perversions of the generally beneficial course of modernization

Post-Modernist Perspective

- Rejects traditional, pre-modern religions
- Allows for "spiritual experiences" to occur without religious constraints
- Considers expressive individualism to be a core value. Globalization brings about the success of expressive individualism breaking up all traditional, local, and governmental structures.

Foreign Policy Research Institute

has at heart an opposition to a certain kind of modernism that is secular, individualistic, and skeptical. Yet, in a curious way, by accepting the modern notion of the nation-state and adopting the technological and financial instruments of modern society, many of these movements of religious nationalism have claimed a kind of modernity on their own behalf.

Religious politics could be regarded as an opportunistic infection that has set in at the present weakened stage of the secular nation-state. Globalization has crippled secular nationalism and the nation-state in several ways. It has weakened them economically, not only through the global reach of transnational businesses, but also by the transnational nature of their labor supply, currency, and financial instruments. Globalization has eroded their sense of national identity and unity through the expansion of media and communications, technology, and popular culture, and through the unchallenged military power of the United States. Some of the most intense movements for ethnic and religious nationalism have arisen in states where local leaders have felt exploited by the global economy, unable to gain military leverage against what they regard as corrupt leaders promoted by the United States, and invaded by images of US popular culture on television, the Internet, and motion pictures.

Other aspects of globalization—the emergence of multicultural societies through global diasporas of peoples and cultures and the suggestion that global military and political control might fashion a "new world order"—has also elicited fear. Bin Laden and other Islamic activists have exploited this specter, and it has caused many concerned citizens in the Islamic world to see the US military response to the September 11 attacks as an imperialistic venture and a bully's crusade, rather than the righteous wrath of an injured victim. When US leaders included the invasion and occupation of Iraq as part of its "war against terror," the operation was commonly portrayed in the Muslim world as a ploy for the United States to expand its global reach.

"BY ADOPTING THE ... INSTRUMENTS OF MODERN SOCIETY, MANY OF THESE MOVEMENTS OF RELIGIOUS NATIONALISM HAVE CLAIMED A KIND OF MODERNITY ON THEIR OWN BEHALF.

This image of a sinister US role in creating a new world order of globalization is also feared in some quarters of the West. Within the United States, for example, the Christian Identity movement and Christian militia organizations have been alarmed over what they imagine to be a massive global conspiracy of liberal US politicians and the United Nations to control the world. Timothy McVeigh's favorite book, *The Turner Diaries*, is based on the premise that the United States has already unwittingly succumbed to a conspiracy of global control from which it needs to be liberated through terrorist actions and guerilla bands. In Japan, a similar conspiracy theory motivated leaders of the Aum Shinrikyo religious movement to predict a catastrophic World War III, and attempted to simulate Armageddon with their 1995 nerve gas attack in a Tokyo subway train.

Identity and Control

As far-fetched as the idea of a "new world order" of global control may be, there is some truth to the notion that the integration of societies and the globalization of culture have brought the world closer together. Although it is unlikely that a cartel of malicious schemers designed this global trend, the effect of globalization on local societies and national identities has nonetheless been profound. It has undermined the modern idea of the state by providing non-national and transnational forms of economic, social, and cultural interaction. The global economic and social ties of the inhabitants of contemporary global cities are intertwined in a way that supercedes the idea of a national social contract—the Enlightenment notion that peoples in particular regions are naturally linked together in a specific country. In a global world, it is hard to say where particular regions begin and end. For that matter, in multicultural societies, it is hard to say how the "people" of a particular nation should be defined.

This is where religion and ethnicity step in to redefine public communities. The decay of the nation-state and disillusionment with old forms of secular nationalism have produced both the opportunity and the need for nationalisms. The opportunity has arisen because the old orders seem so weak, yet the need for national identity persists because no single alternative form of social cohesion and affiliation has yet appeared to dominate public life the way the nation-state did in the 20th century. In a curious way, traditional forms of social identity have helped to rescue one of Western modernity's central themes: the idea of nationhood. In the increasing absence of any other demarcation of national loyalty and commitment, these old staples—religion, ethnicity, and traditional culture—have become resources for national identification.

Consequently, religious and ethnic nationalism has provided a solution in the contemporary political climate to the perceived insufficiencies of Western-style secular politics. As secular ties have begun to unravel in the post- Soviet and post-colonial era, local leaders have searched for new anchors with which to ground their social identities and political loyalties. What is significant about these ethno-religious movements is their creativity—not just their use of technology and mass media, but also their appropriation of national and global networks. Although many of the framers of the new nationalisms have reached back into history for ancient images and concepts that will give them credibility, theirs are not simply efforts to resuscitate old ideas from the past. These are contemporary ideologies that meet present-day social and political needs.

In the context of Western modernism, the notion that indigenous culture can provide the basis for new political institutions, including resuscitated forms of the nation-state, is revolutionary. Movements that support ethno-religious nationalism are therefore often confrontational and sometimes violent. They reject the intervention of outsiders and their ideologies and, at the risk of being intolerant, pander to their indigenous cultural bases and enforce traditional social boundaries. It is thus no surprise that they clash with each other and with defenders of the secular state. Yet even such conflicts serve a purpose for the movements: they help define who they are as a people and who they are not. They are not, for instance, secular modernists.

Understandably, then, these movements of anti-Western modernism are ambivalent about modernity, unsure whether it is necessarily Western and always evil. They are also ambivalent about globalization, the most recent stage of modernity. On one hand, these political movements of anti-modernity are reactions to the globalization of Western culture. They are responses to the insufficiencies of what is often touted as the world's global standard: the elements of secular, Westernized urban society that are found not only in the West but in many parts of the former Third World, seen by their detractors as vestiges of colonialism. On the other hand, these new ethno-religious identities are alternative modernities with international and supernational aspects of their own. This means that in the future, some forms of anti-modernism will be global, some will be virulently antiglobal, and yet others will be content with creating their own alternative modernities in ethno-religious nation-states.

Each of these forms of religious anti-modernism contains a paradoxical relationship between forms of globalization and emerging religious and ethnic nationalisms. One of history's ironies is that the globalism of culture and the emergence of transnational political and economic institutions enhance the need for local identities. They also promote a more localized form of authority and social accountability.

The crucial problems in an era of globalization are identity and control. The two are linked in that a loss of a sense of belonging leads to a feeling of powerlessness. At the same time, what has been perceived as a loss of faith in secular nationalism is experienced as a loss of agency as well as selfhood. For these reasons, the assertion of traditional forms of religious identities are linked to attempts to reclaim personal and cultural power. The vicious outbreaks of antimodernist religious terrorism in the first few years of the 21st century can be seen as tragic attempts to regain social control through acts of violence. Until there is a surer sense of citizenship in a global order, religious visions of moral order will continue to appear as attractive, though often disruptive, solutions to the problems of authority, identity, and belonging in a globalized world.

MARK JUERGENSMEYER is Professor of Sociology and Director of Global and International Studies at the University of California, Santa Barbara.

The Madrassa Scapegoat

Peter Bergen and Swati Pandey

Madrassas have become a potent symbol as terrorist factories since the September 11 attacks, evoking condemnation and fear among Western countries. The word first entered the political lexicon when the largely madrassa-educated Taliban in Afghanistan became the target of a U.S.-led strike in late 2001. Although none of the September 11 terrorists were members of the Taliban, madrassas became linked with terrorism in the months that followed, and the association stuck. For Western politicians, a certain type of education, such as the exclusive and rote learning of the Qur'an that some madrassas offer, seemed to be the only explanation for the inculcation of hate and irrationality in Islamist terrorists.

In October 2003, for example, Secretary of Defense Donald H. Rumsfeld wondered, "Are we capturing, killing or deterring and dissuading more terrorists every day than the madrassas and the radical clerics are recruiting, training and deploying against us?"[1] In the July 2004 report of the National Commission on Terrorist Attacks Upon the United States, commonly known as the 9-11 Commission, madrassas were described as "incubators of violent extremism," despite the fact that the report did not mention whether any of the 19 hijackers had attended a madrassa.[2] In the summer of 2005, Rumsfeld still worried about madrassas that "train people to be suicide killers and extremists, violent extremists."[3] As the United States marked the fourth anniversary of the September 11 attacks in the autumn of 2005, several U.S. publications continued to claim that madrassas produce terrorists, describing them as "hate factories."[4]

Yet, careful examination of the 79 terrorists responsible for five of the worst anti-Western terrorist attacks in recent memory—the World Trade Center bombing in 1993, the Africa embassy bombings in 1998, the September 11 attacks, the Bali nightclub bombings in 2002, and the London bombings on July 7, 2005—reveals that only in rare cases were madrassa graduates involved. All of those credited with masterminding the five terrorist attacks had university degrees, and none of them had attended a madrassa. Within our entire sample, only 11 percent of the terrorists had attended madrassas. (For about one-fifth of the terrorists, educational background could not be determined by examining the public record.) Yet, more than half of the group we assessed attended a university, making them as well educated as the average American: whereas 54 percent of the terrorists were found to have had some college education or to have graduated from university, only 52 percent of Americans can claim similar academic credentials. Two of our sample had doctoral degrees, and two others had begun working toward their doctorates. Significantly, we found that, of those who did attend college and/or graduate school, 48 percent attended schools in the West, and 58 percent attained scientific or technical degrees. Engineering was the most popular subject studied by the terrorists in our sample, followed by medicine.

> **None of the apparent masterminds of the five terrorist attacks had attended a madrassa.**

The data raise questions about what type of education, if any, is actually more likely to contribute to the motivation or skills required to execute a terrorist attack. Researchers such as Dr. Marc Sageman have argued that madrassas are less closely correlated with producing terrorists than are Western colleges, where students from abroad may feel alienated or oppressed and may turn toward militant Islam.[5] Given that 27 percent of the group attended Western schools, nearly three times as many as attended madrassas, our sample seems to confirm this trend. The data also show a strong correlation between technical education and terrorism, suggesting that perpetrating large-scale attacks requires not only a college education but also a facility with technology. This type of education is simply not available at the vast majority of madrassas.

These findings suggest that madrassas should not be a national security concern for Western countries because they do not provide potential terrorists with the language and technical skills necessary to attack Western targets. This is not to say that madrassas do not still pose problems. To the extent that they hinder development by failing properly to educate students in Asian, Arab, and African countries and that they create sectarian violence, particularly in Pakistan, madrassas should remain on policymakers' minds as a regional concern.[6] A national security policy focused on madrassas as a principal source of terrorism, however, is misguided.

The Truth about Terrorist Education

"Madrassa" is a widely used and misused term. In Arabic, the word means simply "school."[7] Madrassas vary from country to country or even from town to town. They can be a day or boarding school, a school with a general curriculum, or a purely religious school attached to a mosque.[8] For the purposes of our study, "madrassa" refers to a school providing a secondary-level education in Islamic religious subjects.[9] We examined information available in U.S., European, Asian, and Middle Eastern newspapers; U.S. government reports; and books about terrorism to determine which of the 79 terrorists responsible for the five major attacks attended such madrassas. In the one instance when a terrorist was found to have attended a madrassa and later a university, we classified him as a graduate of both types of schools. Attacks in which information about the terrorists' education was scant, such as the 2000 attack on the USS *Cole* and the 2004 Madrid train bombings, were excluded from the study.

The 1993 World Trade Center Bombing

On February 26, 1993, a truck bomb exploded in the parking garage beneath the twin towers of the World Trade Center, killing six people and injuring more than 1,000. Ringleader Ramzi Yousef had hoped the bomb would kill tens of thousands by making one tower collapse onto the other. The 12 men, including Yousef, responsible for this first World Trade Center bombing were the best educated of any group we studied. All of them had some college education, with most having studied in universities in the Middle East and North Africa, and two having graduated from Western colleges. The spiritual guide of this terrorist cell, the Egyptian cleric Sheik Omar Abdel Rahman, had a master's degree and had started on his doctoral dissertation at Al Azhar University in Cairo—the Oxford of the Islamic world. Yousef, the mastermind of the plot and the nephew of the operational commander of the September 11 attacks, obtained a degree in engineering from a college in Wales. None of the attackers appeared to have attended a madrassa.

The 1998 Africa Embassy Bombings

On August 7, 1998, nearly simultaneous bombings at the U.S. embassies in Tanzania and Kenya killed 224 people. The attacks were the largest perpetrated by Al Qaeda at the time and catapulted the group and its leader, Osama bin Laden, into the public eye. The 16 attackers who orchestrated the bombings were found largely to be part of a local Al Qaeda cell. This group of men had one Western-born member, Rashed Daoud al-Owhali, who hailed from Liverpool and claimed to have been indoctrinated not at a madrassa but by audio tapes about the Afghan jihad. Recent attacks in Madrid and London have shown that immigrants and the children of immigrants living in Europe may be more dangerous than far-flung madrassa graduates. Sageman, along with scholars such as Olivier Roy and Robert S. Leiken, have noted that living in the West alienates many immigrants and has a strong correlation to Western-based terrorist activity.[10] The Africa group demonstrates another trend that later reappeared in the 2004 Madrid bombings: the formation of ties between the undereducated (and easily influenced or even criminal) and the well educated. Seven of the 16 plotters in the Africa group attended college, with two, Wadih El-Hage and Ali Mohamed, attending Western schools.

The September 11 Attacks

In the most devastating terrorist attack to date, 19 hijackers crashed three planes into the World Trade Center and the Pentagon, with a fourth plane crashing in rural Pennsylvania; nearly 3,000 people were killed. The four pilots who led the September 11 attacks all spent time at universities outside of their home countries, three of them in Germany. As the 9-11 Commission detailed, the plot took shape in Germany as they were completing their degrees.[11] The lead hijacker, Muhammad Atta, had a doctorate in urban preservation and planning from the University of Hamburg-Harburg in Germany. The 15 "muscle hijackers" vary in educational background. Little is known about some of them, including Khalid al-Mihdhar, who met the pilot, Nawaf al Hazmi, while fighting in Bosnia. Indeed, several of the terrorists we studied, if they were not well educated, were so-called career jihadists with experience fighting in regions such as Bosnia and Afghanistan. Still, even of the muscle group, six of the 15 had completed some university studies. Finally, we also examined the so-called secondary planners, who had overall control of the operation, many of whom were longtime Al Qaeda operatives, and found that all of them had attended college in Europe or the United States. Khalid Sheikh Muhammad, the operational commander of the September 11 attacks, had obtained a degree in engineering from a college in North Carolina.

Despite the fact that much of the information about the educational backgrounds of the September 11 planners and pilots has been widely reported, otherwise sophisticated analysts persist in believing that they were the products of madrassas. In his 2005 book *Future Jihad*, Walid Phares, a professor of Middle Eastern studies at Florida Atlantic University and a frequent commentator on U.S. television, explained that Wahhabism "produced the religious schools; the religious schools produced the jihadists. Among them [were] Osama bin Laden and the nineteen perpetrators of September 11."[12] In fact, bin Laden did not attend a religious school when he was growing up in Jeddah, Saudi Arabia, studying instead at the relatively progressive, European-influenced Al Thagr High School and later at King Abdul Aziz University, where he focused on economics.[13] From what is available on the public record, it seems that none of the 19 hijackers attended madrassas.

Madrassas should not be a national security concern for Western countries.

The 2002 Bali Nightclubs Bombing

Islamist terrorists attacked two tourist hotspots in Bali in October 2002, killing more than 200. As was the case with the Africa embassy bombings, the Bali attack was aimed at Western targets, in this case tourists, in a non-Western country. Yet, unlike the previous examples, Bali is the only terrorist attack to have been perpetrated in part by terrorists who attended madrassas. Nine of the 22 perpetrators attended the Al Mukmin, Al Tarbiyah Luqmanul Hakiern, and SMP Pemalang pesantran—Islamic schools of a kind particular to Indonesia. Most Indonesian madrassas are part of the state school system and teach a broad range of subjects. Pesantren, however, such as Al Mukmin, operate outside of this system and are generally boarding schools. The curriculum at pesantren usually focuses on religion and often offers practical courses in farming or small industry.[14] This constellation of schools, particularly Al Mukmin, provided many recruits to Jemaah Islamiyah, the militant Islamist group that seeks to create fundamentalist theocracies in countries across Southeast Asia.

Even in the Bali attacks, however, five of the 22 members of the group had college degrees, particularly the key planners. The mastermind of the Bali plot, Dr. Azahari Husin, who was killed in a shootout with Indonesian police in November 2005, obtained his doctorate from the University of Reading in the United Kingdom prior to becoming a lecturer at a Malaysian university.[15] Noordin Muhammad Top attended the same Malaysian school. The third mastermind, Zulkarnaen, also known as Daud, studied biology at an Indonesian college. Azahari and Top are also suspected of involvement in the 2005 Bali bombings that killed 22 people.

The July 7, 2005, London Bombings

The July 7, 2005, bombings of three subway stations and a bus in London killed 56 people, including the suicide bombers, and were the work of homegrown British terrorists with suspected Al Qaeda ties.[16] The initial news coverage of the attack featured hyperventilating reports that the four men responsible had attended madrassas. One such piece in the *Evening Standard* stated that three of the bombers attended madrassas, which it termed "haven[s] for so-called Islamic warriors."[17] In fact, three of the four suicide attackers had some college education, and none attended a madrassa until adulthood, when their attendance consisted of brief visits lasting for periods from a few weeks to a few months. The suicide bombers made a conscious decision to travel halfway around the globe to attend radical Pakistani madrassas after they had already been radicalized in their hometown of Leeds in the United Kingdom.

Of the four suicide attackers, Hasib Hussain, a man of Pakistani descent, attended Ingram Road Primary School in Holbeck and began his secondary education at South Leeds High School. Although he did not take his postsecondary school subject exams, he held a GNVQ, or vocational degree, in business studies. He visited Pakistan in 2003, when he was likely 16 or 17, after making a pilgrimage to Mecca. It was around this time that, back in England, he began socializing with two of the other bombers, Shehzad Tanweer and Muhammad Sidique Khan, with whom he frequented the Stratford Street mosque and the Hamara Youth Access Point, a teenage center in Leeds. Khan and Tanweer have similar backgrounds of elementary and secondary education in the United Kingdom. Khan later studied child care at Dewsbury College, and Tanweer studied sports science at Leeds Metropolitan University. Similar to Hussain, Tanweer traveled to Pakistan to study briefly at a madrassa in 2004, at the age of 21. The fourth bomber, Jamaican-born Germaine Lindsay, attended Rawthorpe High School in Huddersfield in the United Kingdom and converted to Islam at the age of 15.[18] Local influences appeared to play a far greater role in the radicalization of these young men than did their brief trips to Pakistani madrassas.

Where Are Terrorists Really Educated?

History has taught that terrorism has been a largely bourgeois endeavor, from the Russian anarchists of the late nineteenth century to the German Marxists of the Bader-Meinhof gang of the 1970s to the apocalyptic Japanese terror cult Aum Shinrikyo of the 1990s. Islamist terrorists turn out to be no different. It thus comes as no surprise that missions undertaken by Al Qaeda and its affiliated groups are not the work of impoverished, undereducated madrassa graduates, but rather of relatively prosperous university graduates with technical degrees that were often attained in the West.

Bin Laden, Al Qaeda's leader, is the college-educated son of a billionaire; his deputy, Dr. Ayman al-Zawahiri, is a surgeon from a distinguished Egyptian family. Ali Mohamed, Al Qaeda's long-time military trainer, is a former Egyptian army major with a degree in psychology who started work on a doctorate in Islamic history when he moved to the United States in the mid-1980s. Other Al Qaeda leaders worked in white collar professions such as accounting, the vocation of Rifa'i Taha, a leader of the Egyptian terrorist organization known as the Islamic Group, who signed on to Al Qaeda's declaration of war against the United States in 1998.

Immigrants and their children in Europe may be more dangerous than madrassa grads.

Our findings suggest that policymakers' concerns regarding madrassas are overwrought. More than half of the terrorists that we studied took university-level courses, and nearly half of this group attended Western schools. The majority of the college-educated group had technical degrees, which sometimes provided skills for their later careers as terrorists. Yousef's degree in electrical engineering, for example, served him well when he built the bomb that was detonated underneath the World Trade Center in 1993. It was the rare terrorist who studied exclusively at a madrassa, and only one terrorist managed to transition from madrassa to university, suggesting that madrassas simply should not be part of the profile of a terrorist capable of launching a significant anti-Western attack. Only in Southeast Asia, as seen in the Bali attack in 2002, did madrassas play a role in the terrorists' education. Yet, even in this example, the madrassa graduates paired up with better-educated counterparts to execute the attacks. Masterminding a large-scale attack thus requires technical skills beyond those provided by a madrassa education.

Because madrassas generally cannot produce the skilled terrorists capable of committing or organizing attacks in Western countries, they should not be a national security concern. Conceiving of them as such will lead to ineffective policies, and cracking down on madrassas may even harm the allies that Washington attempts to help. In countries such as Pakistan, where madrassas play a significant role in education, particularly in rural areas, the wholesale closure of madrassas may only damage the educational system and further increase regional tensions. One of Gen. President Pervez Musharraf's plans to reduce extremism, expelling foreign students and dual citizens, may be effective in reducing the number of militant Arabs studying in Pakistan but may also harm neighboring countries such as Afghanistan, which rely on Pakistani madrassas, by leaving thousands of poor Afghans without any education.[19]

Cracking down on madrassas may even harm the allies that Washington attempts to help.

This is not to suggest that Western countries should ignore madrassas entirely. To the extent that they remain a domestic problem because they undermine educational development and spawn sectarian violence, particularly in Pakistan, Western policymakers should remain vigilant about working with local governments to improve madrassas, as well as state schools. Efforts at observation and regulation might be more usefully directed toward European Islamic centers such as the Hamara Youth Access Point, where the London bombers gathered. Armed with a more realistic understanding of religious schools, particularly the differences in the curricula they provide across countries and regions, policymakers can hone their strategy with respect to madrassas. Only by eliminating the assumption that madrassas produce terrorists capable of carrying out major attacks can Western countries shape more effective policies to ensure national security.

Notes

1. "Rumsfeld's War-on-Terror Memo," May 20, 2005, `http://www.usatoday.com/news/washington/executive/rumsfeld-memo.htm` (reproducing memo titled "Global War on Terrorism" to Gen. Dick Myers, Paul Wolfowitz, Gen. Pete Pace, and Doug Feith, dated October 26, 2003).
2. *The 9/11 Commission Report: The Final Report of the National Commission on Terrorist Attacks Upon the United States* (New York: W.W. Norton, 2004), p. 367.
3. Donald Rumsfeld, interview by Charlie Rose, *Charlie Rose Show*, PBS, August 20, 2005.
4. Alex Alexiev, "If We Are to Win the War on Terror, We Must Do Far More," *National Review*, November 7, 2005; Nicholas D. Kristof, "Schoolyard Bully Diplomacy," *New York Times*, October 16, 2005, sec. 4, p. 13.
5. See Marc Sageman, *Understanding Terrorist Networks* (Philadelphia: University of Pennsylvania Press, 2004).
6. For a detailed study of the relationship between madrassas and violence in Pakistan, see Saleem H. Ali, "Islamic Education and Conflict: Understanding the Madrassahs of Pakistan" (draft report, United States Institute of Peace, July 1, 2005).
7. Febe Armanios, "Islamic Religious Schools, *Madrasas:* Background," *CRS Report for Congress*, RS21654 (October 29, 2003), p. 1, `http://fpc.state.gov/documents/organization/26014.pdf`.
8. Ibid., pp. 1–2.
9. Ibid., p. 2.
10. See generally Olivier Roy, *Globalized Islam: The Search for a New Umma* (New York: Columbia University Press, 2004); Robert S. Leiken, "Europe's Angry Muslims," *Foreign Affairs* 84, no. 4 (July/August 2005): 120–135.
11. *9/11 Commission Report*, pp. 160–169.
12. Walid Phares, *Future Jihad: Terrorist Strategies Against America* (New York: Palgrave Macmillan, 2005), p. 63.
13. Peter Bergen, *The Osama bin Laden I Know: An Oral History of Al Qaeda's Leader* (New York: Free Press, 2006), chap. 1.
14. Uzma Anzar, "Islamic Education: A Brief History of Madrassas With Comments on Curricula and Current Pedagogical Practices" (draft report, March 2003).
15. Richard C. Paddok, "Terrorism Suspect Dies in Standoff," *Los Angeles Times*, November 10, 2005, p. A4.
16. See Peter Bergen and Paul Cruickshank, "Clerical Error: The Dangers of Tolerance," *New Republic*, August 8, 2005, pp. 10–12.
17. Richard Edwards, "On Their Way to Terror School," *Evening Standard*, July 18, 2005, p. C1.
18. Ian Herbert, "Portrait of Bomber as a Dupe Fails to Convince Bereaved," *Independent*, September 24, 2005, p. 16.
19. Naveed Ahmad, "Pakistani Madrassas Under Attack," *Security Watch*, October 8, 2005, `http://www.isn.ethz.ch/news/sw/details.cfm?ID=12418`.

Peter Bergen *is a Schwartz fellow at the New America Foundation and an adjunct professor at the School of Advanced International Studies at Johns Hopkins University.* ***Swati Pandey*** *is a researcher and writer at the* Los Angeles Times.

UNIT 8

Women and Terrorism

Unit Selections

Key Points to Consider

- How do cultural and regional differences affect the role of women in terrorism?
- What motivates women to become suicide bombers? Are their motivations different than those of their male counterparts?
- Why are the experiences of girls, used as "weapons of terror," often misunderstood?

Student Website

www.mhcls.com/online

Internet References

Further information regarding these websites may be found in this book's preface or online.

Free Muslims Against Terrorism Jihad
http://www.freemuslims.org/news/article.php?article=140

Foreign Policy Association - Terrorism
http://www.fpa.org/newsletter_info2478/newsletter_info.htm

Israel Ministry of Foreign Affairs—The Exploitation of Palestinian Women for Terrorism
http://www.mfa.gov.il/mfa/go.asp?MFAH0ll10

Women, Militarism, and Violence
http://www.iwpr.org/pdf/terrorism.pdf

Women are more often portrayed as victims than perpetrators of political violence. The fact that women have played a critical role in the evolution of contemporary international terrorism is too frequently ignored. In the 1970s women like Ulrike Meinhof and Gudrun Ensslin of the German Baader-Meinhof Gang, Mara Cagol of the Red Brigades in Italy, Fusako Shigenobu of the Japanese Red Army, and Leila Khaled of the Palestine Liberation Organization held key roles in their organizations and significantly influenced the development of modern terrorism.

Today, while often less visible than their male counterparts, women continue to be actively involved in international terrorism. Women like American Lori Berenson, a former anthropology student at MIT who became involved with the Tupac Amaru Revolutionary Movement (MRTA), and Shinaz Amuri (AKA Wafa Idris), a 28-year-old volunteer medic who became a heroine to young Palestinian women after she killed herself in a suicide bombing, are the role models for a new generation of women. There are indications that the involvement of women in international terrorism may again be on the rise.

In the first article in this unit Karla Cunningham examines the roles of female terrorists in various regions of the world, focusing on why they engage in political violence. She also discusses the future role of women in international terrorism. Next, Terri Toles Patkin examines the role of women in a culture of violence. She argues that the lack of opportunities and a culture of martyrdom motivate young Palestinian women to become suicide bombers. Patkin provides a series of short biographical profiles of what some believe to be a new generation of suicide bombers. Finally, Susan McKay focuses on the role of young girls in rebel movements in Sierra Leone and Northern Uganda. She argues that while the participation of young women in terrorism is often acknowledged, their experiences are "poorly understood."

Cross-Regional Trends in Female Terrorism

Worldwide, women have historically participated in terrorist groups but their low numbers and seemingly passive roles have undermined their credibility as terrorist actors for many observers. This analysis contends that female involvement with terrorist activity is widening ideologically, logistically, and regionally for several reasons: increasing contextual pressures (e.g., domestic/international enforcement, conflict, social dislocation) create a mutually reinforcing process driving terrorist organizations to recruit women at the same time women's motivations to join these groups increases; contextual pressures impact societal controls over women that may facilitate, if not necessitate, more overt political participation up to, and including, political violence; and operational imperatives often make female members highly effective actors for their organizations, inducing leaders toward "actor innovation" to gain strategic advantage against their adversary.

KARLA J. CUNNINGHAM

Department of Political Science
SUNY Geneseo
Geneseo, New York, USA

Although women have historically been participants in terrorist groups[1] in Sri Lanka, Iran, West Germany, Italy, and Japan, to name a few cases, very little scholarly attention has been directed toward the following questions: first, why women join these groups and the types of roles they play; and second, why terrorist organizations recruit and operationalize women and how this process proceeds within societies that are usually highly restrictive of women's public roles. Answering these questions may facilitate the creation of a comprehensive strategy for combating terrorism and limiting political violence. Regardless of region, women's involvement with politically violent organizations and movements highlights several generalizable themes. First, there is a general assumption that most women who become involved with terrorist organizations do so for personal reasons, whether a personal relationship with a man or because of a personal tragedy (e.g., death of a family member, rape). This assumption mirrors theories about female criminal activity in the domestic realm, as well as legitimate political activity by women,[2] and diminishes women's credibility and influence both within and outside organizations.

Second, because women are not considered credible or likely perpetrators of terrorist violence, they can more easily carry out attacks and assist their organizations. Women are able to use their gender to avoid detection on several fronts: first, their "non-threatening" nature may prevent in-depth scrutiny at the most basic level as they are simply not considered important enough to warrant investigation; second, sensitivities regarding more thorough searches, particularly of women's bodies, may hamper stricter scrutiny; and third, a woman's ability to get pregnant and the attendant changes to her body facilitate concealment of weapons and bombs using maternity clothing, as well as further impeding inspection because of impropriety issues. Finally, popular opinion typically considers women as victims of violence, including terrorism, rather than perpetrators, a perspective that is even more entrenched when considering women from states and societies that are believed to be extremely "oppressed" such as those in the Middle East and

North Africa (MENA). Such a perspective is frequently translated into official and operational policy, wherein women are not seriously scrutinized as operational elements within terrorist and guerilla organizations because of limited resources and threat perception.

This analysis contends that female involvement with terrorist activity is widening ideologically, logistically, and regionally for several reasons: first, increasing contextual pressures (e.g., domestic/international enforcement, conflict, social dislocation) creates a mutually reinforcing process driving terrorist organizations to recruit women at the same time women's motivations to join these groups increases; contextual pressures impact societal controls over women thereby facilitating, if not necessitating, more overt political participation up to, and including, political violence; and operational imperatives often make female members highly effective actors for their organizations, inducing leaders toward "actor innovation" to gain strategic advantage against their adversary.[3]

Contextual Pressures and Innovation

Since 11 September 2001 United States law enforcement and national security efforts have been aggressively targeted at identifying current and potential terrorist actors who threaten the country's interests. This activity has largely centered on Muslim males because of the types of terrorist attacks that have threatened the United States over the past decade (e.g., the World Trade Center (1993), the African Embassy bombings (1998), and the USS *Cole* bombing (2000) to name a few). All of the incidents were planned and implemented by Muslim, and predominantly Arab, males residing within the United States or abroad.

Terrorist organizations tend to be highly adaptive and although there are fundamental differences among terrorist groups along ideological lines (e.g., ethnonationalist, religious, MarxistLeninist) that influence the types of ends these organizations seek, they are typically unified in terms of the means (e.g., political violence) they are willing to employ to achieve their goals. The means/goals dichotomy is reflected by the absence of a single definition of terrorism with which all can agree.[4] Nevertheless, an ancient Chinese proverb quickly gets to the heart of terrorism noting that its purpose is "to kill one and frighten 10,000 others."[5]

Problematic, and evidenced by the evolving nature of campaigns in Sri Lanka and Israel/Palestine, as well as historical examples from Ireland and Lebanon, is that terrorist organizations tend to adapt to high levels of external pressure by altering their techniques and targets. Terrorist organizations learn from each other and "[t]he history of terrorism reveals a series of innovations, as terrorists deliberately selected targets considered taboo and locales where violence was unexpected. These innovations were then rapidly diffused, especially in the modem era of instantaneous and global communications."[6] Corresponding to existing terrorism theory, the use of suicide campaigns is an example of one type of tactical adaptation utilized by terrorist organizations, especially in the Arab–Israeli conflict and Sri Lanka, and both cases have also witnessed an evolution in targets (e.g., combatant to civilian).

This analysis suggests that terrorist organizations "innovate" on an additional level, particularly under heavy government pressure or to exploit external conditions, to include new actors or perpetrators.[7] In both Sri Lanka[8] and Palestine, female participation within politically violent organizations has increased and women's roles have expanded to include suicide terrorism. Sri Lanka's "Black Tigers," composed of roughly 50 percent women, is symbolic of this adaptation. In 2002, the Al-Aqsa Martyrs Brigade in the Occupied Territories began actively recruiting women to act as suicide bombers in its campaign against Israeli targets. Other organizations have demonstrated efforts to recruit and employ women. For example, the Algerian-based Islamic Action Group (GIA) operation planned for the Millennium celebration in 1999 reportedly had a woman, Lucia Garofalo, as a central character. The Revolutionary Armed Forces of Colombia (FARC) and Peru's Shining Path have growing levels of female operatives, and even right-wing extremist groups in the United States, such as the World Church of the Creator (WCOTC), are reportedly witnessing high female recruitment levels and one woman associated with the rightist movement, Erica Chase, went on trial in summer 2002 with her boyfriend in an alleged plot to bomb symbolic African-American and Jewish targets.

Women's Political Violence and the Role of Society: The Case of Algeria

Almost universally women have been considered peripheral players by both observers and many terrorist organizations, typically relegated to support functions such as providing safe houses or gathering intelligence. However, women have been central members of some organizations, such as Shigenobu Fusako, founder and leader of the Japanese Red Army (JRA), and Ulrike Meinhof, an influential member of the West German Baader-Meinhof Gang. In Iran, Ashraf Rabi was arrested in 1974 by the SAVAK, the country's secret police, after a bomb accidently detonated in her headquarters. In Sri Lanka, women have been effective suicide bombers for the Liberation Tigers for Tamil Eelam (LTTE) and interestingly, their role is modeled after women's participation in the Indian National Army (INA) during the 1940s war with Britain, which included female suicide bombers.[9] Women have also historically been active, albeit less visible members, of a range of right-wing organizations including the Ku Klux Klan (KKK)[10] and the Third Reich.[11] If women's involvement with political violence is interpreted more broadly to include revolutionary movements, then scholarly discourse clearly demonstrates the importance of women like Joan of Arc and women during the Russian Revolution.[12]

Thus, even a cursory look at history provides numerous examples of a diverse array of cases and roles of women's involvement with political violence. Despite historical evidence though, most observers remain surprised and baffled by women's willingness to engage in political violence, especially within the context of terrorism. Importantly, the "invisibility" of women both within terrorist organizations, and particularly their assumed invisibility within many of the societies that experience terrorism, makes women an attractive actor for these organizations, an advantage that female members also acknowledge. This invisibility also makes scholarly inquiry of the phenomenon more difficult and may lull observers into the false assumption that women are insignificant actors within terrorist organizations.

An analysis of the role of "veiled" and "unveiled" women during the 1950s Algerian resistance against the French provides insights into the process by which women were consciously mobilized into "terrorist" roles within a MENA case by both politically violent organizations and the women who chose to join these organizations. Significantly, "[t]he Algerian woman's entrance into the Revolution as political agent was simultaneous with the deployment of the necessarily violent 'technique of terrorism,' " and the veil became "both a dress and a mask," facilitating women's operational utility during the Revolution.[13] Mirroring scholarly discourse on the "popular upsurge" in transitions against authoritarian rule in which civil society "surges" and then retreats, women's incorporation into the Algerian resistance movement emerged within a process of resistance to international oppression, suggesting that the societal sector may have been momentarily, albeit effectively, mobilized to include women.[14]

The phased mobilization of women into the Algerian resistance movement, and the societal environment that facilitated it, is argued to have had three distinct junctures. Prior to 1956, Algerian resistance led to the "cult of the veil" and women's decisions to veil were an active response to colonial attempts to unveil them and thereby dominate society even further. Only men were involved in armed struggle during this period but French adaptation to resistance tactics prompted male leaders to hesitantly transform their strategy and include women in the "public struggle." This initiated the second phase of women's mobilization, wherein "terrorist tactics are first fully utilized" and the conflict moved to urban areas and women unveiled in order to exploit their opponent. The final phase occurred when "woman ... was transformed into a 'woman-arsenal': Carrying revolvers, grenades, hundreds of false identity cards or bombs, the unveiled Algerian woman move[d] like a fish in the Western waters."[15] "By 1957, the veil reappeared because everyone was a suspected terrorist and the veil facilitated the concealment of weapons." Further, "[r]esistance [wa]s generated through the manipulation, transformation, and reappropriation of the traditional Arab woman's veil into a 'technique of camouflage' for guerilla warfare."[16]

The societal process(es) that facilitated the coalescence of organizational and individualist interests in Algeria is significant. There occurred "[a] transformation of the Muslim notion of femininity, even if only momentarily during decolonization, [which] is central to theorizing the general range of *possibilities* for Algerian women's subjectivity and agency."[17] Significantly, this same type of process has been visible within the Palestinian context for at least three decades. Although women were not actively visible in the earlier periods of the Arab–Israeli conflict, with the creation of the General Union of Palestinian Women (1969) and the spread of education, there was a growing idea among Palestinian leaders "that women constitute half the available manpower resource, one that a small, embattled nation cannot afford to waste. Women began to participate, publicly, in every crisis, from Wahdat camp in the 1970 Amman battles to the latest Israeli invasion in South Lebanon." Although women were willing to participate, and Palestinian leaders were clearly willing to rely on them, Arafat's conception of their role conflicted with societal conceptions of women's roles, thereby making it difficult for women to fully participate in the conflict.[18]

The Algerian case is illustrative of a number of themes that will be developed in this article. First, there was a mutually reinforcing process driving both women and organizations using political violence together. The revolutionary features of Algerian resistance against external, colonial control led to broad political mobilization that included women, a process engendered not only by the promise of socialist "equality" but also by the colonial state's efforts to regulate the veil. Furthermore, the entrenched features of a war for independence inextricably involves virtually every societal segment and ensures that the conflict extends to the household level. Second, the deepening sociopolitical process of the resistance increasingly overlapped with operational imperatives within the all-male resistance movement that indicated the utility of using women against the French.

Third, Algerian men and women generally shared the same political objective—freedom from French colonial domination. Equally significant, however, is that women and men held a secondary, albeit divergent, goal regarding social change; women clearly wished for greater equality, albeit not in the Western feminist sense, whereas men saw social change as asserting more authentic cultural forms (e.g., Islam). The articulation of the latter's vision of social change is captured in *La Charte d'Alger* (1964) wherein women's inferiority under colonialism resulted from poor interpretations of Islam to which women "naturally" reacted. As a result, "[t]he war of liberation enabled the Algerian woman to assert herself by carrying out responsibilities *side by side* with man and taking part in the struggle.... In this sense the charter reveals its unwillingness discursively to allow women's participation in the war to be the product of their chosen activity. Women's historical action is legitimized by their proximity to men ... not by their agency."[19]

This process led to two outcomes that are visible in other cases. First, upon achieving the group's ends, women's participation therein is reinterpreted or reframed as less authentic, which allows women to be legitimately politically peripheralized (i.e., because they were not full and "authentic" participants) and their objectives, particularly with respect to social change, to be dismissed. Second, women's participation is not individually chosen; rather, it is facilitated by relationships with others or structural factors (e.g., poverty) that distance women from the violence they participated in, allowing society (and emergent political leaders) to not only separate women from the citizenship rights inherent in military-type service but also placing women's violence within a more palatable context. Importantly, this process mirrors women's participation in war and even instances of political mobilization within more limited (i.e., less violent) environments of political change, suggesting that it is not unique to terrorist or revolutionary structures and rather reflects more embedded features of female citizenship and political participation.

Patterns of Operational Female Terrorism

Not only have women historically been active in politically violent organizations, the regional and ideological scope of this activity has been equally broad. Women have been operational (e.g., regulars) in virtually every region and there are clear trends toward women becoming more fully incorporated into numerous terrorist organizations. Cases from Colombia, Italy, Sri Lanka, Pakistan, Turkey, Iran, Norway, and the United States suggest that women have not only functioned in support capacities, but have also been leaders in organization, recruitment, and fund-raising, as well as tasked with carrying out the most deadly missions undertaken by terrorist organizations—suicide bombings. Regardless of the region, it is clear that women are choosing to participate in politically violent organizations irrespective of their respective organizational leaders' motives for recruiting them.[20]

European Female Terrorism

European terrorist organizations are among the oldest groups to examine and offer the first insights into women's roles in these organizations. Women have been drawn to leftist and rightist organizations in Europe, and have thus been involved in groups with goals ranging from separatism to Marxist-Leninism. Women have been, and in certain cases continue to be, active members of several terrorist organizations within Europe including the Euskadi Ta Askatasuna (ETA, Basque Homeland and Unity), the Irish Republican Army (IRA), and the Italian Red Brigades (RD), to name a few. Mirroring the Palestinian conflict, which will be discussed later, Irish women, particularly mothers, have been widely active in their conflict with the British, which was waged close to home in their neighborhoods and communities.

One examination of the operational role of women in Italy's various terrorist factions during the 1960s and 1970s identifies several important tendencies. Although women generally accounted for no more than 20 percent of terrorist membership during this period, Italian women who participated in terrorist organizations were overwhelmingly drawn to leftist and nationalist organizations. This corresponded to a general period of social change, evidenced by movement in areas such as divorce, abortion, education, and employment, which allowed the Italian left to recruit and mobilize the country's women.[21] Women within the Italian left had a good chance of functioning as "regulars" and occasionally in leadership roles, particularly during the later stages of the organization's operations.[22]

The Italian experience correlates with a general trend[23] in which leftist organizations tend to attract more female recruits not only because their ideological message for political and social change (e.g., equality) resonates with women, but also because those ideas influence leadership structures within the groups. As a result, "[w]omen tend to be over-represented in positions of leadership in left-wing groups and to be underrepresented in right-wing groups."[24] Conversely, rightist organizations have more limited recruitment of women and they have historically been characterized by an almost uniform absence of female leaders. In Norway, male domination of rightist organizations, and the inability of women to obtain leadership positions, prompted the creation of Valkyria, an all-women rightist organization that allowed members to develop leadership skills and opinions.[25]

North American Female Terrorism

Women's roles North American-based terrorist organizations mirrors the variability of Europe but includes an international element that is distinguishing, at least at this juncture. First, there is an important division between women based on "origination," for lack of a better word. One group of women involved in alleged terrorist organizations are members of, or closely tied with, an expatriate or immigrant community that has links to international terrorism. The other group of women has links to domestic terrorist organizations,[26] and within this category there are three subsets: those belonging to right-wing organizations that include the WCOTC[27] and the Aryan Nation, as well as militia movements and "patriot" organizations; those belonging to "leftist" groups typically linked to Puerto Rican nationalism;[28] and those belonging to "special interest" terrorist groups that range from leftist to rightist including the Animal Liberation Front (ALF), the Earth Liberation Front (ELF), and anti-abortion activists.[29] Second, women's roles in North American terrorist organizations are highly influenced by their organization's target. For most international terrorist organizations, North America is less a theater of operation than an extremely important locus of financial, logistical, and ideological support for operations in other parts of the world. Obviously, for domestic terrorist organizations this is not a limiting factor. Third, readily available social and political freedoms in the region facilitate travel, communication, and organizational advancement that may be unattainable in other states. Finally, most connections involve both Canada and the United States, particularly with respect to legal entry and residency status for international terrorist groups.

Both the Mujahadeen-e-Khalq (MEK) and the Kurdistan Workers' Party (PKK) have attracted official attention in both the United States and Canada since 2000 for incidents involving female members. Mahnaz Samadi became a member of the MEK in 1980 and was an active fighter for the organization against Iranian targets in the 1980s, including alleged terrorist attacks in Tehran in 1982. After becoming leader of the National Liberation Army and the National Council of Resistance (NCR), a MEK civilian front, she replaced Robab Farahi-Mahdavieh in 1993 to head NCR fundraising in North America. Mahdavieh is alleged to have been involved with the 1992 attack against the Iranian embassy in Ottawa, leading to her deportation from Canada in 1993. Samadi was arrested in 2000 by Canadian officials and was deported to the United States where senior officials became involved to prevent her deportation to Iran.[30] A similar fund-raising role was allegedly carried out by Zehra Saygili (a.k.a. Aynur Saygili, Beser Gezer)[31] and Hanan Ahmed Osman (a.k.a. Helin Baran)[32] both with the PKK. Osman allegedly entered Canada in 1984 and was granted refugee status. She then turned to recruiting, fund-raising, and propaganda activities on behalf of the PKK. Saygili arrived in Canada in 1996 and allegedly became active with the Kurdish Cultural Association in Montreal to raise money and support for the PKK.[33]

Another woman suspected of having ties with terrorist networks, but later cleared by the U.S. government under somewhat vague circumstances, was a Montreal woman born in Italy, Lucia Garofalo. Garofalo was allegedly linked to Ahmed Ressam, who was found guilty in U.S. federal court of plotting a terrorist attack within the United States around the Millennium celebrations. Garofalo pled guilty to two counts of illegally transporting individuals into the United States, including her attempt to smuggle Bouabide Chamchi through an unstaffed border crossing in Vermont. She also admitted to providing him with a stolen French passport. Garofalo and Chamchi were arrested after explosive-sniffing dogs positively indicated on the vehicle she was driving. One week before attempting to transport Chamchi into the United States, Garofalo reportedly successfully transported a Pakistani man into the country, raising speculation at the time that she was transporting aliens into the United States. Phone records linking Garofalo with Ressam and other members of the conspiracy, vehicle ownership by a reported member of the Algerian Islamic League, travel records showing numerous trips to Europe, Morocco, and Libya without apparent funding to support such travel, and personal ties linked Garofalo to several individuals indicted in the Millennium operation.[34] Because terrorist charges have been dropped against Ms. Garofalo, her overall role in the Millennium plot is unknown and likely minimal; however, she is reported to have had contact with a large number of individuals linked to the plot and is married to Yamin Rachek who was deported from Canada and has been wanted by both German and British officials for theft and passport fraud. In an effort to secure counsel for her husband, Garofalo allegedly was in contact with one of the individuals linked to the Millennium plot.

Both the Anti-Defamation League (ADL)[35] and the Southern Poverty Law Center (SPLC) have noted an emerging trend in U.S. right-wing movements involving the growing mobilization of female members, particularly on the Internet. According to the SPLC, women now make up 25 percent of right-wing groups in the United States and as much as 50 percent of new recruits, and these young women want a greater role in their organizations, including leadership, than their predecessors have demanded.[36] Considering that domestic terrorism remains the most likely source of terrorist activity in the United States, according to the Federal Bureau of Investigation (FBI), and that right-wing terrorist groups are among the most active domestic terrorists in the country, this trend is noteworthy.

Lisa Turner, founder of the "Women's Frontier" of the WCOTC, provides insights into the perceived role of women for this and other White supremacist organizations.[37] Although acknowledging the role of women in combat and as martyrs for the organization (particularly Vicki Weaver and Kathy Ainsworth[38]), Turner states that "most women are not 'Shining Path' guerilla fighters." She rejects the use of women as suicide bombers ("cannon fodder") by the LTTE not on the basis that such a role is beyond women, but that it emanates from male exploitation of women that appears conjoined with their "non-White" status. Turner concentrates on avoiding a generalized understanding of what is, or is not, a revolutionary and from this argument she asserts that women's roles within the organization should be a function of their unique talents and abilities. This includes leadership positions and she notes that women can become Reverends within the organization as well as Hasta Primus, the second highest position within the organization and the main assistant to the group's leader, the Pontifex Maximus.[39] Female leadership within right-wing groups is not isolated to the WCOTC; Rachel Pendergraft is reportedly a lieutenant in the Ku Klux Klan, an organization that has clearly targeted potential and current members with a women's website.[40] Women have also been associated with potentially more violent activities, such as Erica Chase, who went on trial in summer 2002 for an alleged plot to bomb prominent African-American and Jewish targets.

Women have played a central and important role in the Puerto Rican nationalist movement, particularly the Puerto Rican Armed Forces of National Liberation (FLAN) and *Los Macheteros* (The Machete Wielders or the Puerto Rican Peoples' Army), both designated as terrorist organizations by the FBI. Women such as Blanca Canales and Adelfa Vera were significant leaders in the early nationalist movement and women are significantly represented in nonterrorist, but "supportive" entities like the Puerto Rican New Independence Movement (NMIP) and various demonstrations surrounding U.S. military exercises on Vieques. Additionally, women have been tried and incarcerated in the United States for their affiliation and actions with Puerto Rican nationalist movements. For example, 5 of 15 individuals arrested and tried by the United States between 1980 and 1985 for sedition, conspiracy, and illegal weapons possession were women (Dylcia Pagan, Alejandrina Torres, Carmen Valentin, and Alicia and Ida Luz Rodriguez) with Pagan (a.k.a. Dylcia Pagan Morales) considered the leader of the group by the government.[41]

For both the ALF and ELF, the most visible members are male, as evidenced by spokespeople (e.g., Craig Rosebraugh of the ELF) and arrests. However, this surface impression is likely not indicative of the actual rosters of these organizations and the individuals who take part in their operations. According to the FBI, the ELF and ALF have committed 600 criminal acts since 1996 amounting to more than US$42 million in damages.[42] Neither the ALF nor ELF disseminate lists of their members and members' names tend to only come to the surface based on arrest records.[43] In a report documenting actions undertaken by the ELF and ALF in 2001, of the 23 individuals associated with various legal actions ranging from arrest, imprisonment, and subpoenas only 3 were women.[44] However, this should not be construed as totally representative of female participation rates within these organizations, based upon historical trends that clearly demonstrate higher female participation rates within leftist organizations.

To date women affiliated with designated terrorist organizations in North America, both international and domestic, have played mixed roles in their respective organizations. International organizations appear to have incorporated women into more important structures, particularly those associated with fund-raising and recruitment, although cases remain few and far between. Domestic terrorist groups are increasingly targeting females for recruitment, and are attracting a diverse occupational and generational group of women. However, their roles within the leadership structures of their respective organizations is either minimal (right wing) or unknown (special interest). Leftist organizations have traditionally centered on Puerto Rican independence and have frequently involved women in a variety of capacities, including leadership positions, mirroring trends visible in Latin America and other "nationalist" settings.

Latin American Female Terrorism

Women have historically been involved in numerous revolutionary movements in Latin America (e.g., Cuba, El Salvador, Nicaragua, Mexico) so their more visible role in groups like the FARC and Shining Path is not surprising.[45] Within Latin America, two of the most notable terrorist organizations designated by the U.S. Department of State, Colombia's FARC and the Shining Path of Peru, have increasingly incorporated women into their organizations. Figures on total female membership within the FARC vary from 20 to 40 percent, with a general average of 30 percent.[46] Although the FARC's senior leadership structure, particularly the Secretariat, remains all male, women have been ascending throughout the group's ranks, with women now reportedly bearing the title "Commandante." Like Shining Path, the FARC has recruited and retained women for more than a dozen years. Unlike the FARC, the Shining Path's senior leadership structure, the Central Committee, is composed of 8 women (out of 19).[47] The Latin American phenomenon of "machismo" is noted as responsible for the continuation of senior male leadership for the FARC and the "cult of personality" that is said to surround the Shining Path's former leader, Abimael Guzman. As with the LTTE, women of both groups experience the same types of training and expectations as their male counterparts and women have been increasingly used in intelligence roles by the FARC.[48]

In Latin America, female activism in politically violent organizations remains concentrated within leftist movements, corresponding to themes seen in Europe and North America. In both Colombia and Peru, the revolutionary features of the respective movements is significant, mirroring processes in Palestine and Sri Lanka, as well as Iran, South Africa, and Eritrea. For the most part, women join the FARC and Shining Path while young, engage in all facets of the organization, and often remain members for life, although activism rates may alter with age, as is true with their male counterparts. Also noteworthy is that cases drawn from three regions (South Asia, Middle East, Latin America) confront more generalized poverty and "youth bulges" than is true of North America and Europe. Between 1983 and 2000, the percentage of the population living on less than US$2 per day was 45.4 for Sri Lanka, 36 for Colombia, and 41.4 for Peru. Data released by the Palestinian Central Bureau of Statistics in early 2002 showed that 57.8 percent of those living in the West Bank and 84.6 percent of those living in the Gaza Strip were living below the poverty line. In addition to poverty, each of these states is confronting some form of "youth bulge" evidenced by the percentage of their populations between 0 and 14 as reported in 2001. These figures ranged from 25.9 percent in Sri Lanka, 31.88 percent in Colombia, 34.41 percent in Peru, 49.89 percent in the Gaza Strip, and 44.61 percent in the West Bank.[49] Thus, the fact that poor, young individuals are frequently drawn to terrorist organizations and politically violent groups is neither regionally limited nor gendered.

South Asian Female Terrorism

The Sri Lankan case shares some parallels with MENA terrorist organizations, including the structural imperatives that favor the use of women as suicide bombers, the intersection of political and sociocultural goals of liberation, and sociocultural norms that idealize sacrifice.[50] As of 2000, roughly half of the LTTE's membership[51] were females, who are frequently recruited as children into the Black Tigers, an elite bomb squad composed of women and men.[52] Women enjoy equivalent training and combat experience with their male counterparts and are fully incorporated into the extant structure of the LTTE. Women's utility as suicide bombers derives from their general exclusion from the established "profile" of such actors employed by many police and security forces (e.g., young males), allowing them to better avoid scrutiny and reach their targets. The 1991 assassination of Rajiv Gandhi, then leader of India, by a young Tamil woman who garlanded him, bowed at his feet, and then detonated a bomb that killed them both, provides proof of the power of this terrorist weapon. However, that woman, identified as Dhanu (a.k.a. Tanu), suggests some of the contradictory themes that arise when considering women's roles in the LTTE.

Reportedly prompted to join the LTTE because she was gang-raped by Indian peacekeeping forces who also killed her brothers,[53] Dhanu has become an important mythical force utilized for further recruitment as rape has been identified as one of the primary reasons motivating young women to join the LTTE. The goal of *eelam* (freedom) pursued by the LTTE is said to be conjoined with the pursuit of similar personal, and perhaps even societal, freedom for female recruits as "[f]ighting for Tamil freedom is often the only way a woman has to redeem herself."[54] Also inherent in the struggle is the idea(l) of sacrifice, particularly for Tamil rape victims who are said to be socially prohibited from marriage and childbearing. Equating the sacrifice of the female bomber as an extension of motherhood, suicide bombings become an acceptable "offering" for women who can never be mothers, a process that is reportedly encouraged by their families.[55] "As a rule, women are represented as the core symbols of the nation's identity" and the "Tamil political movements have used women's identity as a core element in their nationalism."[56]

According to *Jane's Intelligence Review*, "suicide terrorism is the readiness to sacrifice one's life in the process of destroying or attempting to destroy a target to advance a political goal. The aim of the psychologically and physically war-trained terrorist is to die while destroying the enemy target.[57] It is also on the increase. Aside from the LTTE, the main groups that employ suicide terrorism in pursuit of their objectives are located in, or linked to, the Middle East, such as Hamas, Hizballah, and Islamic Jihad. The LTTE is the only current example of a terrorist organization that has permanently adopted "suicide terrorism as a legitimate and permanent strategy."[58] Suicide terrorism in this context is the result of a "cult of personality" rather than a religious cult, demonstrating that "under certain extreme political and psychological circumstances secular volunteers are fully capable of martyrdom."[59]

Sri Lanka is not the only place in South Asia, however, where women are, or have been, allegedly involved with terrorism. Among Sikh militants, women have participated in an array of roles including armed combat. Importantly, Sikhism does not distinguish between male and female equality forming a religio-societal grounding that neither precludes female combat nor categorizes that role as uniquely masculine (or "unfeminine"). Rather, societal resistance to female combat roles is fostered by well-founded fears of sexual abuse, rape, and sexual torture of women if captured. Within the Sikh case, women's "support" roles are not viewed as peripheralized or indicative of women's marginalization within the political sphere. Instead, women's support of their husbands and sons is seen to critically enable their ability to fight and die for the nation, and women's roles as mothers producing future fighters for that nation is also recognized. As a result, "[w]hile it is obvious that the celebrated virtues of courage, bold action, and strong speech are consonant with masculinity as understood in the West, among Sikhs these qualities are treated as neither masculine nor feminine, but simply as Sikh, values. Women may be bound to the kitchen and may have babies in their arms, but they are still fully *expected* to behave as soldiers, if necessary."[60]

Additional examples of women's participation with politically violent organizations relate to the Indian–Pakistan confrontation over Kashmir. According to Indian sources, Shamshad Begum was arrested by Indian security forces in October 2001 for allegedly acting as a guide responsible for identifying safe travel routes for members of Hizbul Mujahadeen.[61] Another female member of the same organization was reportedly killed by Lashker-e-Taiyaba members. Indian sources claim that women are drawn to the organization for financial motives, and women's roles as couriers have been improved by a "requirement" to wear a *burqa*.[62] Reports of female involvement in terrorist groups expanded by December 2001 as the Indian press reported female bomb squads were being prepared by Pakistan-supported groups in Kashmir for attacks against senior officials during the Republic Day Parade.[63]

Several themes arise from the South Asian context that provide additional insight into female terrorists, particularly suicide terrorism. First, personal motives (e.g., family, rape, financial) are argued to greatly influence women to join organizations like the LTTE and, even more importantly, into becoming suicide bombers (e.g., rape). Second, freedom and liberation are key themes at both the collective and individualistic levels. Collectively, freedom and liberation capture the legitimating ideology of the LTTE vis-á-vis the Sinhalese and the Indian governments, the mujahadeen in Kashmir vis-á-vis India, and the Sikhs vis-á-vis India for Khalistan. Liberation also appears to be conceptualized individualistically as, according to one Tamil Tiger, "the use of women in war is part of a larger vision of the guerrilla leadership to liberate Tamil women from the bonds of tradition."[64] However, this has led to accusations that women are less committed to *eelam* as their primary motivation for participating in the LTTE, joining instead for personal vengeance.[65] The idea of sacrifice as an ideal is the third theme and it centers both on the role of women within society as a whole (e.g., motherhood) as well as for suicide bombers more particularly. Female sacrifice for her family, and particularly for her male children, is seen as a generalized cultural norm that is usefully extended to female self-sacrifice for her community and family, particularly if she is unable (e.g., because of rape), to undertake her role as wife and mother within the society. In both the Sikh and Sri Lankan examples, female martyrdom is viewed as necessary to overcome the individual and—more importantly—collective shame of dishonor caused by rape. Fourth, the personalism of women's motives that arguably drive them to join organizations like the LTTE is both responsible for somehow diminishing the overall "authenticness" of women's roles in these organizations, particularly for outside observers, and allowing for charges of LTTE exploitation of its female cadre who are used as "throw-aways" or "as artillery."[66]

Middle East and North African Female Terrorism

From the earliest days of the Palestinian resistance, women have been involved in both the leftist and rightist sides of the Palestinian struggle against Israel.[67] The events of 2002 suggest that this pattern remains intact. Through April 2002 four Palestinian women have become suicide bombers on behalf of the Al-Aqsa Martyrs Brigade, an offshoot of Fatah, prompting, in part, a major Israeli military offensive against the Palestinians begun in March 2002. However, although these attacks have shocked Israeli security analysts, there is a sustained, and varied, history of Palestinian women who have been involved with terrorist organizations, particularly since the nationalist-based movements began to increasingly carry out violent activities in the 1960s. One of the most well-known female terrorists is Leila Khaled, affiliated with the Popular Front for the Liberation of Palestine (PFLP), who hijacked a plane in 1969. Another woman convicted of planting a bomb in a Jerusalem supermarket during 1969, Randa Nabulsi, was sentenced to 10 years imprisonment.[68] Although there has been a low probability that women will be used by Islamist terrorist groups, continuing the trend of lower female representation among rightist organizations, there is precedent for such inclusion in Palestine. Etaf Aliyan, a Palestinian woman who is also a member of Islamic Jihad, was scheduled to drive a car loaded with explosives into a Jerusalem police station in 1987 but was apprehended before the attack could take place. If the attack had occurred, it would have represented "the first suicide vehicle bombing in Israel"[69] and significantly, it would have been implemented by a woman.

Women's roles were increasing among secular and Islamist Palestinian organizations before 2002, suggesting a warning sign of the impending escalation of Palestinian violence against Israeli targets. In particular, there was an apparent trend in women's growing roles within the Palestinian resistance that was initiated with examples of male/female collaboration (e.g., suggesting female training by more experienced males), followed by individual women planting explosive devices but not detonating them, to the culmination wherein women were tasked with actually detonating bombs on their own persons. Thus, in hindsight suicide bombing by women appeared to be a logical progression in women's operations within various organizations, and suggests that women may be tasked with tandem suicide bombing and other operations in the future.

For example, Ahlam Al-Tamimi was arrested by Israel's Shabak in 2001, charged with extending logistical support to the Hamas cell that attacked the Sbarro pizzeria in West Jerusalem. She reportedly worked with Mohamed Daghles, a member of the Palestinian Authority security body. The two are linked to at least two incidents in summer 2001. In July, Al-Tamimi reportedly carried a bomb disguised as a beer can into a West Jerusalem supermarket that detonated but did not injure anyone. In August, Al-Tamimi was linked to a Hamas bomber who carried a bomb in a guitar case into a Sbarro pizzeria that killed the bomber and 15 others.[70] In another instance, on 3 August 2001, Ayman Razawi (a.k.a. Imman Ghazawi, Iman Ghazawi, Immam Ghazawi), 23,[71] a mother of 2, was caught before she could plant an 11-pound bomb packed with nails and screws hidden in a laundry detergent box in a Tel Aviv bus station. However, despite the escalating role of women in the *intifada*, the prospect of a female suicide bomber remained remote through the first weeks of 2002 because "[t]here have been very few cases of Arab women found infiltrating Israel on a mission to murder civilians."[72]

That perception changed dramatically in the wake of 28 January 2002 when Wafa Idris (a.k.a. Wafa Idrees, Shahanaz Al Amouri),[73] 28, detonated a 22-pound bomb in Jerusalem that killed her, an 81-year-old Israeli man, and injured more than 100 others. Confusion punctuated the immediate aftermath of the attack given that heretofore women had only helped plant bombs and it was not clear whether Idris had intended to detonate the explosive or whether the explosion was accidental. Equally unclear was whether she was acting on behalf of some group or how she had obtained the explosives. This confusion made the Israelis reticent to confirm that the attack constituted the first "official" case of a female suicide bomber related to the Arab–Israeli conflict and, therefore, a significant shift in the security framework within which the Israelis would have to operate. As Steve Emerson is quoted as stating in the wake of Idris's attack, if true the bombing "opens a whole new demographic pool of potential bombers."[74] By early February the Israelis declared that Wafa Idris was a suicide bomber[75]—a first. The Fatah-linked Al-Aqsa Martyrs Brigade (a.k.a. Al Aqsa Brigades) claimed responsibility for her attack and described Idris as a "martyr."

Idris's motivation to commit a suicide operation was arguably prompted by a sense of hopelessness under occupation and rage, not heaven as promised to her male counterparts.[76] As a result, her action is seen "to have been motivated more by nationalist than religious fervor,"[77] a motivation that is frequently attributed to her male counterparts. In addition to not being a "known" member of a terrorist organization, and therefore more likely to be identified as a potential suicide bomber, Idris did not carry out the attack in the "normal" fashion. She carried the bomb in a backpack, rather than strapped to her waist, raising widespread speculation that she did not intend to detonate the bomb and the explosion was accidental.[78] Another cause for skepticism about Idris's role in the attack arose from the lack of a note and martyr's video, which are typically left behind by one engaging in a "martyr's operation."

The response by secular and Islamist Palestinian leaders to the attack is important. Although the Al-Aqsa Martyrs Brigade claimed responsibility for the attack, it did not do so immediately. The strong reaction by the "Arab street" to the attack, and the heightened sense of insecurity noted by Israeli officials, provide two excellent reasons why women's operational utility increased for Al-Aqsa's leaders. First, Idris's action resonated strongly throughout the Arab world. Egypt's weekly *Al-Sha'ab* published an editorial on 1 February 2002 entitled "It's a Woman!" that is reflective of the general tone that emanated throughout the Arab press regarding the attack. The editorial stated, in part, "It is a woman who teaches you today a lesson in heroism, who teaches you the meaning of Jihad, and the way to die a martyr's death.... It is a woman who has shocked the enemy, with her thin, meager, and weak body.... It is a woman who blew herself up, and with her exploded all the myths about women's weakness, submissiveness, and enslavement."[79]

The profound reaction to her attack by the masses both within and outside Palestine created a turning point for the Al-Aqsa Martyrs Brigade that had two effects: first, a willingness to use both men and women in terrorist attacks and second, an acknowledgment of the utility of using suicide bombers against civilian targets within Israel to undermine Israeli security and force Israel to negotiate from a position of weakness. Within days of the attack, Abu-Ahman, founder and leader of Al-Aqsa, showed signs of a tactical shift, asserting that there would be a "qualitative military operation by Al-Aqsa Battalions (*sic*) against Israeli targets," within a short period of time[80] that was clearly designed to take advantage of the psychological and tactical significance of female members of the organization. By the end of February, the Al-Aqsa Martyrs Brigade reportedly confirmed it had created a "special women's unit" named after Wafa Idris[81] to carry out attacks. Subsequent attacks by female suicide bombers over the next three months did not confirm the existence of a "special unit," but it did signify the group's willingness to utilize female members for suicide operations was not a fluke.

Reactions by Islamists were more mixed and muted, but not rejective in the immediate aftermath of the attack. Sheikh Ahmad Yassin, spiritual leader of Hamas, initially opposed Idris's action citing personnel imperatives, stating that "in this phase (of the uprising), the participation of women is not needed in martyr operations, like men." He went on to note that "[w]e can't meet the growing demands of young men who wish to carry out martyr operations," and "women form the second line of defence (*sic*) in the resistence to the occupation." However, he later qualified his objection when he added that if a Hamas woman wanted to carry out a "martyr operation," she should be accompanied by a man if the operation required her to be away more than a day and a night. Hamas leaders Sheikh Hassan Yusef and Isma'eel Abu Shanab noted that there was no *fatwa* (religious decree) that prevented a woman from being a martyr, ostensibly against Israeli occupation in the Palestinian territories.[82] Sheikh Abdullah Nimr Darwish, spiritual leader of Arabs in Israel, was more forceful in advocating the new role for women, driven in large part by the extension of the occupation to the home. He stated "the women will fight. Now the Palestinians prefer to be killed at the front rather than wait and be killed at home.... Israel has the Dimona nuclear plant, but we Palestinians have a stronger Dimona—the suiciders. We can use them on a daily basis. He also pointed, with pride, at the sight of Palestinian women in white shrouds at funerals—a sign of their readiness to become shuhada, or martyr." Further, women lined up to become martyrs, shouting "make a bomb of me, please!"[83] All of these reactions were in keeping with the August 2001 *fatwa* issued by the High Islamic Council in Saudi Arabia urging women to join the fight against Israel as martyrs.

Despite Israeli assertions that Idris's attack was a planned suicide bombing, significant uncertainty surrounds the authenticity of this attack as an Al-Aqsa Martyrs Brigade—planned suicide attack. More likely is that Idris was to plant the bomb as Al-Tamimi and Razawi were to have done in 2001. Nevertheless, Al-Aqsa learned an important lesson about the utility of female suicide bombers, and the uncertainties of the Idris case were addressed in subsequent attacks by female martyrs: Darin Abu Aysheh (a.k.a. Dareen Abu Ashai, dareen Abu Eishi), 21, detonated an explosive device on 27 February 2002 at an Israeli checkpoint in the West Bank;[84] Ayat Akhras, 18, blew herself up on 29 March 2002 at a Jerusalem neighborhood grocery store in a wave of Passover attacks that followed Israeli attacks against Arafat's headquarters;[85] and Andalib Takafka blew herself up in a crowded Jerusalem market, killing 6 and wounding more than 50 people on 12 April 2002, undermining efforts by the U.S. Secretary of State Colin Powell to move ahead with peace talks.[86]

Historical and recent cases of female Palestinian terrorism suggest several trends. First, female activism has tended to be more active within the secularist context (e.g., leftist) rather than among Islamists (e.g., rightist), reflecting a general global trend. However, although women have been more active with the nationalist/secular side of the Palestinian movement, women have been linked to Islamist groups either directly or in terms of their overall support. Second, as the conflict with Israel deepened, the scope of activism widened to include women in an increasing array of activities, up to and including suicide bombing, and women pushed for these expanded roles. Third, women activists have tended to be young, with one or more politically active family members (male), and exposed to some form of loss (e.g., within their family or immediate community) that arguably contributed to their mobilization. Importantly, marital, educational, and maternal status were not uniform factors. Also, these factors are not radically divergent from males who undertake suicide operations within this context. Fourth, Palestinian secular leaders' willingness to include women in martyr operations was influenced by security assessments (e.g., an ability to evade security scrutiny and travel more deeply into Israel), operational constraints (e.g., growing Israeli pressure on male operatives), and publicity. Female suicide bombers represent one way to overcome Israeli security pressures, heighten Israeli insecurity, and exhaust Israeli security resources by significantly increasing the operational range and available pool for suicide operations. Akhras is an exemplary case as witnesses noted she looked "European," and dressed like any Israeli schoolgirl.[87] Trying to protect against that type of terrorist represents a fundamental challenge for any security apparatus.

Conclusion: Preliminary Trends and Themes

Although there is a tendency to dismiss the overall threat of women suicide bombers, or female terrorists more broadly, because they have historically engaged in such a small percentage of terrorist activities, contextual pressures are creating a convergence between individual women, terrorist organization leaders, and society that is not only increasing the rate of female activity within terrorist and politically violent organizations, but is also expanding their operational range. The tactical advantage of this convergence is apparent particularly with respect to female suicide bombers, a tactic designed to attract attention and instill widespread fear in the target audience, because as one observer noted in the wake of Idris's attack, "it's the women we remember."[88] Because suicide terrorism is designed to attract attention and precipitate fear, in an increasingly charged atmosphere it takes more and more to attract attention, increasing the utility of female suicide bombers. Female suicide bombers also fundamentally challenge existing security assessments and socially derived norms regarding women's behavior, heightening the fear factor. Finally, and more significantly, the small number of women who have, to date, been used in such operations suggests that they will be able to better evade detection than their male counterparts.

Leftist organizations may be more likely to initially recruit or attract women because their goals tend to conform more easily to general processes of social change in society. Nevertheless, security, operational, and publicity assessments inducing secular organizations to recruit and operationalize women in a variety of roles, including as suicide bombers, may spread to rightist organizations including Islamist groups and right-wing organizations. This process may first be visible in Palestine if violence is prolonged or deepens for four reasons: first, women are operationally significant to achieve the over-all goals of the *intifada* in a manner that at least immediately overrides potential social costs of their mobilization; second, given the nascent "public" political roles of women in the region, and sociocultural factors that facilitate this role, women could very well be "demobilized" back into the private realm with little effort; third, nothing in Islam precludes women from serving in this function; and fourth, as the conflict has progressed the lines between the secular/nationalists (e.g., Fatah) and the Islamists (e.g., Hamas) has blurred.[89] Furthermore, it is also possible that groups like Al Qaeda may see women as operationally useful, as enabling conditions abound, including: the horizontal structure and loose affiliations of these organizations, the "war on terrorism" and its escalating enforcement efforts, and no overt religious prohibition against women's activity.

There is a real fascination for many observers with why women join and participate in groups like the FARC, Shining Path, the LTTE, and even the Palestinian groups, perhaps in part because this membership is fairly visible and sizable within their respective organizations. This focus is not overly surprising "[b]ecause politics, and especially revolutionary politics, has traditionally been regarded as a male affair... [and as a result] the historian has never really had to 'explain' why an individual man chose to enter political activity."[90] Ergo, trying to "explain" why an individual woman engages in not just political activity but violent political activity becomes quite necessary because there is something not quite "natural" about it.

Both women and men join politically violent organizations, and engage in an array of activities within those organizations, for similar reasons. Most frequently, individuals want to achieve some form of political change, whether revolutionary or more limited in nature. At the most basic level, groups that use political violence as a tactic have as their end-goal a right to draw up and implement new rules of the political game. In revolutionary or nationalist contexts like Palestine, Colombia, and Sri Lanka, the potential political change is far-reaching and typically involves replacing some form of external (or externally linked) leadership. However, political institutions do not arise in a cultural vacuum and often necessitate some form of social change.

Typically women are said to have engaged in political violence for personal (private) reasons, whether because of a male family member, poverty, rape, or similar factors. Importantly, this argument suggests women do not choose their participation consciously, but are rather drawn in as reluctant, if not victimized, participants. Even women who join for ideological (public) reasons are suspect, especially in revolutionary contexts. Here, women's motivations for "freedom" are viewed dualistically as both collective (e.g., independence) and individualistic (e.g., equality) or their ideological motivations are not fully developed, making them "helpers" to men rather than ideologues in their own right. The Algerian case suggests dualistic goals for both men and women, differing only with respect to their conceptualization of social change, the secondary goal, not political change, the primary goal. Nevertheless, there remains an entrenched belief that women's motives are more personal (and private), leaving behind an impression of insincerity and shallowness that prevents women from having any fundamental voice in creating new structures.

In addition to determining why women join organizations there is an equal effort to untangle what women do once they join. Although women have historically been involved with politically violent organizations, most of their activities have been in "support" capacities; thus, their presence has been seen as passive. Usually, this support has come from mothers, who have moral authority, a certain degree of safety vis-á-vis the adversary, and fairly clear boundaries within which they operate (both with respect to their own societies and the adversary).[91] Such action is typically viewed as initiated by the women themselves, and while resistance or terrorist leaders may exploit this activity through propaganda, the role is so natural, if not expected in a highly conflictual context, that very few find this type of activity threatening (e.g., Ireland, Sri Lanka, and Palestine).

This is not true of the "warrior" women. Even a cursory review of interviews with these women demonstrates that women are pushing for expanded roles within their respective organizations, from leadership to combat, and that a growing number of younger women are joining organizations and staying. However, what is equally clear is that for most observers (e.g., academic, journalistic, policymakers) this choice seems so foreign and unnatural to women that there must be an explanation beyond simply that women want to fight for their respective causes.[92] As a result, women are duped into being "cannon fodder" as they are tasked with the most dangerous missions because they are expendable to their leaders. Additionally, history is replete with cases where women's support and service have not produced extended political freedom. But here is the rub, and the significance of the expanding roles of women in various organizations on both the left and right. As female "warriors," women are able to carve out roles themselves both within their respective organizations and with the hope of doing so in the structures that result from the struggle. Significantly, the women who are being drawn to these movements may be attracted by political opportunities implied by combatant (public) roles regarding citizenship that were denied their mothers who remained altogether private during earlier conflicts. Although it is safer and easier to simply dismiss "warrior" women as pawns of male leaders, as "dupes," and as misguided women who have lost sight of their femininity, this obscures the more interesting issue of why and how women have concluded that political violence will help them achieve desired political (and perhaps social) ends.

In evaluating the roles of women in terrorist and politically violent organizations, it remains prudent to be cognizant of the following: first, the implications of limited data; second, the possibility of denial and deception; third, that invisibility does not necessarily equate with passivity or powerlessness; and fourth, organizational versus societal imperatives. The secretiveness of many of the groups addressed in this study underscores the difficulties with obtaining reliable information related to both male and female recruitment, leadership, and operational roles. Furthermore, group leaders may mislead observers regarding the depth and breadth of female participation in their groups, either through inflation or under-inflation, to gain strategic advantages vis-á-vis their adversaries. Relatedly, just because women are not necessarily visible participants within organizations does not correlate with their absence or passivity within said organizations. Women's operational strengths and tactical advantages may induce leaders to keep female participants well-hidden until contextual pressures necessitate the group show its hand. Finally, it should not be immediately concluded that societal structures that traditionally limit female public roles will hold under tremendous conflict, nor that such structures will necessarily dictate women's roles once within politically violent organizations.

As a result, academic and policy observers must be extremely cautious in how they approach and frame female activism within terrorist or politically violent organizations.

Women have been, and will continue to be, willing to serve in a variety of groups, including right wing/religious and, significantly, they may very well be tasked in combatant roles. Terrorist or politically violent organizations are extremely aware of the potential utility of female members because this actor allows them to play on established biases and assumptions in their adversary. Terrorist organizations engage in "actor innovation" because women are able to penetrate more deeply into their targets to gather intelligence or carry out violent operations than many of their male counterparts. These organizations are interested in immediate results; the system that results will be dealt with later. This same imperative drives both female members of these organizations and the societies within which this process occurs. Societies that are under extreme strain due to occupation or conflict will often loosen their constraints on women to facilitate the convergence of individual and terrorist organizational interests. The aftermath of this process remains generally uncertain, as many of the cases discussed herein remain unresolved.

Address correspondence to Karla J. Cunningham, PhD, Department of Political Science, SUNY Geneseo, 1 College Circle, Geneseo, NY 14454, USA. Email: cunningh@geneseo.edu

Notes

1. Organizations labeled as "terrorist" are derived from the United States Department of State listing of designated terrorist organizations through either support or operational activities (see *Patterns of Global Terrorism 2000*, available at (`http://www.state.gov/s/ct/rls/pgtrpt/2000/2450.htm`). This analysis will utilize this designation for the sake of simplicity.
2. Because a woman's place is "naturally" private her motivation to become "public" would have to be personal. This suggests as well that once this personal reason has been resolved she will willingly and naturally return to her normal, private, role.
3. The common belief that women's participation in political violence is quite limited is not supported by even a cursory examination of history. However, what is clear from that cursory look is that women's experiences with political violence have not received sustained attention, and what examination has occurred has often been heavily influenced by established Western norms of appropriate female behavior. Given the constraints of any article-length analysis, certain limitations were necessary in approaching the subject matter. As a result, this work should not be construed as an exhaustive inventory of women's participation in politically violent or terrorist organizations, past or present, but rather a selective examination of primarily current critical cases.
4. Several of the most oft-quoted terrorism definitions include those used by the United States Federal Bureau of Investigation (FBI), the United States Department of State, and the United States Department of Defense (DoD). The FBI defines terrorism as "the unlawful use of force and violence against persons or property to intimidate or coerce a government, the civilian population, or any segment thereof, in furtherance of political or social objectives" (28 Code of Federal Regulations Section 0.85). The State Department defines terrorism as "premeditated, politically motivated violence perpetrated against noncombatant targets by subnational groups or clandestine agents, usually intended to influence an audience" (United States Department of State, *Patterns of Global Terrorism 2000*, available at (`http://www.state.gov/s/ct/rls/pgtrpt/2000/`). 13 April 2001). Problematic with both definitions, however, is that they fail to capture organizations motivated by religious or economic motives, such as Islamist organizations in the Middle East and North Africa (MENA) or narcoterrorist organizations such as the Revolutionary Armed Forces of Columbia (FARC) and National Liberation Army (ELN) in Colombia. The DoD partially overcomes this deficiency by widening the goal orientation of terrorist organizations as it defines terrorism as "the calculated use of violence or the threat of violence to inculcate fear; intended to coerce or to intimidate governments or societies in the pursuit of goals that are generally political, religious, or ideological" (Department of Defense, "DoD Combating Terrorism Program," Directive Number 2000.12, available at (`http://www.defenselink.mil/pubs/downing_rpt/annx_e.html`), 15 September 1996).
5. Jamie L. Rhee, "Comment: Rational and Constitutional Approaches to Airline Safety in the Face of Terrorist Threats," *DePaul Law Review* 49(847) Lexis/Nexis (Spring 2000).
6. Martha Crenshaw, "The Logic of Terrorism: Terrorist Behavior as a Product of Strategic Choice," in Walter Reich, ed., *Origins of Terrorism: Psychologies, Ideologies, Theologies, States of Mind* (Washington, DC: Woodrow Wilson Center Press, 1998), p. 15.
7. If we examine state behavior with respect to military recruitment, we see a similar process. Samarasinghe notes "most nations have increased women's military roles only when there has been a shortage of qualified men and a pressing need for more warriors.... The decision to permit women into combat is made by men.... [And] the allowable space within which women could operate in military units is also determined by them." (Vidyamali Samarasinghe, "Soldiers, Housewives and Peace Makers: Ethnic Conflict and Gender in Sri Lanka," *Ethnic Studies Report* XIV(2) (July 1996), p. 213).
8. As of early 2002, a cease-fire deal was secured between the Tamil Tigers and the government of Sri Lanka, halting the type of violence that will be discussed in this article. However, even if this activity is now a matter of historical record, rather than a current phenomenon, it offers important insights into how women were (are) mobilized into a politically violent movement.
9. Peter Schalk, "Women Fighters of the Liberation Tigers in Tamil Ilam. The Martial Feminism of Atel Palacinkam," *South Asia Research* 14(2) (Autumn 1994), pp. 174–175.
10. See Kathleen M. Blee's *Women of the Klan: Racism and Gender in the 1920s* (Berkeley: University of California

Press, 1991) for an interesting study of this widely overlooked phenomenon.

11. See Claudia Koonz, "Women in Nazi Germany," in Renate Bridenthal and Claudia Koonz, eds., *Becoming Visible, Women in European History* (Boston: Houghton Mifflin, 1977).
12. Marie Marmo Mullaney, "Women and the Theory of the 'Revolutionary Personality': Comments, Criticisms, and Suggestions for Further Study," *The Social Science Journal* 21(2) (April 1984), pp. 49–70.
13. Jeffrey Louis Decker, "Terrorism (Un)Veiled: Frantz Fanon and the Women of Algiers," *Cultural Critique* 17 (Winter 1990), pp. 180–181.
14. Although O'Donnell and Schmitter's argument centers around mobilization against domestic authoritarian rule, there are parallels in the decolonization process that makes this comparison useful. See Guillermo O'Donnell and Philippe C. Schmitter, *Transitions from Authoritarian Rule: Tentative Conclusions About Uncertain Democracies* (Baltimore, MD: The Johns Hopkins University Press, 1986). With respect to civil society, O'Donnell and Schmitter argue that "private" civil society mobilizes only temporarily to become "public" to achieve its goal (transition from authoritarianism). Once that goal is achieved, civil society willingly returns to its "natural" private sphere. This conceptualization bears striking parallels to the role of women wherein Algeria's "private" women (a role physically visible through the veil) are temporarily mobilized into "public" action to achieve independence. However, once the aim of the mobilization is completed (e.g., independence) they are assumed to willingly and naturally return to their private role. However, not all scholars (Karla J. Cunningham, "Regime and Society in Jordan: An Analysis of Jordanian Liberalization," Dissertation, University at Buffalo, 1997; Peter P. Ekeh, "Historical and Cross-Cultural Contexts of Civil Society in Africa," Paper presented at the United States Agency for International Development (USAID)—Hosted Workshop on *Civil Society, Democracy, and Development in Africa*, 9–10 June 1994) are convinced that the rigid conceptual boundaries between "public" and "private" in transition are meaningful, with important ramifications for transition.
15. The account of these phases are taken from Decker, "Terrorism (Un)Veiled," pp. 190–192. The first quotation is located on p. 191, the second is on p. 192.
16. Ibid., p. 193.
17. Ibid., p. 183, emphasis in original.
18. This account of the conflicting interests of Palestinian women, leaders, and society was discussed by Soraya Antonius, "Fighting on Two Fronts: Conversations with Palestinian Women," *Journal of Palestine Studies* 5 (October 1979), pp. 28–30.
19. Marnia Lazreg, "Citizenship and Gender in Algeria," in Saud Joseph, ed., *Gender and Citizenship in the Middle East* (Syracuse, NY: Syracuse University Press, 2000), p. 62, emphasis in original.
20. The regional cases that are discussed later are utilized to demonstrate these developments given the constraints of an article. However, it should be understood that this is not, and is not intended to be, an exhaustive inventory of cases in which women have engaged in political violence or terrorism. Cases from Africa (Eritrea, South Africa) and East Asia (Japan, Korea, Vietnam) are also worth investigating.
21. Leonard Weinberg and William Lee Eubank, "Italian Women Terrorists," *Terrorism: An International Journal* 9(3) (1987), p. 247.
22. Weinberg and Eubank, 1987, pp. 250–252. The authors' conclusions are based on biographical reviews of female terrorists reported in two major Italian newspapers. Concentrating on individuals identified and arrested by the Italian government, the authors admit that their information "does not represent a sample of terrorists" (Ibid., p. 248). A point to consider is that women's roles and representation may remain somewhat skewed, even in this worthwhile study, because one of the apparent operational advantages of female members to terrorist organizations, at least in other contexts, is that they tend to go unnoticed by officials. As a result, relying on official recognition of key women may not provide the fullest picture of women's roles in varying terrorist organizations.
23. For a good analysis of female participation in left- and right-wing organizations within the United States during the 1960s and 1970s please see Jeffrey S. Handler, "Socioeconomic Profile of an American Terrorist: 1960s and 1970s," *Terrorism* 13(3) (May–June 1990), pp. 195–213.
24. Ibid., 1990, p. 204.
25. Katrine Fangen, "Separate or Equal? The Emergence of an All-Female Group in Norway's Rightist Underground," *Terrorism and Political Violence* 9(3) (Autumn 1997), pp. 122–164. In contrast, leftist women tend to organize their own organizations to pursue a particular objective (Ibid., p. 122).
26. According to the FBI, domestic terrorism is "the unlawful use, or threatened use, of force or violence by a group or individual based and operating entirely within the United States or Puerto Rico without foreign direction committed against persons or property to intimidate or coerce a government, the civilian population, or any segment thereof in furtherance of political or social objections" (United States Department of Justice Federal Bureau of Investigation, Terrorism in the United States 1999, Counterterrorism Threat Assessment and Warning Unit, Counterterrorism Division, available at `http://www.fbi.gov/publications/terror/terror99.pdf`, 1999). For the purposes of this study, FBI official designations of domestic terrorist status will be utilized in characterizing a group as terrorist. Between 1980–1999 there were 327 incidents or suspected incidents of terrorism within the United States, of which 239 were attributed to domestic terrorism (Ibid.). The analysis of domestic terrorism offered in this article is focused on groups or categories the FBI deems as generally active. As a result, historical examples of female participation may not be included, particularly if the group is no longer actively identified by the FBI as a terrorist threat.
27. According to the FBI, the WCOTC has been linked to acts of domestic terrorism including the July 1999 shootings of several racial minorities by Benjamin Nathaniel Smith in Il-

linois and Indiana (United States Department of Justice Federal Bureau of Investigation, 1999).

28. Women's participation with left-wing movements is longstanding, with prominent examples from the 1960s and 1970s including the Weathermen, the Black Panthers, and the Symbionese Liberation Army. The discussion of left-wing terrorism in this article does not focus on these examples because they have diminished or disappeared, at least with respect to FBI reporting of left-wing terrorism.
29. Please note that these three generalized categorizes have been created to facilitate discussion within the limited confines of this article. There is tremendous variation within the three categories that such grouping tends to obscure.
30. Background on Samadi and Mahdavieh were drawn from: Aaron Sands, "Secret Arrest of Saddam Ally," Ottawa Citizen 1 February 2000, Lexis/Nexis, 3 March 2002; Moira Farrow, "Woman Ordered Deported Not a Terrorist Lawyer Says," *The Vancouver Sun*, 8 April 1993 Lexis/Nexis, 3 March 2002.
31. See *Zehra Saygili v. The Minister of Citizenship & Immigration and Solicitor General for Canada*, Court No. DES-6-96, available at (`http://decisions.fct-cf.gc.ca/cf/1997/des-6-96.html`).
32. Tom Godfrey, "Lax Security Screening Has Allowed 'Sleeper' Terrorists to Infiltrate Canada for Years," *Toronto Sun*, 7 October 2001, available at (`http://www.canoe.ca/TorontoNews/04n1.html`).
33. The PKK reportedly used women as suicide bombers in Turkey during 1998, but ended the tactic thereafter, suggesting that suicide terrorism was used temporarily to achieve a specific objective (Ehud Sprinzak, "Rational Fanatics," *Foreign Policy* 120 (September/October 2000), ProQuest, 25 March 2002, pp. 4–5). For more information on the PKK's use of suicide bombing during 1998 please see "Female Separatist Rebel Captured in Southeastern Turkey," *BBC Worldwide Monitoring*, 15 August 1998, Lexis/Nexis, 31 January 2002; "Female 'Terrorist' Reportedly Carries Out Suicide Bombing," *BBC Worldwide Monitoring*, 24 December 1998, Lexis/Nexis, 31 January 2002; and "Child Wounded in Female Suicide Bombers' Attack in Southeastern Turkey," *BBC Worldwide Monitoring*, 17 November 1998, Lexis/Nexis, 31 January 2002.
34. For the ups and downs of this particular case see Neil MacFarquhar, "Woman Freed After Pleading in Border Case," *The New York Times*, 16 February 2000, Lexis/Nexis, 25 March 2002; Michael G. Crawford, "MILNET: The Algerian Y2K Bomb Case," 2001, available at (`http://www.milnet.com/milnet/y2kbomb/y2kbomb.htm`), 7 March 2002; Cindy Rodriguez, "Stress Line US Tries to Tighten Security on Canadian Border," *The Boston Globe*, 7 November 2001, Lexis/Nexis, 6 March 2002; Lloyd Robertson, "Lucia Garofalo Pleaded Guilty to Immigration Charges Today But Was Cleared of Terrorism Charges," *CTV Television, Inc.*, 15 February 2000, Lexis/Nexis, 6 March 2002; David Arnold, "Garofalo Might Go Free: U.S. to Recommend Release of Montrealer Suspected of Terrorism Link," *The Gazette* (Montreal), 15 February 2000, Lexis/Nexis, 6 March 2002; "Special Report: The Future of Terror: On Guard: America is the Dominant Nation Entering the New Century—and the Top Target for Extremists," *Newsweek International*, 10 January 2000, Lexis/Nexis, 6 March 2002; "Canadian Police Search Apartment of Accomplice of Terrorism Suspect," *Agence France Presse*, 24 December 1999, Lexis/Nexis, 6 March 2002; Butler T. Gray, "U.S. Prosecutors Link Arrests in Vermont and Washington State," *Washington File, United States Department of State International Information Programs*, 1999, available at (`http://usinfo.state.gov/topical/pol/terror/99123004.htm`), 15 March 2002; and "Canadian Woman Has Ties to Washington Bomb Suspect, 2 Algerian Terrorist Groups," *CNN.com*, 30 December 1999, available at (`http://www.cnn.com/2000/US/01/12/border.arrest.02`), 15 March 2002.
35. "Feminism Perverted: Extremist Women on the World Wide Web," Anti-Defamation League, 2000, available at (`http://www.adl.org/special_reports/extremist_women_on_web/print.html`), 18 February 2002.
36. "All in the Family," Southern Poverty Law Center, n.d., available at (`http://www.splcenter.org/intelligenceproject/ip-4k2.html`), 28 March 2002. Also see Jim Nesbitt, "The American Scene: White Supremacist Women Push for Greater Role in Movement," Newhouse News Service, 1999, available at (`http://www.newhousenews.com/archive/story1a1022.html`) accessed 24 July 2002.
37. Turner's efforts to create a women's organization within the larger movement is noteworthy and parallels Norwegian experiences (see Fangen, "Separate or Equal?," especially pp. 124–127, 128–140, 144–155).
38. Vicki Weaver, wife of Randy Weaver, was shot by an FBI sniper in August 1992. Randy Weaver, a white separatist, was accused by the government of illegal weapons sales. Kathy Ainsworth was killed by the FBI in 1968 when she and another man tried to plant a bomb at the house of an ADL leader in Mississippi, allegedly on behalf of the Ku Klux Klan. She is one of the only known women affiliated with the white supremacy movement in the United States to be tasked with this type of mission. Interestingly, an additional woman often noted as a "martyr" is Hanna Reitsch who was reportedly a leading proponent of suicide plane missions on behalf of the Nazis during World War II (see `http://www.sigrdrifa.com/sigrdrifa/67hanna.html` for a sample biography). For additional information on "martyrs" identified by the white supremacist movement (see `http://www.volksfrontusa.org/martyrs.shtml`).
39. Turner's argument regarding women's roles in the WCOTC were taken from Sister Lisa Turner, "The Women of the Creativity Revolution," ChurchFliers.com, n.d., available at (`http://www.churchfliers.com/sub_articles/women.html`). 2 April 2002. In looking at the WCOTC site over a period of several months, there have been clear changes in the positioning of women's sites. In April 2002, women's issues were clearly not a priority but there was a direct link on the main page directing women to four white

women's movement sites: Elisha Strom: A Woman's Voice, available at (http://www.elishastrom.com), Free Our Women Campaign (FOW), available at (http://www.midhnottsol.org/fow/index.html). Mothers of the Movement (MOTM), available at (http://www.sigrdrifa.com/motm), and Sigrdrifa.com—Premier Voice of the Proud White Women, available at (http://www.sigrdrifa.com). Sigdrifa publishes a journal that addresses a wide range of issues important to women in the movement including feminism, women's roles in the organizations, recruitment, and prison outreach. Elisha Strom's "Angry White Woman" site covers an array of issues clearly central to women in the movement, including debates over feminism and the importance of motherhood. She is also extremely critical of Kathleen M. Blee's works on the white power movement (see Blee, Women of the Clan, 1991 and *Inside Organized Racism: Women in the Hate Movement*, Berkeley: University of California Press, 2002). The WCOTC links, however, are not fully representative of the websites oriented toward women in the white power movement. Stormfront has a women's page as well which links into a variety of profiles of women who have joined the white power movement (see http://www.stormfront.org). Through their links page Women for Aryan Unity can be assessed at (http://www.waul4.cjb.net/), which features a picture of a white woman holding her baby that acts as the site's gateway to a site dedicated to the more pagan side of the white power movement, pictures of the Aryan sisterhood including tatoos, childrearing tips, and similar features. By July 2002, the WCOTC had removed the linkage to women's sites from their main page for unknown reasons, although this author speculates that the growing outside scholarly and activist scrutiny of these women is unwelcome by the organization for various reasons, including operational. Attempts to find the Women's Frontier using the WCOTC search engine as of July 2002 were ineffectual, bringing up only four articles apparently targeted to women, including the aforementioned article, none of which was accessible.

40. See (http://www.kukluxklan.org/lady4.htm) for the KKK's "Woman to Woman" website, which covers a range of issues including children, attacks against the feminist movement, and even women's roles in combat.
41. In 1999 President Clinton offered the individuals arrested and convicted during this time, known by many Puerto Rican activists as the "independentistas," clemency. All but two accepted the offer.
42. Dale L. Watson, "The Terrorist Threat Confronting the United States," Statement before the Senate Select Committee on Intelligence, Washington, D.C., 6 February 2002, available at (http://www.fbi.gov/congress/congress02/watson020602.htm). 18 February 2002.
43. See "What is the Earth Liberation Front (ELF)?" available at (http://www.animalliberation.net/library/facts/elf.html) for details on organizational features of the group.
44. "2001 Year End Direct Action Report Released by ALF Press Office," 2001, available at (http://www.earthliberationfront.com/library/2001DirectActions.pdf), 30 March 2002.
45. For a useful examination of women's roles in Latin American guerilla movements please see Linda M. Lobao, "Women in Revolutionary Movements: Changing Patterns of Latin American Guerilla Struggle," *Dialectical Anthropology* 15 (1999), pp. 211–232.
46. For varying figures see Jeremy McDermott, "Girl Guerillas Fight Their Way to the Top of Revolutionary Ranks," *Scotland on Sunday*, 23 December 2001, Lexis/Nexis, 2 April 2002; Karl Penhaul, "Battle of the Sexes: Female Rebels Battle Colombian Troops in the Field and Machismo in Guerilla Ranks," *San Francisco Chronicle* 11 January 2001, Lexis/Nexis, 2 April 2002; and Martin Hodgson, "Girls Swap Diapers for Rebel Life," The Christian Science Monitor, 6 October 2000, available at (http://www.csmonitor.com/durable/2000/10/06/p6s1.htm). 2 April 2002. Aside from a fascination with the makeup habits of the female FARC members, these articles offer some insights into the motivations driving women into the FARC's ranks.
47. M. Elaine Mar, "Shining Path Women," n.d., *Harvard Magazine*, available at (http://www.harvardmagazine.com/issues/mj96/right.violence.html). 2 April 2002. During the late 1980s, "approximately 35 percent of the military leaders of … [the Shining Path], primarily at the level of underground cells … [were] also women" (Juan Lazaro, "Women and Political Violence in Contemporary Peru," *Dialectical Anthropology* 15(2–3) (1990), p. 234). Additionally, by 1987 roughly 1,000 women had been arrested on suspicion of terrorism in Peru including four senior Shining Path female leaders: Laura Zambrano ("Camarada Meche"), Fiorella Montano ("Lucia"), Margie Clavo Peralta, and Edith Lagos (Ibid., p. 243).
48. This position is advanced by McDermott, "Girl Guerillas Fight Their Way to the Top."
49. This data was drawn from several sources. Poverty rates for Colombia, Peru, and Sri Lanka were taken from the United Nations Development Programme, *Human Development Report 2002*, available at (http://www.undp.org/hdr2002/) whereas the data for the Gaza Strip and the West Bank were found in "More Than Two Thirds of Palestinian Children Living on Less than US$1.90/day," 21 May 2002, available at (http://www.iap.org/newsmay213.htm). The demographic data can be found in the Central Intelligence Agency's *The World Factbook 2001*, available at (http://www.cia.gov/cia/publications/factbook/index.html).
50. Interestingly, the LTTE's creation of an organized squad of female suicide bombers is said to be mirrored after the Indian National Army's (INA) activities against the British during the early to mid-1940s (see Schalk, "Women Fighters of the Liberation," p. 174).
51. United States Department of State, *Patterns of Global Terrorism 2000*, "Asia Overview," 30 April 2001, available at

(http://www.state.gov/s/ct/rls/pgtrpt/2000/2432.htm). 2 April 2002.

52. Some observers further identify the female cadre of the Black Tigers as the "Birds of Freedom." See, for example, Charu Lata Joshi, "Sri Lanka: Suicide Bombers," *Far Eastern Economic Review*, 1 June 2000, available at (http://www.feer.com/_0006_01/p64currents.html), 11 March 2002. The idea of a bird carrying the soul of the martyr to paradise is a theme seen in Islamist discourse on martyr operations.
53. Ana Cutter, "Tamil Tigresses: Hindu Martyrs," n.d., available at (http://www.columbia.edu/cu/sipa/PUBS/SLANT/SPRING98/article5.html), 11 March 2002. Also see Frederica Jansz, "Why Do They Blow Themselves Up?" *The Sunday Times*, 15 March 1998, available at (http://www.lacnet.org/suntimes/980315/plus4.html), 3 April 2002.
54. Cutter, "Tamil Tigresses."
55. Ibid.
56. Joke Schrijvers, "Fighters, Victims and Survivors: Constructions of Ethnicity, Gender and Refugeeness among Tamils in Sri Lanka," *Journal of Refugee Studies* 12 (3 September 1999). The quotation on women as core national symbols is on p. 308; the quote on Tamil use of women's identity is on p. 311; and the quote on purity and suicide bombing is on p. 319 with emphasis in the original.
57. "Suicide Terrorism: A Global Threat," *Jane's Intelligence Review*, 20 October 2000, available at (http://www.janes.com/security/regional_security/news/usscole/jir001020_1_n.shtml), 11 November 2001.
58. Sprinzak, "Rational Fanatics," p. 6.
59. Ibid.
60. The discussion of the role of Sikh women was drawn from Cynthia Keppley Mahmood, *Fighting for Faith and Nation: Dialogues with Sikh Militants* (Philadelphia: University of Pennsylvania Press, 1996), pp. 213–234. The quotation is located on pp. 230–231, emphasis added.
61. "Veiled Women Show the Way to Terrorists in the Kashmir," *The Statesman* (India), 20 October 2001, Lexis/Nexis, 31 January 2002.
62. Ibid. This line of reasoning is very reminiscent of Decker's discussion of Algerian women during the Resistance.
63. For example see "Indian Intelligence Agencies Warn of Possible Female Suicide Squad Attacks," *BBC Worldwide Monitoring*, (originally published in *The Asian Age*, Delhi), 14 December 2001, Lexis/Nexis, 31 January 2002. Although no attacks occurred during the 26 January 2002 festivities, security was reportedly tight.
64. "Female Fighters Push on for Tamil Victory," *Michigan Daily.com* CX (93) 10 March 2000, available at (http://www.pub.umich.edu/daily/2000/mar/03-10-2000/news/09.html), 2 April 2002.
65. Jansz, "Why Do They Blow Themselves Up?"
66. The artillery reference was reportedly made by a Sri Lankan military source ("Female Fighters Push on for Tamil Victory").
67. For two good studies on the role of women in Palestinian resistance both before and during the first *intifada* see Antonius, "Fighting on Two Fronts," pp. 26–45 and Graham Usher, "Palestinian Women, the Intifada and the State of Independence," *Race & Class* 34(3) (January–March 1993), pp. 31–43.
68. Majeda Al-Batsh, "Mystery Surrounds Palestinian Woman Suicide Bomber," *Agence France Presse*, 28 January 2002, Lexis/Nexis, 6 February 2002.
69. David Sharrock, "Women: The Suicide Bomber's Story," *The Guardian*, 5 May 1998, Lexis/Nexis, 30 March 2002.
70. For more information on Al-Tamimi and the Summer 2001 incidents that appear linked to her see Wafa Amr, "Palestinian Women Play Role in Fighting Occupation," *Jordan Times*, 29 January 2002, available at (http://www.jordantimes.com/tue/news/news6.htm). 3 February 2002; also see "Shabak Accuses Young Palestinian Woman of Assisting Hamas Cell," The Palestinian Information Center, 17 September 2001, available at (http://www.palestineinfo.com/daily_news/prev_editions/2001/ep01/17sep01.htm), 3 February 2002. As of 12 February 2002, Al-Tamimi remains in Israeli custody awaiting trial (see http://www.palestinemirror.org/Other%20Updates/palestinian_women_political_prisoners.htm).
71. Ghazawi's age has been quoted as either 23 or 24 (see Majeda Al-Batsh, "Palestinian Mother, 24, Is Among Loners Mounting Attacks On Israel," *Agence France Presse*, 6 September 2001, Lexis/Nexis, 30 March 2002; David Rudge, "Alert Security Guard Foils TA Bombing," *The Jerusalem Post*, 5 August 2001, Lexis/Nexis, 30 March 2002; "Palestinians' New Weapon: Women Suicide Bombers," *The Straits Times (Singapore)*, 6 August 2001, Lexis/Nexis, 30 March 2002; Uzi Mahnaimi, "Israeli Fear As Women Join Suicide Squad," *Sunday Times (London)*, 5 August 2001, Lexis/Nexis, 30 March 2002; and Douglas Davis, "Women Warriors," *Jewish World Review*, 9 August 2001, available at (http://www.jewishworldreview.com/0801/women.warriors.asp), 30 March 2002).
72. Phil Reeves, "The Paramedic Who Became Another 'Martyr' for Palestine," *The Independent*, 31 January 2002, available at (http://www.ccmep.org/hotnews/parameic013102.html), 6 March 2002.
73. Hizbollah television identified the bomber as Shahanaz Al Amouri following the attack. See Imigo Gilmore, "Woman Suicide Bomber Shakes Israelis," *The Daily Telegraph* (London), 28 January 2002, Lexis/Nexis, 6 March 2002.
74. William Neuman, "Femmes Fatales Herald New Terror Era," *The New York Post*, 28 January 2002, Lexis/Nexis, 11 March 2002.
75. James Bennet, "Israelis Declare Arab Woman Was In Fact a Suicide Bomber," *The New York Times*, 9 February 2002, Lexis/Nexis, 11 March 2002.
76. Larnis Andoni, "Wafa Idrees: A Symbol of a Generation," *Arabic Media Internet Network* (AMIN), 23 February 2002, available at (http://www.amin.org/eng/uncat/2002/feb/feb23.html), 6 March 2002.

77. Reeves, "The Paramedic Who Became Another 'Martyr' "; James Bennet, "Filling in the Blanks on Palestinian Bomber," *The New York Times*, 31 January 2002, Lexis/Nexis, 6 March 2002; and Wafa Amr, "Palestinian Woman Bomber Yearned for Martyrdom," *The Jordan Times*, 31 January 2002, available at (http://www.jordantimes.com). 31 January 2002.

78. Peter Beaumont, "From an Angel of Mercy to Angel of Death," *The Guardian*, 31 January 2002, available at (http://www.guardian.co.uk/Print/0,3858,4346503,00.html). 6 March 2002.

79. Quoted in "Inquiry and Analysis No. 84: Jihad and Terrorism Studies Wafa Idris: The Celebration of the First Female Palestinian Suicide Bomber—Part II," *The Middle East Media and Research Institute*, 13 February 2002, available at (http://www.memri.org). 6 March 2002. Also see James Bennet, "Arab Press Glorifies Bomber as Heroine," *The New York Times*, 11 February 2002, Lexis/Nexis, 6 March 2002.

80. "Militant Palestinian Leader on Imminent Operations with 15-km Rockets," *BBC Monitoring Middle East*, 4 February 2002, Lexis/Nexis, 4 March 2002.

81. Sophie Claudet, "More Palestinian Women Suicide Bombers Could Be On the Way: Analysts," *Agence France Presse*, 28 February 2002, Lexis/Nexis, 16 March 2002.

82. Yassin and Yusef's points were taken from "We Don't Need Women Suicide Bombers: Hamas Spiritual Leader," *Agence France Presse*, 2 February 2002, Lexis/Nexis, 6 March 2002; "Islam Not (sic) Forbid Women From Carrying Out Suicide Attack," *Xinhua*, 28 February 2002, Lexis/Nexis, 31 January 2002. For further accounts of the range of religious responses to Idris' action please see "Inquiry and Analysis No. 83: Jihad and Terrorism Studies—Wafa Idris: The Celebration of the First Female Palestinian Suicide Bomber—Part I," *The Middle East Media and Research Institute*, 12 February 2002, available at (http://www.memri.org), 6 March 2002.

83. Darwish's statements were taken from "Palestinians' New Weapon: Women Suicide Bombers," *The Straits Times (Singapore)*, 6 August 2001, Lexis/Nexis, 31 January 2002.

84. Mohammed Daraghmeh, "Woman Suicide Bomber Rejected by Hamas," *The Independent*, 1 March 2002, Lexis/Nexis, 6 March 2002; Mohammad Daraghmeh, "Woman Bomber Wanted to Carry Out Sbarro-Like Attack," *The Jerusalem Post*, 1 March 2002, Lexis/Nexis, 6 March 2002; "Woman Suicide Bomber was 21-Year Old Palestinian Student," *Agence France Presse*, 28 February 2002, Lexis/Nexis, 6 March 2002; Sandro Contenta, "Student 'Had a Wish to Become a Martyr,' " *Toronto Star*, 1 March 2002, Lexis/Nexis, 13 March 2002; Stephen Farrell, "Daughter's Dedication Was Beyond Doubt," *The Times* (London), 1 March 2002, Lexis/Nexis, 13 March 2002.

85. See "Deadly Secret of Quiet High School Girl Who Became a Suicide Bomber," *The Herald* (Glasgow) 30 March 2002, Lexis/Nexis, 1 April 2002; Anton La Guardia, "The Girl Who Brought Terror to the Supermarket," *The Daily Telegraph* (London), 30 March 2002, Lexis/Nexis, 1 April 2002; and Cameron W. Barr, "Why a Palestinian Girl Now Wants to Be a Suicide Bomber," *The Christian Science Monitor*, 1 April 2002, Lexis/Nexis, 1 April 2002; Eric Silver, "Middle East Crisis: Schoolgirl Suicide Bomber Kills Two in Supermarket," *The Independent* (London), 30 March 2002, 1 April 2002; Philip Jacobson, "Terror of the Girl Martyrs," *Sunday Mirror*, 31 March 2002, Lexis/Nexis, 1 April 2002. The reference to the militia linked to Arafat is a thinly disguised reference to the Al-Aqsa Martyrs Brigade.

86. David Lamb, "The World; Gruesome Change from the Ordinary; Conflict: A Quiet, Young Seamstress Further Widened the Mideast Breach When She Joined the Ranks of Palestinian Suicide Bombers," *The Los Angeles Times*, 14 April 2002, ProQuest, 3 June 2002;

"Jerusalem Shocked by Suicide Bomb; Woman Bomber Kills Six in Attempt to Derail Powell Peace Talks," Belfast News Letter, 13 April 2002, Lexis/Nexis, 3 June 2002.

87. Jacobson, "Terror of the Girl Martyrs."

88. Melanie Reid, "Myth That Women Are the Most Deadly Killers of All," *The Herald (Glasgow)* 29 January 2002, Lexis/Nexis, 6 February 2002.

89. This last point is reinforced by reports emanating from the territories that suggest at least a temporary "alignment" between the two sides. For example, in Jenin Hamas and Fatah reportedly joined together to distribute "explosive belts" and hand grenades to individuals in the camp for self defense. A woman, Ilham Dosuki, reportedly blew herself up on 6 April 2002 as soldiers approached the door to her home ("Fierce Battles in Jenin, Nablus: Unconfirmed Reports: Scores of Palestinians Killed and Injured in Jenin Refugee Camp," 2002, *Al-Bawaba*, 6 April 2002, available at (http://www.albawaba.com/). 6 April 2002.

90. Mullaney, "Women and the Theory of the 'Revolutionary Personality'," p. 54.

91. See Antonius, "Fighting on Two Fronts," pp. 26–45 and Juliane Hammer, "Prayer, *Hijab* and the *Intifada*: The Influence of the Islamic Movement on Palestinian Women," *Islam and Christian-Muslim Relations* 11(3) (October 2000), pp. 299–320 for additional information on the role of mothers in the Palestinian resistance to Israeli occupation.

92. This argumentation is directed from a number of sources, including feminist scholars who view violent women as "unnatural" because women are naturally peaceful, a feminine attribute that is superior and morally virtuous. Thus, violent women are either duped by male leaders or have internalized masculine (violent) traits in lieu of female traits (nonviolence). This reasoning is shared, interestingly enough, by many conservative thinkers.

Explosive Baggage: Female Palestinian Suicide Bombers and the Rhetoric of Emotion

Abstract: *This paper examines the rhetoric of emotion surrounding the first female Palestinian suicide bombers. The influence of gender in recruitment, training and compensation by the terrorist organization are considered within the context of the tension between gender equality and tradition in Palestinian culture. The carefully-edited discourse of the bombers themselves is juxtaposed with the discounting of those statements by friends, family and the media in an attempt to understand the motivations for engaging in terror. Media coverage, particularly in the West, appears to actively search for alternate explanations behind women's participation in terror in a way that does not seem paralleled in the coverage of male suicide bombers, whose official ideological statements appear to be taken at face value.*

Terri Toles Patkin

There is a powerful psychological effect associated with being prepared to die for a cause. A *suicide bombing* is a bomb attack on people or property, delivered by a person who knows the explosion will cause his or her own death. Although the concept predates the label (suicide attacks occurred in the ancient world, kamikaze pilots in World War II chose to die for their country), the term became popularized in 1983 after an explosives-laden pickup track crashed into a Beirut, Lebanon, facility housing U.S. Marines. However, the use of suicide operatives in nationalist terror organizations in recent decades marks a change from the 1960s and 1970s practice of conserving manpower by carrying out attacks while keeping operatives at a safe distance (Lewis, 2003). Suicide bombings—inexpensive, effective, media-friendly and with a built-in intelligent guidance and delivery system—are chillingly effective as psychological warfare (Hoffman, 2003). Suicide bombing redefines basic cultural relationships and merges private, psychological motivations with public, ideologically-charged actions. Killing oneself is no longer an act of self-destruction (*intihar*), but rather divinely commanded martyrdom (*istishad*) in defense of the faith (Stern, 2003).

Today, the Arab press generally refer to a suicide bomber as a *human bomb*. The Bush administration briefly tried to get journalists to use the term *homicide bombing*, but it did not gain currency (Suicide Bomber, 2003; Suicide Bombing, 2003). Suicide bombers are not suffering from clinical depression or emotional difficulties; they perceive themselves as fulfilling a holy mission that will make them martyrs. The action is not "suicide" but rather "martyrdom" and thus does not violate religious prohibitions against killing oneself (Atran, 2003; Lewis, 2003; Reuter, 2004; Schweitzer, 2000).

The tactic was introduced into Palestinian areas gradually starting in the late 1980s. Hezbollah pioneered the use of suicide bombing, claiming responsibility for attacks on the U.S. Marine barracks in Beirut (1983), the hijacking of TWA flight 847 (1985) and a series of lethal attacks on Israeli targets. Like many other Islamist organizations, Hezbollah engages in both guerrilla warfare against Israeli military targets and terrorism targeting the civilian population, as well as sponsoring social programs for the Palestinian population (Byman, 2003).

By the mid-1990s, Hamas, Islamic Jihad, and Hezbollah had all used suicide bombings as a means to derail the Oslo peace process. Palestinian terrorist groups in Israel during this period also included Palestinian Islamic Jihad, Islamic Resistance Movement (Hamas), Umar al-Mukhtar Forces, Al-Aqsa Martyrs Brigade, and Salah al-Din Battalions (Office of the Coordinator for Counterterrorism, 2000). The Islamist agenda shared by many of these organizations has led to the "second intifada," during which suicide bombings have escalated. There have been more volunteers for suicide attacks (including women) and planning for each attack has been less rigorous than in the past (Atran, 2003).

As the Palestinian point of view shifted from negotiation about specific tracts of land to a no-compromise drive toward a final victory, the psychology of terrorism shifted from martyrdom as a means to martyrdom as an end (Brooks, 2002). The Palestine Liberation Organization's (PLO) goal is the creation of a secular state; Hamas and similar organizations merge religion with political and social activism and add terrorist activities to the mix (Reuter, 2004). For example, "Hamas calls all Muslims to give up their secular culture and lifestyles and return to religious observance: prayer, fasting, Islamic dress, moral and social values to re-create a proper Islamic society so that Muslim society can again become strong and wage a successful jihad to liberate Palestine from Israeli control" (Esposito, 2002: 95–96).

Terrorist activity in Israel is organized; suicide bombings do not represent the actions of lone, crazed individuals. Suicide terrorism is perceived by the sponsoring organizations as most painful way to inflict damage on the Israeli occupation forces, and a way to make the cost of the conflict unbearable. The movement recognizes that it does not have a nuclear arsenal, tanks or rockets, but says that its "exploding Islamic human bombs" are far superior (Hassan, 2001). Two-thirds of all suicide bombings in Israel have occurred in the past three years (Burns, 2003; Hoffman, 2003; IDF Spokesperson, 2002). The attacks take place in shopping malls, on buses, in supermarkets, in restaurants and cafes, on street corners—places where the fabric of everyday life is suddenly rent by an explosion, blood and terror.

What Makes Terrorists Tick: Motivations Sacred and Secular

The dynamics of the terrorist group shape individual behavior, giving many members a strong sense of belonging, of importance, and of personal significance (Post, 1990). Suicide bombers often articulate a sense of personal, sacred mission. When Hezbollah introduced suicide bombing as a tactic in the mid-1980s, it soon became clear that the religious fervor of the bombers could help the organization compensate for its small numbers and inadequate military capabilities (Kramer, 1990).

Resentment and self-righteousness are often considered to be the underlying motivators for engaging in terrorism. Perceiving themselves as victims, the terrorists hone a hypersensitive awareness of slights and humiliations inflicted upon themselves or their particular group, and picture themselves as part of an elite heroically struggling to right the injustices of an unfair world. Terrorists share several characteristics: "oversimplification of issues, frustration about an inability to change society, a sense of self-righteousness, a utopian belief in the world, a feeling of social isolation, a need to assert his own existence, and a cold-blooded willingness to kill" (Davis, 2001: n.pag.). According to the Palestinian Authority, the typical suicide bomber (prior to the Second Intifada and the September 11 attacks) fit a standard profile: young, male, unemployed, with few prospects economically or socially, mildly religious. He is persuaded to join the movement because of both pragmatic and ideological reasons; the allure of martyrdom may in fact take second place to the very tangible economic and social benefits his family will receive after his action (Reuter, 2004; Stern, 2003). Post September 11, the profile has become less clear, with men, women and even children being included. All strata of society are represented, all marital statuses, all educational levels (Hoffman, 2003).

The myth that suicide bombers are driven to their actions by the frustration stemming from poverty and ignorance is exploded by the actuality that today's Palestinian bombers tend to be well educated and relatively economically stable (Atran, 2003; Brooks, 2002; Hassan, 2001; Stern, 2003). While cash payments from abroad to families of suicide bombers continue, now all levels of the economic and educational spectrum are represented (Stern, 2003; Tierney, 2002). Despite well-publicized photos of families holding checks for as much as $25,000, the bomber's family may receive little direct financial incentive (Reuter, 2004). Often, the bomber has a close friend or family member who has been killed by Israeli soldiers or has spent time in Israeli custody, but the most crucial factor appears to be loyalty to the terrorist organization, about which members speak in family metaphors (Atran, 2003; Brooks, 2002; Hassan, 2001; Stern, 2003).

Religious terrorists believe their goals and activities are sanctioned by divine authority. Martyrdom, the voluntary acceptance of death as a demonstration of religious truth, is a concept central to Islam. Suicide bombers view themselves as martyrs fighting a *jihad* against their heretic, apostate opponents (Rapoport, 1990). Transforming oneself into a living bomb is perceived as the equivalent of using a gun against one's enemies. The struggle is much the same, the only difference being one of chronology: the bomber dies *while* killing several enemies rather than *after* doing so (Kramer, 1990). Being "ready to die" is not the same as "seeking to die." Further, some whose death is interpreted as suicide may have been tricked into the action by those controlling their activities ("remote-control martyrs") (Esposito, 2002; Lewis, 2003; Merari, 1990).

Suicide bombing to date always occurs within the context of a terrorist organization; bombers do not act individually (Hassan, 2001; Hoffman, 2003). Personal revenge is not the primary motivator for Palestinian suicide bombers, and such an impetus in fact would negate the promise of martyrdom. There is not one instance of a lone, crazed Palestinian who has gotten hold of a bomb and set off to kill Israelis or even of an independent suicide bomber acting without the support of an established organization (Atran, 2003; Brooks, 2002; Victor, 2003).

It is, of course, difficult to ascertain what terrorists are "really" thinking or what "really" motivates them, especially considering the tendency of terror organizations to maintain high levels of secrecy and the contextual situation of long-standing sociocultural conflicts (Hassan, 2001). Similarly, it is easy to misinterpret the happy expressions often seen on the faces of suicide bombers. A smile may mean contemplation of eternal paradise or it may represent satisfaction that the individual has helped the organization advance their goals one step forward (Kramer, 1990). Of necessity, this analysis relies on secondary sources. It is important to note that, with few exceptions, we do not hear the voices of the women involved in terrorist organizations themselves, except in the officially sponsored and edited video testaments that the martyrs leave behind. Perhaps inevitably, we cannot know with certainty the extent of their ideological fervor, nor can we pinpoint their emotional and cognitive responses to engaging in terror. We are left only with observations of behavior in public, i.e. the actual suicide bombing or attempt, and the post-detonation interpretations of family and friends, and so must extrapolate all manner of important background as we reconstruct the influences leading up to the terrorist act.

The Care and Feeding of Terrorist Trainees

The terrorist recruitment process is complex. Hamas and Islamic Jihad do not accept all the volunteers for martyrdom who approach them; in fact, leaders call fending off the crowds insisting on retaliatory human bombing missions their biggest problem lately (Hassan, 2001). Until recently, all recruits have been male. The sponsoring organizations usually (but not always) reject those under 18, and those who are married or the sole wage earners in their families. Siblings are not accepted together. Pious youths who can be discreet among friends and family are preferred as are those who could "pass" as an Israeli Jew for long enough to infiltrate into the targeted area.

Training for *jihad* includes instruction in small arms practice, cartography, targeting, mines, and demolitions and poisons, as well as religious instruction and prepackaged justifications for killing Americans and Jews (Olcott and Babajanov, 2003). New recruits are asked to provide their reasons for volunteering and assure the trainer of the seriousness of their intentions. At the same time, the organization interviews the recruit's friends and family members in order to ensure that the newcomer is not an Israeli spy (Victor, 2003). Appropriate disguise and demeanor are discussed and practiced during training sessions (Reuter, 2004) and recruits are sometimes placed in situations where their lives are in danger in order to assess their responses (Victor, 2003).

Recruits view suicide missions as the shortest path to heaven. If their joy at attaining paradise ever falters, the "assistants" who constantly accompany the trainees remind them of the pain associated with sickness and old age, encourage them to re-enact previous terror operations, and assure them that death will be swift and painless and that the doors to paradise beckon. In actuality, fear of impending death is not an issue for recruits as much as awe: the prospective martyr expects to attain paradise imminently and is anxious that something might go wrong and keep him from the presence of Allah (Brooks, 2002; Hassan, 2001).

Heaven is conceptualized as a place of perfection, a lovely garden containing trees, fruit orchards, animals, exquisite foods, beverages, clothing and scenery (Stern, 2003). One lives in a beautiful home with a pleasant smell of perfume, with servants attending to one's every need, and family members from this life and the next close at hand (Rewards Promised to Suicide Bombers in Paradise, 2002). The martyr achieves atonement for all of his sins with the first drop of blood shed, and ten minutes after martyrdom weds 72 beautiful dark-eyed virgins whose home is in heaven. The 72 virgins are actually the reward for every believer admitted to paradise, according to mainstream Islamic theology, and the pleasures they offer are not sensual. But that doesn't make the prospect any less appealing to teenage boys. Indeed, the "Israeli Defense Forces report that one of the suicide bombers whose attack they managed to prevent had wrapped toilet paper around his genitals, apparently to protect them for later use in paradise" (Stern, 2003: 55). (As it turns out (Stern, 2003), the promise of 72 virgins (*houri*) may result from a mistranslation of the word *hur* (white raisins, an ancient regional delicacy).

Trainees pray and read sections of the Koran dealing with themes of jihad, war, Allah's favors and the importance of faith. They attend religious lectures for two to four hours a day, fast, pay off their debts, and ask for forgiveness for actual or perceived offenses. After studying for many months, the candidate is titled *al shaheed al hayy*, the "living martyr" or "one who is waiting for martyrdom." In the days preceding an operation, the candidate prepares a will (on paper, audiotape or video), emphasizing the voluntary nature of the mission and exhorting others to imitate him. He repeatedly watches his own video as well as those of others, growing more comfortable with the idea of death (Hoffman, 2003). He may also view videos showing Israeli attacks on Palestinians to bolster his resolve (Victor, 2003).

On the designated day, he completes a ritual bath, puts on clean clothes and tucks a Koran in the left breast pocket above the heart, prays, and straps on the explosives or picks up the briefcase or bag containing the bomb. The trainer wishes him success so that he will attain paradise and the trainee responds that they will meet in paradise. As he pushes the detonator, he says "*Allahu akbar*" ("Allah is great. All praise to him.") (Hassan, 2001). Afterwards, the sponsoring organization pays for the *shaheed*'s (martyr) memorial service and burial as well as making financial contributions to the bomber's family (Reuter, 2004).

The Changing Role of Female Recruits in Terrorist Organizations

Women have participated in terrorist groups worldwide, but their relatively low numbers and roles often centering on support of their male colleagues have diminished onlooker perceptions of their importance. Women tend to be more actively involved in nationalist/secular terror organizations rather than Islamist/religious groups. Women in Palestinian groups are often enthusiastic about their increased roles, especially as the conflict with Israel deepens (Cunningham, 2003).

The arguments for women to join armed forces have been well-developed: the defense of one's country is the duty and right of all citizens, women need to participate in the military if they are to have real equality with men, it will give women more self-confidence and might benefit the armed forces. The counter-arguments have been equally well-developed: engaging in violence does not serve women's (or anyone else's) ultimate interests, women have too many other burdens (such as childcare) in society, women as givers of life should not be involved in taking life (Brock-Utne, 1985). Female involvement in historical terrorist or revolutionary uprisings has also been well documented (Schweitzer, 2000). Women have been responsible for significant numbers of the suicide bombings carried out by the Liberation Tigers of Tamil Eelam in Sri Lanka and the Kurdish Workers' Party PKK.

In some cases, women accounted for as many as 66% of the suicide bombings completed by the organization (Stern, 2003).

Still, most terrorist organizations are androcentric, and many are rooted in fundamentalist religious ideologies which require the exclusion of women from public life, especially in Islamic contexts. Although women are often motivated to join terrorist movements by the same political and economic concerns as men, they also join or are encouraged to join revolutionary struggles on behalf of their practical and strategic gender interests (Peterson and Sisson, 1999). Women may also play a supporting role in terrorism in their traditional roles as wives and mothers by nurturing families committed to terrorist causes, willingly sacrificing their children to militarist actions and engaging in peripheral activities such as carrying supplies or messages for terrorist groups. While men may support terrorism from a desire to bring about social justice, women are more often found to articulate "private" concerns such as using terrorism as a means to protect their families, homes and communities (Caiazza, 2001). Although patriarchal norms often preclude women from militaristic actions and limit their public roles, some women have taken part in terrorism when there are few perceived outlets for gender equality (Caiazza, 2001; Elshtain, 2003). Indeed, the rationale offered for engaging in terrorism may center on defending the "purity" of women in traditional roles. However, there is increasing evidence that women in terrorist organizations are moving away from the traditional "support" role (FBI Warns of Female Terror Recruits, 2003; Lewis, 2003).

Women who join terrorist groups tend to be older and better educated than their male counterparts. And yet, perceptions of women's motivations for terrorism continue to be colored by the notion that women are emotional and irrational, perhaps even driven by hormonal imbalances; rarely have their actions been interpreted as intelligent, rational decisions. "The average depiction of women terrorists draws on notions that they are (a) extremist feminists; (b) only bound into terrorism via a relationship with a man; (c) only acting in supporting roles within terrorist organizations; (d) mentally inept; (e) unfeminine in some way; or any combination of the above.... She is seldom the highly reasoned, non-emotive, political animal that is the picture of her male counterpart; in short, she rarely escapes her sex" (Talbot, 2001: n.pag.). But when one asks women themselves about their terrorist activity, they do not perceive their involvement as passive; they regard themselves as empowered political actors, not as auxiliaries to their more self-aware male counterparts (Talbot, 2001).

Al-Qaeda is reported to have begun recruiting women for terrorist attacks following successful Chechen and Palestinian operations (FBI Warns of Female Terror Recruits, 2003). This would, of course, expand their personnel, but the concept is ironic, given Al-Qaeda's historical link with the socially repressive policies towards women enacted by the Taliban in Afghanistan. Women drawn to terrorist organizations for support of societal and ideological change may also hold a parallel desire for change in private role behaviors. Political changes on the societal level are often reflected in the home. Men in terrorist groups often report a desire to return to "authentic" traditional role relationships where women have a stronger interest in attaining social equality (Cunningham, 2003). The emotional baggage surrounding cultural gender roles cannot be jettisoned as easily as a bomb-filled suitcase on a crowded city bus.

Although women have been involved since the beginning of the struggle between the Israelis and the Palestinians (in 1960, one woman hijacked a plane, others have successfully planted bombs in various locations), recently women's roles have escalated. Suicide bombing represents the next step after completing assignments to plant and detonate bombs without injuring oneself (Cunningham, 2003). Until recently, female suicide bombers were extremely rare among Muslims, and some fundamentalist Islamic terror groups do not even now permit women to take part in terrorist activities, particularly not suicide operations. Historically, Hamas and Islamic Jihad were adamant that women should not participate in violent demonstrations but rather remain at home in their established roles as mothers and homemakers, donning traditional dress and head coverings (Victor, 2003).

But in 2002, Yasser Arafat gave his "army of roses" speech in which he called upon women to join as equals in the struggle against Israel, coining the term *shaheeda*, the feminine of the Arabic word for martyr (Victor, 2003). That same afternoon, Wafa Idris became the first female Palestinian suicide bomber (Reynolds, 2002; Tierney, 2002). Soon afterward, Al-Aqsa Brigades actively began recruiting women as suicide bombers, opening a woman's suicide unit in Idris' honor (Victor, 2003). Palestinian women are recruited by men (brothers, uncles, teachers or religious leaders), not by other women, although young girls now look to the mediated images of the first female suicide bombers as role models (Victor, 2003). The men persuade the female recruits that the most valuable thing they can do with their life is to end it; a suicide bombing often provides a dual function as an attack against Israel and a redemption of personal or family honor, a highly salient value in Palestinian culture (Victor, 2003).

If anything, women may be seen as holding a deeper commitment to the "cause" than men, due to the emotive soul-searching that shapes their decision to participate. While not all women who apply are accepted, those who are believe it is their duty to volunteer for their country. "I could die at any time, so I will die for my people" one trainee says (Tierney, 2002: n.pag.). Women terrorists are likened to a lioness protecting her cubs; it is said that the woman views her cause as a surrogate child. Indeed the presence of women as terrorist actors may play on cultural images of victimization just as much as retaliatory strikes do (Talbot, 2001).

Ironically, the perception of female weakness can increase a woman's effectiveness in terror operations. Many female terrorists have exploited male assumptions about the "innocent woman" as a way to evade search and detection by predominantly male military forces, sometimes reverting to voluminous traditional dress, other times using fictive pregnancy or even real infants to hide explosive equipment. Other stereotypes, such as age, are also utilized. Soldiers may ignore an old woman egging on stone throwers, feeling it more important to capture the young boys following her instructions (Talbot, 2001). Israeli Security Sources admit (2002, 2003b) that, especially when dressed in Western clothes with modern hairstyles or maternity clothes, women can exploit the presumption of innocence, and soldiers may be hesitant to perform thorough body searches of women passing through checkpoints.

However, this strategy can also backfire for the sponsoring terrorist organization. For example, Thawiya Hamour, 26, decided to abort her suicide mission at the last moment, "claiming her operators directed her to dress provocatively like an Israeli woman, such as wearing her hair down, using heavy makeup, and donning tight pants. During media interviews Hamour stated, 'I wasn't afraid. I'm not afraid to die. I went for personal reasons. However, I did not want to arrive 'upstairs' for impure reasons. I did not want to dress that way, because it is against my religion.'" (Israeli Security Sources, 2003b: n.pag.). Not surprisingly, checkpoint security guidelines have evolved in response to terrorists' use of gender expectations (Victor, 2003).

Media Images and the Rhetoric of Domesticity

Communication serves as a dynamic foundation for interaction among individuals, a systemic process through which meanings are created and relationships formed. The connection between gender roles and the structures, vocabularies and styles of using language has been well documented. Women are more concerned with inner feelings, relational issues, nurturance and emotional support; men are more concerned with sharing activities than sharing feelings. Women's relational perspective contrasts with men's contextual perspective (Gilligan, 1982; Lakoff, 1976; McConnell-Ginet, 1980; O'Barr and Atkins, 1980; Tannen, 1990; Wardhaugh, 1986: Wood, 1994). This difference is intuitively exploited in media reports about terrorist motivations.

Terrorist organizations typically plan their activities to achieve the greatest media coverage and most positive spin for their story. The rhetoric surrounding suicide bombers makes good use of the metaphors of domesticity. Small training cells operate in isolation from one another (Reuter, 2004), and one's training cell of three to eight other terrorists becomes a "family of fictive kin for whom they are willing to die as a mother for her child or a soldier for his buddies" (Atran, 2003: 11). Indeed, the very point of suicide terrorism is an attack on the opposition's comfortable domesticity; the killer's goal is to disrupt everyday life to the greatest degree possible. The warm image of family distracts from the cold reality of carefully premeditated mass murder.

Following a bombing, even in the midst of personal grief, the terrorist's "family and sponsoring organization celebrate his martyrdom with festivities, as if it were a wedding. Hundreds of guests congregate at the house to offer congratulations. The hosts serve the juices and sweets that the young man specified in his will. Often, the mother will ululate in joy over the honor that Allah has bestowed upon her family" (Hassan, 2001, n.pag.). After a "successful" suicide bombing, the bomber's parents and family members are often interviewed on television, proudly expressing their joy at their child's martyrdom and even their readiness to send another child off to the afterlife should the opportunity present itself. Parents may distribute candy to neighborhood children in celebration of their child's martyrdom (Victor, 2003). "We are receiving congratulations from people.... Why should we cry? It is like her wedding today, the happiest day for her," (Murphy, 2003: n.pag.) said a brother following Hanadi Tayseer Jaradat's death.

The image is of parents so wronged and humiliated by the Israelis that they would rather sacrifice their children than continue to endure. Both mothers and fathers present this public face, downplaying the material and status benefits associated with having a martyr in the family and utilizing their personal tragedy as an illustration of the organization's ideological stance. While some parents assure the Western media that they would have stopped their child from committing suicide if they have known of the child's plan, they simultaneously express pride in the child's final act (Copeland, 2002). However, there is some evidence that parents, especially mothers, recognize this dissonance and experience a delayed grief reaction that must of necessity be masked from the Palestinian media (Victor, 2003). Privately, some parents will admit that they had "other plans" for their child than martyrdom (Reuter, 2004). These "other plans" may reflect traditional values: one mother, whose daughter did not complete her mission, expresses relief: she would have been "proud" of a son who martyred himself, because that is "normal," but her unmarried daughter should aspire to marriage and children, not martyrdom (Victor, 2003).

Following the attack, the organization distributes copies of the martyr's audio or video testament to the media and to local organizations in addition to the posed photographs, often posted on billboards, that the terrorist organizations use for recruiting purposes. "The video testaments, which are shot against a background of the sponsoring organization's banner and slogans, show the living martyr reciting the Koran, posing with guns and bombs, exhorting his comrades to follow his example, and extolling the virtues of jihad" (Hassan, 2001, n.pag.). These, of course, also make it that much more difficult for the "living martyr" to back out of his commitment to be used by the organization in a "sacred explosion" (Brooks, 2002; Hoffman, 2003; Stern, 2003).

The videos follow a standard format, although the *shaheed* or *shaheeda* is permitted to choose from a variety of common backdrops such as a plaster model of the Al-Aqsa mosque, various organizational flags, weapons and the like. Typically, the bomber stands before a flag, holding a Koran and an automatic weapon, and reads a script that may have been written by the bomber or by the handler. The speech discusses motivations for the bombing and leaves messages of hope and inspiration for surviving family members (Reuter, 2004; Victor, 2003). There may also be a private testament left for family only, and while this appears to reflect a more authentic view of the individual's motivations, the private testaments are for obvious reasons more difficult to obtain for analysis (Reuter, 2004).

Posters and calendars glorifying the "martyr of the month" reinforce the culture of martyrdom along with chants, slogans, graffitti and victory gestures. These not only legitimate the organization's success in the specific attack, but provide encouragement to other young people, background for sermons in mosques, and material for posters, videos and demonstrations (Brooks, 2002). Suicide bombing is particularly well-suited to the television age—from the compelling footage of the bomber's last farewells to the graphic images of death and destruction to the marches and celebrations after the attacks, from the newspaper announcements of the weddings between bombers and the dark-eyed virgins in paradise to the displays of material wealth the family acquires from the cash awards, a suicide bombing makes for gripping media drama (Dickey, 2002; Hassan, 2001). Names of terrorists may even appear in crossword puzzles as answers to clues such as "famous Palestinian martyr" (Marcus, 2002). Martyrdom is emerging as the short road to celebrity for impoverished teens with few prospects.

For instance, the Palestinian Authority immediately turned Wafa Idris, the first female suicide bomber, into a heroine, holding a demonstration in her honor with young girls carrying posters illustrated with her picture and eulogizing her with "great pride" (Marcus, 2003). Music videos morphed images of a woman singing into a uniformed female warrior proclaiming her willingness to die as a martyr, a concert honoring Idris has been broadcast repeatedly, and summer camps for Palestinian girls were named to honor Idris and other female suicide bombers (Marcus, 2003).

To the terrorist, the identity of the victims is incidental to the larger purpose of gaining publicity for the terrorist agenda and instilling fear in the public at large. Moral justification can be offered through theological gymnastics that diffuse and displace responsibility for murder, minimizing the consequences and dehumanizing the victims. One such psychosocial strategy centers on the use of euphemistic labeling: terrorists become "freedom fighters" engaging in "operations" that cause "collateral damage" to an "occupying force." Such sanitizing language cleans up audience evaluations of the terrorists' actions (Bandura, 1990). Sanitized language allows participants to divorce themselves from accountability and perhaps to tame the uncontrollable forces associated with dangerous technologies. Just as imagery that domesticates and humanizes weapons make it possible for use to think about the unthinkable precisely because that language makes domesticity, the warm and playful, even sexuality, part of the technological world (Cohn, 1987a; Cohn, 1987b; Cohn, 1990), so too can terrorism be presented in a positive light.

Potential bombers who complete the training but have a change of heart before the final operation appear to take a broader view of the consequences and implications of the act. Tauriya Hamamra, who rejected her orders to dress in modern clothing, said "I began to think about killing people—babies, women, sick people, and to imagine my family sitting in a restaurant and someone coming in and blowing them up ... God would not see it as a good reason for committing suicide and therefore would not accept me as a *shaheed*" (Shin Bet, 2004: n.pag.). Hamamra accused her operators of "making a business out of the blood of *shaheeds*" (International Christian Embassy Jerusalem, 2002: n.pag.). Similarly, Arin Ahmed decided not to follow through on a bombing when she was unable to depersonalize her victims, but instead began to see them as people who looked like a friend, an aging grandmother, etc. "I suddenly understood what I was about to do and I said to myself, 'How can I do such a thing?'" (Fields, 2002: n.pag.).

How *can* suicide bombers do such a thing? How does "Suha" come to the conclusion that "you don't think about the explosive belt or about your body being ripped into pieces. We are suffering. We are dying while we are still alive ... I am prepared to sacrifice my life for the cause" (Zoroya, 2002: n. pag.), while others pull back from the brink of self- and other-destruction?

What little we know of the motivations of female Palestinian suicide bombers emerges from the juxtaposition of the carefully edited video testaments with their practiced statements and the public interpretations of the bombers' motives offered by family and friends to reporters. We are left with a jarring disconnect between the discourse of the bombers themselves and the discounting of those statements by friends, family and the media as we attempt to understand the action. Chesler (2004) asks whether Palestinian female suicide bombers are willing participants in terror or victims of indoctrination, force or clinical depression. Are they victims of honor killings, atoning for cheating on their husbands or becoming pregnant or being raped? While the male terrorist is pictured as a "living weapon" the ultimate in macho potency (Morgan, 2002), the female terrorist is often suspected of joining the movement for emotional or social reasons.

Palestinian women have historically been among the least bound by traditional roles in Arab society, and some may see martyrdom as a way to achieve equality and fight powerlessness (Copeland, 2002). However, women in Palestinian culture embody the honor of their family, and any hint of impropriety may have serious consequences (Elshtain, 2003; Hassan, 2001). "The unmarried Palestinian woman today lives under a stringent set of social and religious rules: if she is too educated, she is considered abnormal; if she looks at a man, she risks exclusion; if she sleeps with a man, and especially if she gets pregnant, she disgraces the family and risks death at the hands of her male relatives" (Victor, 2003: 193). Terrorism may be a means of rehabilitating one's personal or family status (Israeli Security Sources, 2003b). Palestinian women voluntarily engaging in terrorism are described in media accounts as having a large amount of "personal baggage." They are portrayed as divorced, barren, influenced by brothers, uncles or other male family members, grief-stricken from the death of a friend or relative, wanting to clear the family's name following a drop in status (either personal or that of a family member), romantically attached to other terrorists, or suicidal following a broken love affair and exploited by the organization (Israeli Security Sources, 2003b). While there are individual differences, the attribution of personal and social motives appears to dominate.

While each of the female Palestinian suicide bombers to date arrived at the decision to self-detonate by a unique path, their official statements share the same general tone as the officially-sponsored discourse as their male counterparts. Media coverage, particularly in the West, appears to actively search for alternate explanations behind women's participation in terror in a way that does not seem paralleled in the coverage of male suicide bombers, whose official ideological statements appear to be taken at face value. In the case of the relatively few female terrorists, media coverage profoundly emphasizes the emotional over the ideological in an effort to provide comprehensible explanations.

For example, media coverage of trend-setting female suicide bomber Wafa Idris (detonation 1/27/02) focused on her roles as a good friend and a loving daughter who volunteered with the Palestinian Red Crescent and had twice been hit by plastic-coated bullets in the line of duty. Friends say she was haunted by the terrible things she'd seen, but still wondered if she chose to die because her marriage had broken up (Beaumont, 2002). Although her sister reported that Idris used to say that she wanted to die as a martyr, her family expressed surprise at learning of her terrorist links. They said she was a cheerful if sometimes hot-tempered young woman. She and her husband had divorced when it became clear she could not have children after a miscarriage (Victor, 2003). Idris may have been depressed, stating (Victor, 2003: 196): "I have become a burden on my family. They tell me they love me and want me, but I know from their gestures and expressions that they wish I didn't exist."

Rather than exploring her ideological motivations, however, reporters struggled to uncover a domestic explanation for her actions: "She moved back to the family home, where her corner of one room was dark and simple—a battered teddy bear sat on a table. On another sat a can of hair foam and a brush" (Female Suicide Bomber Wanted to the a Martyr, 2002: n.pag.). Her attack resonated throughout the Arab world, sparking editorials celebrating her heroism and saying she also exploded myths about women's weakness. Even though there was some ambiguity about whether Idris was meant to merely plant the bomb or perform a suicide mission—she never made the customary suicide video but a video with very poor production values featuring a fully-veiled women speaking in a muffled voice has been attributed to her (Victor, 2003)—Al-Aqsa quickly realized the efficacy of using women as suicide bombers (Cunningham, 2003).

The second female suicide bomber, Dareen Abu Aysheh, 21 (detonation 2/27/02), was a student who highlighted the role of women in the struggle against the Israelis in her video (Palestinian Women Martyrs Against the Israeli Occupation, 2004). An independent-minded scholar and a feminist who planned to become a university professor of English literature, she had been resisting strong family pressure to marry and bear children for some time. During a humiliating encounter at an Israeli checkpoint, her honor was stained when she was forced by soldiers to kiss a male cousin in order to save a baby's life. However, she rejected the cousin's later offer of marriage in order to preserve her reputation. The event seems to have crystallized her rage at the occupation, and she accepted the cousin's offer of an alternative plan to avoid family disgrace, i.e. becoming a *shaheeda* (Victor, 2003). Dareen Abu Aysheh said in her suicide video, "Let Sharon [the Israeli Prime Minister] the coward know that every Palestinian woman will give birth to an army of martyrs, and her role will not only be confined to weeping over a son, brother or husband instead (sic), she will become a martyr herself" (Palestinian Women Martyrs Against the Israeli Occupation, 2004).

Ayat Akhras, 18 (detonation 3/29/02), the youngest female suicide bomber to date, had the previous day sat with her fiancé and talked about getting married after graduating from high school (Palestinian Women Martyrs Against the Israeli Occupation, 2004). In her video, Akhras stated "I am going to fight instead of the sleeping Arab armies who are watching Palestinian girls fighting alone" (Copeland, 2002: C01). Although she was known for her intense interest in political matters, observers indicate that she may have been motivated in large part by the disgrace her family had faced in the Palestinian community when her father refused to quit his job working for Israelis (Victor, 2003). Again the words of the bomber herself are discounted by observers in favor of an emotion-based interpretation.

Andaleeb Takafka, 20 (detonation 4/12/02), was concerned with the suffering of the Palestinian people (Palestinian Women Martyrs Against the Israeli Occupation, 2004). Her latent motives have been identified as apolitical, however. She had long collected movie magazines and posters of celebrities, but in the months before her bombing she replaced those with posters of martyrs, especially Wafa Idris. Takatka indicated to her handlers that she viewed martyrdom as a road to celebrity (Victor, 2003).

Hiba Daraghmah, 19 (detonation 5/19/03), was a student of English literature who showed the world her unveiled face for the first time on the Islamic Jihad poster released after her death (Palestinian Women Martyrs Against the Israeli Occupation, 2004). She became very religious and began wearing traditional dress after being raped by an uncle at the age of fourteen, and her decision to kill herself may have resulted from long-term psychological trauma from the episode (Victor, 2003).

Hanadi Tayseer Jaradat, 29 (detonation 10/4/03), was an attorney who may have been motivated by revenge for the killing of her younger brother and cousin by Israeli forces in the raid on Jenin (Palestinian Women Martyrs Against the Israeli Occupation, 2004). "By the will of God I decided to be the sixth martyr who makes her body full with splinters in order to enter every Zionist heart who occupied our country. We are not the only ones who will taste death from their occupation. As they sow so will they reap," she said in her video (Toolis, 2003: n.pag.). Her family reports that Jaradat was "inconsolable" following her brother's death and that her religiosity had strengthened in recent weeks (Murphy, 2003). The only way to release her emotions was, apparently, to become a suicide bomber.

Reem Salih al-Rayasha, 21 (detonation 1/14/04), was a university student from a wealthy family who was said to love her two children dearly (Palestinian Women Martyrs Against the Israeli Occupation, 2004). She was photographed with her two small children prior to her attack but noted that motherhood does not compare to the ability to "turn my body into deadly shrapnel against the Zionists and to knock on the doors of heaven with the skulls of Zionists" (Myre, 2004: n.pag.). She said she always wanted "to carry out a martyr attack, where parts of my body can fly all over" (Suicide bombing kills 4 Israelis, 2004: n.pag.) and continues (Myre, 2004: n.pag.), "God gave me the ability to be a mother of two children who I love so. But my wish to meet God in paradise is greater, so I decided to be a martyr for the sake of my people. I am convinced God will help and take care of my children." Observers struggled to find a motive behind al-Rayasha's attack: she was a wealthy woman, married with children, and had no close friends or family members to avenge (El-Haddad, 2004). However, a number of press reports indicated that she may have carried out the attack in order to atone for an affair she was having with a married man (O'Loughlin, 2004).

Women's Role in the Culture of Martyrdom

To some extent, it does not matter whether these seven women decided to commit acts of terror because of external social pressures or internal ideology, or whether the media spin simply reflects the outsider's need to make sense of the apparent contradiction between "nurturing female" and "calculating killer". In either case, the point remains: the women believe. This belief may uphold the rhetoric of martyrdom or it may sustain the cultural values that encourage individuals to sacrifice themselves for personal or family honor. But in either case, the decision to become a suicide bomber reflects a lifetime of immersion in a culture that regards terrorism as an acceptable behavioral choice, and is as voluntary as any culturally-influenced choice may be.

Today's Palestinian children experience anticipatory socialization into terror from early childhood; infants have even been proudly dressed as mini-bombers by their parents (Atran, 2003). The innocent look of these young people combined with their susceptibility to persuasion makes them a likely target for terrorist influence. Recruitment of young people in schools, camps and through pervasive cultural support has grown stronger over time (Israeli Security Sources, 2003a). Polls show that 70 to 80 percent of Palestinians now support this postmodern culture of terror—far more than ever supported the peace process (Brooks, 2002; Stern, 2003). Young girls join boys at playing at suicide missions, and an eight-year-old girl may calmly sit at the dinner table and announce her intention to become a *shaheeda* (Reuter, 2004). Six year old girls in class offer their reasons for wanting to become martyrs: "to have everything in Paradise ... to kill the Jewish ... to live near our God ... we never die" (Victor, 2003: 185).

Twelve year old girls are even more articulate. They hope to become martyrs in order "to follow my brother ... to my country everything I can ... to free my people from occupation ... there is no hope for peace" (Victor, 2003: 188–189). A "good" Palestinian girl may ask for an automatic rifle as a wedding gift, as did Jasmeen, who said, "I do not want gold, or a diamond ring, or jewelry, but rather a M-16, and if only I can acquire this I will wish for no more to be paid by my fiancé" (Marcus, 2002: n.pag.). But it is not clear that young children really understand the meaning behind the rhetoric about "travelling to Paradise." Shireen Rabiya, 15, who was captured by the Israelis before she could complete her suicide mission, says "It sounded like fun. It sounded exciting and so many others had done it or tried that I thought, why not me?" (Victor, 2003: 261).

To some extent, all suicide operatives are victims, not only of the terrorist organizations, but of the cultural conditioning that lures them into believing that their ultimate life purpose lies in an untimely death. Military commanders for Hamas and Islamic Jihad see the human bomb—male or female—as an inexpensive, easily targetable weapon that is uniquely capable of striking fear in Israeli hearts. "The more training a soldier receives, the more skilled he is at avoiding death, whereas the opposite is true for a suicide bomber" (Stern, 2003: 52). The routinization of suicide means that operational planning has become less intensive in recent months as the numbers of volunteers increase (Reuter, 2004). The only needed supplies are readily available and inexpensive: gunpowder, nails, a light switch, a battery, mercury, acetone, a wide belt, and transportation to the target site (Reuter, 2004; Victor, 2003). Another cost-effective reason for scheduling suicide bombings is that they eliminate the need to arrange an escape plan—often the most challenging part of a terrorist operation (Hoffman, 2003). The total cost of a suicide operation is approximately $150—and the bomber's life (Atran, 2003; Hassan, 2001). The bombings are simultaneously simple and sophisticated, the ultimate poor person's smart bomb.

From an economic point of view, the female suicide bomber is a much better investment than even her male counterpart. A volunteer female suicide bomber typically trains for a period of only two to eight weeks, depending on the woman involved, far less than the months-long male course (Cunningham, 2003). Women require less persuasion (they are considerably less inclined to be swayed by promises of virgins in paradise) and the simplicity of their missions demands little technical expertise (Tierney, 2002). Women have already made a long ideological journey before they set foot in the door of the terrorist organization; they arrive ready to take that final step. They are paid less, too: the organization that takes responsibility for the suicide attack typically gives a lifetime stipend to the family of $400 a month for male suicide bombers but only $200 a month for females (Victor, 2003).

In radical Islam, women's status as subordinate is fundamental: women are considered unclean, they must be kept hidden and their bodies covered, they must be made subordinate to men (Elshtain, 2003). Just as lower status laborers do the menial work of society—cleaning, taking out the garbage—so too can they be expended in the task of removing the enemy a few at a time. "Indeed, up till now, the children of leaders have not been involved in suicide missions, but are usually sent off to Amman, Europe or the United States to study, far from the trauma and danger of the Intifada" (Victor, 2003:114–115).

The female suicide bomber turns into a victim in the midst of what she may consider the most empowered act of her life. Her complex mix of ideological, psychological and sociological motivations is reduced by the media to a poignant struggle with her feelings as an outsider in a warm, traditional community. And if the terrorist organizations can convince women that killing themselves for the "cause" not only incorporates political and religious benefits but also serves as a way to bring personal honor to themselves and their families, they manage to isolate and eliminate the most dangerous women of all in a traditional society—those courageous enough to independently take on a previously male role and perform an ideological act in a public setting.

Understanding female terrorists as socially vulnerable victims of calculated emotional blackmail by male-dominated terrorist organizations is obvious sexism. But the refusal to admit that both men and women absorb the lessons of the Palestinian culture of martyrdom is equally limiting. In a society with restricted options and opportunities, especially for women, where children are socialized into terror from their earliest years and where martyrs attain the status of celebrities, where daily life provides endless examples of humiliation and deprivation in a culture where honor has historically been among the most salient values, where religious leaders provide elaborate theological justifications for martyrdom, is it any surprise that young people, female and male, eagerly line up for a one-way ticket to Paradise?

References

Atran, S. (2003). *Genesis and Future of Suicide Terrorism.* Retrieved 9/29/03 from www.interdisciplines.org/terrorism/papers/1/12/printable/paper

Bandura, A. (1990). Mechanisms of Moral Disengagement. In W. Reich (Ed.), *Origins of Terrorism: Psychologies, ideologies, theologies, states of mind* (pp. 161–191). Cambridge: Cambridge University Press.

Beaumont, P. (2002). Suicide Notes. *The Observer.* Retrieved 3/28/04 from http://observer.guardian.co.uk/2002review/story/0,12715,862850,00.html.

Brock-Utne, B.(1985). *Educating for Peace: A Feminist Perspective.* New York: Pergamon Press.

Brooks, D. (2002). The Culture of Martyrdom. *Atlantic Monthly,* 289(6): 18–24.

Burns, J.F. (2003, October 7). Bomber Left Her Family With a Smile and a Life. *The New York Times,* p. A13.

Byman, D. (2003). Should Hezbollah Be Next? *Foreign Affairs* 82(6):54–67.

Caiazza, A. (2001). *Why Gender Matters in Understanding September 11: Women, Militarism, and Violence.* Washington, D.C.: Institute for Women's Policy Research.

Chesler, P. (2004, January 22). Forced Female Suicide. FrontPageMagazine.com. Retrieved 6/29/04 from www.frontpagemag.com/Articles/Printable.asp?ID=11855

Cohn, C. (1987a). Sex and Death in the Rational World of Defense Intellectuals. Signs: *Journal of Women in Culture and Society* 12(4): 687–718.

Cohn, C. (1987b). Slick'ems, Glick'ems, Christmas Trees, and Cookie Cutters: Nuclear Language and how we learned to pat the bomb. *Bulletin of the Atomic Scientists* 43(5): 17–24.

Cohn, C. (1990). "Clean Bombs" and Clean Language. In J.B. Elshtain and S. Tobias (Eds.), *Women, Militarism & War: Essays in History, Politics and Social Theory* (pp. 33–55). Savage, MD: Rowman and Littlefield.

Copeland, L. (2002, April 27). Female Suicide Bombers: The New Factor in Mideast's Deadly Equation. *Washington Post*: C01.

Cunningham, K.J. (2003). Cross-Regional Trends in Female Terrorism. *Studies in Conflict & Terrorism* 26:171–195.

Davis, P.B. (2001). The Terrorist Mentality. *Cerebrum: The Dana Forum on Brain Science* 3(3). Retrieved 10/4/03 from www.dushkin.com/powerweb/0072551054/article.mhtml?Article=30849

Dickey, C. (2002, April 15). Inside Suicide, Inc. *Newsweek,* pp. 26–32.

El-Haddad, L. (2004, January 23). A Palestinian mother becomes a human bomb. Retrieved 7/8/04 from http://english.aljazeera.net/NR/exeres/554FAF3A-B267-427A-B9EC-54881BDE0A2E.htm

Elshtain, J.B. (2003). *Just War Against Terror.* New York: Basic Books.

Esposito, J.L. (2002). *Unholy War: Terror in the Name of Islam.* New York: Oxford University Press.

FBI Warns of Female Terror Recruits. (2003, April 1). Retrieved 3/28/04 from www.girlswithguns.org/news/news0005.htm

Female suicide bomber wanted to die a martyr. (2002, January 31). *Irish Examiner.* Retrieved 9/29/03 from http://archives.tcm.ie/irishexaminer/2002/01/31/

Fields, S. (2002, July 1). When a suicide bomber fails. *Jewish World Review.* Retrieved 6/29/04 from www.jewishworkreview.com/cols/fields070102.asia

Gilligan, C. (1982). *In A Different Voice.* Cambridge: Harvard University Press.

Hassan, N. (2001, November 19). An Arsenal of Believers. *The New Yorker.* Retrieved 11/11/03 from www.newyorker.com/printable/?fact/011119fa_FACT1

Hoffman, B. (2003, June). The Logic of Suicide Terrorism. *Atlantic Monthly.* Retrieved 9/29/03 from http://www.theatlantic.com/issues/2003/06/hoffman.htm

IDF Spokesperson. (2002). Suicide Terror: Its use and rationalization. Retrieved 9/29/03 from www.mfa.gov.il/mfa.go.asp?MFAH0m6k0

International Christian Embassy Jerusalem. (2002, May 31). Female Bomber Disgusted By Fatah Handlers. Retrieved 6/29/04 from http://truthnews.com/world/2002060103.htm

Israeli Security Sources. (2002). Blackmailing Young Women into Suicide Terrorism. Retrieved 9/29/03 from www.mfa.gov.il/mfa.go.asp?MFAH0n2a0

Israeli Security Sources. (2003a). Participation of Children and Teenagers in Terrorist Activity during the "Al-Aqsa" Intifada. Retrieved 9/29/03 from www.mfa.gov.il/mfa.go.asp?MFAH0n100

Israeli Security Sources. (2003b). The Role of Palestinian Women in Suicide Terrorism. Retrieved 9/29/03 from www.mfa.gov.il/mfa.go.asp?MFAH0n210

Kramer, M. (1990). The Moral Logic of Hizballah. In W. Reich (Ed.), *Origins of Terrorism: Psychologies, ideologies, theologies, states of mind* (pp. 131–157). Cambridge: Cambridge University Press, pp. 131–157.

Lakoff, R. (1976). *Language and Women's Place.* New York: Octagon Books.

Lewis, B. (2003). *The Crisis of Islam: Holy War and Unholy Terror.* New York: Modern Library.

Marcus, I. (2002, March 12). Encouraging Women Terrorists. *Palestinian Media Watch Bulletin.* Retrieved 6/29/04 from www.science.co.il/Arab-lsraeli-conflict/Articles/Marcus-2002-03-12.asp

Marcus, I. (2003, October 9). Promoting Women Terrorists. *Palestinian Media Watch Bulletin.* Retrieved 6/29/04 from www.israel-wat.com/idris_eng2.htm

McConnell-Ginet, S. (1980). Linguistics and the Feminist Challenge. In S. McConnell-Ginet, R. Borker and N. Furman (Eds.), *Women and Language in Literature and Society* (pp. 3–25). New York: Praeger Publishers.

Merari, A. (1990). The readiness to kill and die: Suicidal terrorism in the Middle East. In W. Reich (Ed.), *Origins of Terrorism: Psychologies, ideologies, theologies, states of mind* (pp. 192–207). Cambridge: Cambridge University Press.

Morgan, R. (2002). Demon Lover. Ms (Dec.2001-Jan.2002). Retrieved 10/4/03 from www.dushkin.com/powerweb/0072551054/article.mhtml?article=34717

Murphy, V. (2003, October 15). Mid-East cycle of vengeance. BBC News Online. Retrieved 7/8/04 from http://news.bbc.co.uk/2/hi/middleeast/3165604.stm

Myre, G. (2004, January 15). Gaza Mother, 22, Kills Four Israelis in Suicide Bombing. *New York Times.* Retrieved 6/29/04 from www.nytimes.com/2004/01/15/international/middleeast/15MIDE.html

O'Barr, W.M. and Atkins, B.K. (1980). "Women's Language" or "Powerless Language"? In S. McConnell-Ginet, R. Borker and N. Furman (Eds.), *Women and Language in Literature and Society* (pp. 93–110). New York: Praeger Publishers.

Office of the Coordinator for Counterterrorism, U.S. Department of State. (2001, April 30). Background Information on Terrorist Groups. Retrieved 9/29/03 from http://www.state.gov/s/ct/rls/pgtrpt/2000/2450.htm

Olcott, M.B. and Babajanov, B. (2003). The Terrorist Notebooks. *Foreign Policy* (March/April): 30–40.

O'Loughlin, E. (2004, January 27). As deadly as the male. *Sydney Morning Herald.* Retrieved 7/8/04 from www.smh.com.au/articles/2004/01/26/1075087955902.html?from=storyrhs

Palestinian Women Martyrs Against the Israeli Occupation. (2004). Retrieved 6/29/04 from www.911review.org/Wget/aztlan.net/women_martyrs.htm

Peterson, V. S. and Runyan, A.S. (1999). The Politics of Resistance: Women as Nonstate, Antistate, and Transstate Actors. In V.S. Peterson and A.S. Runyan (Eds.), *Global Gender Issues* (pp. 163–211). Boulder, CO: Westview Press.

Post, J.M. (1990). Terrorist psycho-logic: Terrorist behavior as a product of psychological forces. In W. Reich (Ed.), *Origins of Terrorism." Psychologies, ideologies, theologies, states of mind* (pp. 25–40). Cambridge: Cambridge University Press.

Rapoport, D.C. (1990). Sacred Terror: A contemporary example from Islam. In W. Reich (Ed.), *Origins of Terrorism: Psychologies, ideologies, theologies, states of mind* (pp. 103–130). Cambridge: Cambridge University Press.

Reuter, C. (2004). *My Life Is A Weapon: A Modern History of Suicide Bombing.* Princeton: Princeton University Press.

Rewards Promised to Suicide Bombers in Heaven. (2002) Retrieved 10/1/03 from www.idf.il/newsite/english/1201-3.stm

Reynolds, J. (2002). Mystery over female 'suicide bomber.' Retrieved 9/29/03 from http://news.bbc.co.uk/1/hi/world/middle_east/1788694.stm

Schweitzer, Y. (2000). *Suicide Terrorism: Development and Characteristics.* Herzliya, Israel: International Policy Institute for Counter-Terrorism.

Shin Bet, IDF nab reluctant female suicide bombers. (2004, June 29). *Ha'aretz.* Retrieved 6/29/04 from www.haaretzdaily.com/hasen/pages/ShArt.jhtml?itemNo=170364

Stern, J. (2003). *Terror in the Name of God: Why Religious Militants Kill.* New York: HarperCollins.

Suicide Bomber. (2003). Retrieved 9/29/03 from www.wordspy.com/words/suicidebomber.asp

Suicide Bombing. (2003). Retrieved 9/29/03 from www.wikipedia.org/w/wiki/phtml?Title=Suicide_bombing

Suicide bombing kills 4 Israelis. (2004, January 14). Retrieved 6/29/04 from www.cbc.ca/storyview/MSN/2004/01/14/israelbomb040114

Talbot, R. (2001). Myths in the Representation of Women Terrorists. *Eire-Ireland.* Retrieved 10/4/03 from www.dushkin.com/powerweb/0072551054/article.mhtml?article=31680

Tannen, D. (1990). *You Just Don't Understand. Women and Men in Conversation.* New York: Ballantine Books.

Tierney, M. (2002, August 2). Young, Gifted, and Ready to Kill. *The Glasgow Herald.* Retrieved 10/4/03 from www.dushkin.com/powerweb/0072551054/article.mhtml?artiele=34716

Toolis, K. (2003, October 12). Why Women Turn To Suicide Bombing. *The Observer.* Retrieved 6/29/04 from www.countercurrents.org/pa-toolis121003.htm

Victor, B. (2003). *Army of Roses: Inside the Worm of Palestinian Women Suicide Bombers.* New York: St. Martin's Press.

Wardhaugh, R. (1986). *An Introduction to Sociolinguistics.* Oxford: Basil Blackwell. Wood, J.T. (1994). *Gendered Lives: Communication, Gender and Culture.* Belmont, CA: Wadsworth Publishing Company.

Zoroya, G. (2002, April 22). Woman described the mentality of a suicide bomber. *USA Today.* Retrieved 6/29/04 from www.usatoday.com/news/world/2002/04/22/cover.htm

---

Terri Toles Patkin is Professor of Communication at Eastern Connecticut State University. Her research interests include the intersection between mass and interpersonal communication, the influence of communication on culture, organizational communication, and persuasion. Please direct all correspondence to patkin@easternct.edu.

Girls as "Weapons of Terror" in Northern Uganda and Sierra Leonean Rebel Fighting Forces

Girls—both willingly and unwillingly—participate in terrorist acts within the context of contemporary wars. These acts range from targeting civilians for torture and killing to destroying community infrastructures so that people's physical and psychological health and survival are affected. Girls witness or participate in acts such as mutilation, human sacrifice, forced cannibalism, drug use, and physical and psychological deprivation. This article focuses upon girls in two fighting forces: the Lord's Resistance Army (LRA) in Northern Uganda and the Revolutionary United Front (RUF) in Sierra Leone and their roles as combatants whose primary strategy is perpetrating terrorist acts against civilians. In analyses of gender and terrorism, girls are typically subsumed under the larger category of female, which marginalizes their experiences and fails to recognize that they possess agency and power.

SUSAN McKAY

Women's and International Studies and Nursing
University of Wyoming
Laramie, Wyoming, USA

In the majority of contemporary wars, conflicts are internal to a nation, although often with regional and sub-regional involvement. Terrorist acts are implicit strategies used in fighting within rebel and opposition movements although all sides of an armed conflict will perpetrate forms of terrorism. Such acts include torture and killing of parents, siblings, neighbors, and teachers, looting and burning property, amputations of limbs, disfigurement of body parts such as nose, lips, and ears, and gender-specific acts of rape and sexual mutilation that can be directed toward either sex but are usually female-focused.

During intra-state wars, civilians are the most frequent targets of such terror tactics. Civilian casualties, particularly women and children, are estimated to be as high as 90 percent.[1] Targeting civilians for horrific and capricious acts of terror conveys powerful political and psychological messages and creates widespread fear that is characteristic of terrorism. Such actions during armed conflicts are consistent with Deborah Galvin's definition of terrorism as "those acts and events systematically protagonized for the purpose of instilling massive fear in individuals and/or the public at large, and which are deliberately used for coercive purposes. Terrorists are those who engage in these activities, whatever form they take. Terrorism is never accidental ... [but] is deliberately aimed at the human mind through the calculated infliction of pain or loss or the threat of the same.... Terrorism is something done by people to other people."[2] In addition, purposeful destruction of the public health infrastructure—such as through damaging agricultural lands and water systems, looting health care clinics, and destroying highways and electrical sources—jeopardizes civilians. This article argues that destroying such infrastructures also constitutes terrorist acts because of their powerful effects on people's physical and psychological health and survival. Also increasingly common is the terrorist practice of targeting humanitarian workers who, under threat of injury and death, are prevented from providing assistance in the form of food, water, and medical care.[3] Thus, terrorism as it occurs during civil wars is directed against people and also occurs indirectly by targeting their community infrastructures and those who work in humanitarian relief operations to make continued civilian existence possible.

Girls and Terrorist Acts

Although women and "females" are now more often identified as participants in terrorism,[4] girls' experiences are poorly understood and only occasionally acknowledged. This is true regardless of whether they are socialized, volunteered, or coerced to participate in such acts. Also, girls' efficacy, actions, resistance, and survival skills within fighting forces are inadequately appreciated.[5]

This article focuses on girls as actors within two rebel forces known for directing terrorism at civilians. Within the context of rebel forces, girls typically are characterized as victims who lack agency although recent research indicates that girls in these forces, willingly or otherwise, also participate in terrorist acts.[6] As members of rebel forces, many witness and participate in terrorist mutilation, ritualistic murder (human sacrifice), forced cannibalism and drug use, and physical and psychological deprivation.[7] After situating girls' involvement as child soldiers in fighting forces as a global phenomenon, including their recruitment and roles, this article details girls' participation, with a focus on their agency as fighters within rebel groups in two African countries—Northern Uganda and Sierra Leone. It draws on data gathered between September 2001 and June 2002 when the author conducted field work in both countries.[8]

Girl Child Soldiers

Throughout the world, participants in armed conflicts involve children under 18 who are internationally referred to as child soldiers. These children, boys in particular, have been a focus of international attention and advocacy on their behalf, largely because of the efforts of a child advocacy consortium, the International Coalition to Stop the Use of Child Soldiers [Coalition]. This group has systematically identified the use of girls and boys in fighting forces, published, and publicized these data.[9] As a result of the Coalition's documentation, accurate estimates of the use of child soldiers are increasingly possible. Its *2004 Global Reports* provides the most relevant and in-depth information on the worldwide use of children in fighting forces.

Just as women's war experiences have been overlooked until recently,[10] girls' presence in fighting forces has received even less exposure. A long history exists of women's participation in fighting forces, some of whom were girls—such as Joan of Arc who was 16 when her military career began in 1428.[11] Also, until recently and largely because of the emphasis placed on the girl child at the 1995 UN Fourth World Conference on Women,[12] girls have been subsumed under the larger categories of "women" or "females" so that their presence in fighting forces has been shrouded, and girls have been widely perceived as lacking agency in perpetrating acts of terror.

Some of girls' invisibility can be related to culturally specific definitions of who is a girl and who is a woman. For example, in some African countries where the Western cultural notion and rite of passage of being a teenager do not exist, pubescent girls are considered to be women after initiation rites. In contrast, in contemporary Western societies, females are normally thought of as girls younger than 18 years of age. The definition, found in the Cape Town Principles, commonly accepted by the international community, is used in this article for defining a child soldier.

> ... any person under 18 years of age who is part of any kind of regular or irregular armed force in any capacity, including but not limited to cooks, porters, messengers, and those accompanying such groups, other than purely as family members. Girls recruited for sexual purposes and forced marriage are included in this definition. It does not, therefore, only refer to a child who is carrying or has carried arms.[13]

Global Involvement

Throughout the world, from Colombia to Kosovo/a to Chechnya to Israel and Africa, girls are actors within fighting groups.[14] Between 1990 and 2003, girls were part of fighting forces in 55 countries. They were present in 38 armed conflicts in 13 African countries, 7 countries in the Americas, 8 countries in Asia, 5 countries in Europe, and 5 countries in the Middle East.[15] Most of these armed conflicts were internal to the country although, in some cases such as in Macedonia, Lebanon, Uganda, and Sudan, girls also fought in international conflicts.

Country-specific cases provide examples of girls' involvement in some of these fighting forces. During Cambodia's civil war, girls were used by both governmental forces and the Khmer Rough. A 17-year-old girl taken as an orphan into the Khmer Rough when she was 2 years old reported that, together with a group of 300 to 500 girls, she was given military training from the age of 5. Provided with guns and uniforms, they became active soldiers when they reached 14 years of age.[16] In the PKK, the Kurdistan Worker's Party, a 14-year-Syrian national fought as a female guerrilla against the Turkish army. She received military and political training in Iraq. According to a report of the Coalition to Stop the Use of Child Soldiers, "In 1998 ... more than 10 percent of the PKK's total number of child soldiers were said to be girls."[17] In Asia, approximately 900 to 1,000 girls fought in the northeastern state of Manipur, India, constituting 6–7 percent of the total number of child soldiers fighting there. In Nepal, Maoist insurgents have used girls extensively in what they call "the People's War."[18] Similarly, in Sri Lanka, Tamil girls have been recruited into the Liberation Tigers of Tamil Eelam (LTTE) since the mid-1980s. In LTTE, girl fighters participate in grueling training and in fierce fighting.[19] According to government sources, because girls are less suspect than boys and less often subjected to body searches, girls in Sri Lanka have been chosen to become or forced to be suicide bombers at as young as 10 years old.[20] Of the LTTE fighting forces, 40–60 percent are estimated to be under 18 years of age, most being girls and boys ages 10–16.[21] However, these data lack precision and have been critiqued as needing reliable, field-based estimation.[22] In Colombia, 6,000 boys and girls are estimated to be involved in armed groups. Again, although reliable statistics are unavailable, girls in Colombia are thought to constitute approximately 20 percent of children in guerrilla groups and 15 percent of children in paramilitary groups.[23]

In military forces in Ethiopia, Israel, the Philippines, Sri Lanka, and Colombia, girls have been highly respected and regarded as fighters. In Eritrea, where females comprised one-third of fighters, Veale[24] studied 11 former female participants. The girls' ages when recruited into the Tigrean People's Liberation Front (TPLF) averaged 12.68 years, with the youngest 5 years old and the oldest 17. On the average, these girls spent 11.6 years as fighters, with a range of 4 to 18 years. In Liberia, older girls and young women in Liberians United for Reconciliation and Democracy (LURD) were reported as particularly fierce fighters who commanded respect from their male peers. A female commander reported that her unit entered combat clad only in undergarments due to beliefs that their appearance would intimidate enemies and strengthen their magical protection.[25] The terrorist elements of surprise and invoking fear as well as violating cultural taboos are strongly operational in this strategy, which has also been used by women as a form of nonviolent protest.[26] Ellen S., a fighter and commander of girls for LURD, said that during attacks, girls and boys were captured for the force. She described how they would enter battle wearing yellow or brown T-shirts inscribed with the initials LURD. Ellen S. also wore ammunition around her chest and carried an automatic weapon. She related how she terrorized captured enemies, "if my heart was there, I would bring them to the base for training. But if my heart was bad lucky [sic], then I would kill them right there."[27]

These data indicate that girls globally are actors in fighting forces, often because they were forced to participate, but they also volunteer for ideological or pragmatic reasons.[28] Regardless of their rank and situations in a force, girls participate in acts that terrorize civilians in countries where they fight. Perpetrating violence and torture become normal and routine within a culture of violence that pervades every aspect of daily routines and activities.[29]

Recruitment

Although the idea of children freely choosing to join a force is a contested one, girls may volunteer—meaning that they were not physically forced, abducted, or otherwise coerced. Some girls volunteered or were coerced into a force. Other entry points included being born of an abducted mother or captured by another fighting force. They may enter a force for ideological reasons, to fulfill a compulsory obligation, escape poverty, and/or seek opportunities such as employment or sponsorship in school. They also join because of untenable family situations such as sexual abuse and overload of domestic work, and to find protection, join with other family members, and seek adventures.[30] Some girls find new freedoms and capabilities, with fewer gender restrictions and opportunities to exert authority that have not been previously possible.

Girls may be gang-pressed, meaning they are physically coerced into a force when they are in places such as schools, discotheques, and markets or simply walking along a road or snatched from their homes. For example, during the war in Mozambique (1976–1992), the Frelimo government force recruited and gang-pressed girls to fight in the war against Renamo rebel forces. Frelimo recruiters arrived with buses at schools where they asked girls to volunteer for the military. When few agreed, girls were forced onto buses and taken to a military base where they met with other "recruited" girls and began military training.[31] Girls also joined Frelimo because of the promise of new and emancipatory roles, to escape rural areas and expand traditional gender roles, and in hopes of improving their educational and career opportunities. In the Renamo force, most girls were abducted but some were recruited. Others joined because they felt discontent over Frelimo socialist policies, wanted to be with family members, or because they were lured into the force with the promise of educational opportunities.

In many rebel forces, notably in Africa, girls have entered rebel forces involuntarily, usually by abduction. Also, cross-border abductions have occurred in both Sierra Leone and Northern Uganda and other countries. Between 1990 and 2003, girls were abducted in 12 African countries, 4 countries in the Americas, 8 Asian countries, 3 European countries, and 2 Middle Eastern countries.[32] In some countries, boys and girls are taken from orphanages, as reported in Sri Lanka where the LTTE purportedly runs its own orphanages and uses these children as fighters.[33]

Roles

Girls' and women's participation within fighting forces are key because they carry on supportive tasks that maintain the fighting force. Also, they are fighters, which can mean being sent to the frontlines as cannon fodder, sometimes with their babies drugged and strapped on their backs. Girls also conduct suicide missions, provide medical care, and serve as mine sweeps.[34]

Colombian ex–girl soldiers who joined a fighting force as teenagers were taught how to care for and use guns, conduct military maneuvers and communications operations, and serve as bodyguards for commanders.[35] As such, girls' roles are multi-faceted and vary according to the force in which they are enrolled, their ages, and how gender is constructed within the force, such as whether girls are viewed as "equal" to boys (even though power differentials inevitably exist) or are treated as slaves and servants.

In some forces where girls serve primarily as combatants, sex is consensual or forbidden, and severe punishment directed to sexual perpetrators was reported in Colombia, the Philippines, and Sri Lanka.[36] In many African fighting forces such as in Angola, Mozambique, Northern Uganda, and Sierra Leone, primary roles for pubescent girls are providing sex and being "wives" who give birth to children who are raised within a rebel force to be fighters.[37]

Regardless of whether they primarily are fighters or serve as spies, porters, or "wives" of rebel-captor "husbands," girls typically hide their involvement in terrorist acts when they come out of a force because they are reluctant to acknowledge roles that violate broader community and gender norms. They often feel shame, even though they acknowledge that they would have been killed had they refused to participate. Therefore, a veil of secrecy continues to surround their acts and experiences within a force. Only recently have researchers focused on deconstructing their experiences and expanding the scope of their inquiry about child soldiers to include girls as both perpetrators of terror as well as terrorism's victims.

Northern Uganda and Sierra Leone

The Lord's Resistance Army (LRA) in Northern Uganda

Since 1986, the LRA, which is led by Joseph Kony, has waged a war of terror in Northern Uganda and Southern Sudan against the governmental Ugandan People's Defence Force (UPDF). The primary victims have been the Acholi people whose community infrastructures have been shattered. Thousands of people have been displaced, often in camps for internally displaced people (IDPs). Today, they continue to live in terror of surprise LRA attacks in these highly vulnerable camps and in villages throughout Northern Uganda. In the main, the LRA force consists of abducted children from Northern Uganda and Southern Sudan with 80 percent of the force estimated to be children. Girls are thought to comprise one-fourth or more of child soldiers although actual numbers of children abducted into the Lord's Resistance Army (LRA) are imprecise.[38] Children are also born into the LRA, fathered by rebel commanders. These children may grow up in the LRA to become fighters. When they are taken into a force, girls (and boys) are immediately subject to intensive abuse and torture and many are killed or die because they are unable to cope with the harsh circumstances of rebel existence

After training in military tactics and use of weaponry, girls participate in front-line combat, with some holding command positions within the LRA. They engage in terrorist acts that create widespread fear, such as attacking their own families and neighbors, abducting other children, and killing civilians. Girls also perform support roles within the military bases such as raising crops, selling goods, preparing food, carrying loot, moving weapons, and stealing food, livestock, and seed stock.[39] They fetch firewood and water, cook food, climb trees to spy, transport ammunition, participate in guard duty, and fight during ambushes.[40] Younger girls are servants to commanders and their "wives," and they work continuously.[41]

The Revolutionary United Front (RUF) in Sierra Leone

Sierra Leone's 11-year war began in 1991 and officially ended in January 2002. The war pitted the Sierra Leone Army (SLA) and pro-government civilian militias such as the Civil Defense Forces (CDFs) against the rebel RUF force. Gross human rights violations were committed by all sides. The RUF was especially culpable because of its extensive abduction of children and adults and of terror tactics that resulted in countrywide fear. These included attacking villages, destroying community infrastructures such as schools, homes, and health facilities, and perpetrating atrocities such as severing hands, arms, feet, and legs, cannibalism, and ritual murders.[42] The war was waged throughout the country, including trans-border regions. Girls fought on all sides and comprised an estimated 25 percent of the child soldiers in all forces, with child soldiers estimated to constitute one-third of all fighters. An estimated 8 percent of the total forces, both adults and children, during the Sierra Leonean war consisted of girls although their use and numbers varied between forces. For example, within the Revolutionary Unit Front (RUF) rebel group, girls are estimated to have constituted at least one-third of all child soldiers and approximately 16 percent of the total RUF force.[43]

In Sierra Leone, girls' roles within the RUF were similar to those within the LRA. They were fighters, cooks, domestic laborers, and also porters, "wives," and food producers. They cared for the sick and wounded, passed messages between rebel camps, served as spies, and some worked in diamond mining for their commanders or rebel-captor "husbands."[44] Ramata Y. was taken into the RUF when her mother and father were killed by the RUF. In the rebel force, she fetched water, cooked, and was a "wife." Ramata Y. reported that girls were trained to use guns. They killed people, stole property, and looted and burned houses.[45]

Victim, Perpetrator, or Both?

In the two rebel forces, the LRA and the RUF, most girls entered because they were abducted. Researchers studying 32 girls who were in the RUF force in Sierra Leone found that all were abducted, often by children their own age who threatened them with death.[46] Sophia R.'s story is a typical one. At age 11, she was captured at school and spent the next 9 years in the RUF. She was immediately "disvirginalized" by many men. Her leg was tattooed with the letters "RUF." In the RUF Sophia R. was a "wife," but she also learned to use a gun and was given combat clothes. She was introduced to cocaine, which emboldened her to fight. She explained that the cocaine enabled her to destroy and "cause bad havoc."[47] Dorothy G. was 13 years old when she was abducted by the LRA. She was taken to Sudan and taught how to work a gun and to be a spy. She climbed trees to see when the Ugandan army was coming. She was also given as a "wife" and used for sex. She said that if a girl refused sex, she was beaten or killed.[48]

Given the realities of the almost-ubiquitous experience of girls being abducted into the rebel forces in Northern Uganda and Sierra Leone, a tension arises in explaining the paradox whereby victims of terrorist violence subsequently become perpetrators of similar violence. However, discrete categorization as "victim" or "perpetrator" fails to underscore the complexities of shifting roles and experiences such as the seeming paradox of girls becoming allies with individuals who were responsible for abducting and victimizing them and who continue to sexually abuse them. Or, a girl who has never felt herself to be efficacious might experience the lure that can occur from the power of carrying a gun and defying traditional gender roles.[49] This dialectic has parallels with the 1970s abduction of Patricia Hearst who later became an actor in the Symbionese Liberation Army (SLA) and then came to be viewed as a perpetrator instead of a kidnapped and terrorized victim. Similarly, differentiating victim and perpetrator roles of girls is problematic because of the fluid roles and situations within a force and changes that can occur over time for impressionable children who are socialized, often for many years, into a culture of violence that encouraged perpetration of terrorist acts as a fighting strategy.

Girls in the Northern Ugandan and Sierra Leonean rebel forces have been victimized because they have been forced, at the threat of their lives, to participate in terrorist acts such as killing friends or family members and torching homes. Yet they also demonstrate resiliency, agency, and ability to resist—although usually not successfully—their oppressors Over time, as they continue to participate in terrorist acts, some become combatants, spies, and communications personnel who hold key responsibilities within the force. Nevertheless, they remain relatively powerless within a force and coerced to participate. Except for the most powerful girls who hold commander status or are commander's "wives," girls are subjected to abuse from men and boys and, in some cases women, because of their low status and traditional gender roles.

Boys' experiences are both similar to girls and also differ as an effect of gender. As young children taken into a force, boys may carry out domestic tasks, be porters, and participate in terrorist acts. Although some boys are thought to experience sexual abuse, little is known about the extent of sexual violence perpetrated against boys by male and female commanders in a force; its occurrence is thought to be much less widespread for boys than for girls. Boys also may be forbidden to sexually approach girls and women until they attain rank, such as a commander, within the rebel force.[50]

Girls As Fighters and Resisters

Children abducted into the LRA before 2002 who spent time in Sudan were given long and formalized military training. Since 2002, training has been sporadic, and some of the youngest abductees are not trained at all. Others are trained but not given uniforms or weapons.[51] In 2003, ex-LRA children told Human Rights Watch researchers that they were forced to participate in beatings or tramplings of other abductees. Susan A. told of being forced, along with three other girls, to beat and kill civilians in villages and internally displaced people camps.[52] Elizabeth B. was 12 when she was abducted into the LRA and was in the force for 2 years. Her father was killed trying to protect her from abduction. She described how another girl in her group was asked to beat somebody they [the LRA] wanted to kill. When the girl refused, she was killed. Elizabeth B. now becomes annoyed very quickly; when she's angry, she feels like killing somebody.[53] Another girl, Alice R., was abducted into the LRA when she was 17 years old. When Alice R. crossed into Sudan with the LRA, she carried guns and was subsequently trained to be a soldier. Although allocated to an army commander to be his "wife," she was also a fighter[54] Janet M. was 15 when abducted into the LRA and spent most of two and one-half years in southern Sudan. Her story is one of resistance to participating in violence and terrorist acts although ultimately she became a fighter. Initially, she received training in Sudan after which time she was to be "allocated" to a commander to be his "wife." Because she was young and feared sexual abuse, she deceived her captors by saying she was [sexually] infected. Janet M. was next taken for a medical examination. When a report was given that she was not infected, angry commanders ordered her killed. Kony, the LRA commander, intervened to spare her life because, he said, her actions showed she was wise and tricked people. She was then given to another commander but refused sex with him. Beaten for six months as punishment, she was sent to fight the Dinkas in southern Sudan. While fighting the Dinkas, she looted property and foodstuff; even minerals were taken. When she escaped the LRA, she was in advanced pregnancy, evidence that despite her resistance she was subjected to forced sex.[55]

In Sierra Leone, girls also received military training. Many girls who were with the RUF reported that although they learned to cock and load a gun, they did not participate in combat.[56] It is possible that girls do not view themselves as combatants, and few would self identify as perpetrators of terror unless they possessed and used guns or held commander rank within a force. As in Northern Uganda, training could be intense and lengthy and consisted of how to use guns, engage in physical training, and to kill.[57] Grace J. said that she was trained "on barbed wire." Trainees were put into a kind of cage and told that if they managed to escape, they were perfectly trained. Grace J. escaped whereas many others died during training exercises.[58] Arlene N. was trained to hold onto a gun but did not handle [carry] a gun. She explained that most girls were trained, and some participated in terrorist acts such as shooting and killing, stealing properties, and looting and burning houses.[59] Christine P. was young when she was abducted. She stayed with the RUF throughout most of the war. When she was abducted, the RUF force encountered her mother, in advanced pregnancy at the time, and her father walking along the road. The rebels tied her father's hands behind his back. Her mother was given a heavy load to carry that she threw away because of its weight. The rebels then caught her mother, slit open her abdomen, took out and killed the unborn child, and killed both her father and brother.[60] Her story provides insights into the types of terrorist acts perpetrated by the RUF, and in which girls participated.

Margaret C. recounted that she was trained and took part in fighting. Within the force they would "kill and eat." In the jungle they ate humans and reptiles; if she refused to eat, she would have been killed. She described a terrorist ambush when a woman was killed, and they [commanders] told them to eat this woman. So they opened the upper part of the body, cooked it [presumably the heart] and threw away the other parts of the body.[61]

As noted by Denov and Maclure,[62] routinization of violence through training and everyday experiences, such as these girls had, helps those who perpetrate terrorist acts to see themselves as effectively performing a job. Mary J., an ex-combatant in the RUF, explained that over time she came to view her role of killing as normal and to understand that overcoming the enemy was part of her job. Killing without a reason showed commitment and willingness to work with other rebels. She was not allowed to show remorse, sadness, or shame whereas brutal acts of torture and violence were encouraged and celebrated.[63] Girls also have reported that by carrying small arms, they gained power, status, and control and they felt pride, self confidence, and a sense of belonging.[64] Also, when girls hold positions of power, such as being a military commander, feelings of pride may become more salient than identifying oneself as a victim.

Disarmament, Demobilization, and Reintegration (DDR)

Despite recent and increasingly robust data detailing girls in fighting forces,[65] the international community, governments, and militaries continue to ignore and deny the extent of girls' involvement and offer inaccurate and reductionistic explanations for their presence. Pervasive gender discrimination in war-affected countries such as that existing in Northern Uganda and Sierra Leone, perpetuates the notion of girls solely as victims, most notably "sex slaves," and as having lesser agency in perpetrating violence and terror than boys. Girls, therefore, are not thought of as ex-combatants or as having held responsible positions within the rebel force. A consequence of "not seeing" girls as actors and perpetrators is that girls are seldom included in disarmament, demobilization, and reintegration (DDR) programs. Instead, boys and men are privileged in receiving DDR benefits, which typically include opportunities to enroll in skills training, attend school, or participate in rehabilitation programs. For example, in Angola, despite recognition that large numbers of girls were abducted into Angolan fighting forces, thousands of boys were formally demobilized although no girls were.[66] In Sierra Leone, 6,052 boys passed through DDR whereas only 506 girls did.[67] Reflecting the reality of their situations and without negating the knowledge that they also experienced sexual and other forms of gender-specific violence, girls should be recognized as serving in capacities that parallel or are complementary to those of boys.[68]

Community Responses to Girls' Return

Community members often react with hostility and fear to girls coming back from a rebel fighting force. This is understandable because these girls were among those who either witnessed or perpetrated acts of terror against community members and profoundly violated community norms of behavior.[69] Consequently, returning girls are often provoked, stigmatized, and poorly accepted by community members and at school, if they attend.

Girls returning with children conceived and born in a rebel force are especially stigmatized. The presence of these children makes explicit, regardless of forced maternity, that they have violated traditional gender norms that mandate girls should be virgins before they marry. Further, their children often are of unknown paternity or their fathers are rebel-captor "husbands." One effect is that some girls do not marry. Others eschew marriage because of the horrific sexual and other violence they have experienced from boys and men and their lack of trust of them. This resistance to marriage can be construed as a radical act in some African societies, where marriage is perceived as mandatory to avoid being viewed as a social outcast.

Consequently, many girls leave their communities because they are poorly accepted, unable to adjust to community life, cannot marry, or find no way to secure an economic livelihood. Additionally, gender discrimination, such as the long hours girls must work at home or notions such as education, is more important for boys and affect returning girls' ability to attend school or learn skills such as a trade. Returning girls also often experience gender-specific effects that reduce their life choices and span, such as sexually transmitted diseases resulting from forced sex within a force.[70]

Thinking about Girls in Rebel Forces and Terrorism

Terrorism, as construed in Western minds, is usually closely equated with 9/11 and similar episodic and unexpected acts that rivet the world's attention because of the magnitude of their effects. Girls and women are only occasionally seen as actors in these scenarios. Yet, history says that women, too, are involved although usually with limited visibility. Within the context of rebel wars in Northern Uganda and Sierra Leone, as discussed in this article, girls' presence in these forces is pervasive and, as such, they routinely witness and experience violence and participate in terrorist acts. Their involvement is not isolated but occurs throughout the world, although with distinct experiences according to the force.

For some girls, often taken into the LRA at young ages and for relatively short periods, if they survive they may be young enough to avoid full participation as terrorists in the force; instead they are first assigned to domestic work and serving as porters. Yet, they are inevitable witnesses, which has its own traumatic effects. In both Northern Uganda and Sierra Leone, large numbers of girls spent years in a force, and many grew into motherhood and adulthood within this context. They have literally lost their childhoods and are socialized into a culture of violence where terrorist acts become normal. Further, their children are socialized into this same violent culture. When these girls, often now women, escape or the war ends, they are ignored, stigmatized, and refused DDR benefits. Most go directly back to their communities, if these still exist, find relatives to stay with, or migrate to urban areas. Life does not continue as "normal" despite the changed circumstances of their lives. All are traumatized and often display inappropriate social responses or behavioral deficits. For example, community members in a village in Sierra Leone told about how returning girls would steal and kill neighbors' chickens and that some girls emotionally withdrew; others were belligerent and hostile—behavior that is consistent with the culture of violence in which they spent so much time. Yet, these same girls have been victimized and forced to participate in terrorist acts. This understanding must permeate initiatives to help them.

These girls can never "go back" to being innocents because they have experienced what most people cannot imagine, and they also have gained strengths from their survival. The challenge, therefore, is to empower them to use these strengths and to expand cultural definitions of gender to enable these strengths to be harnessed. Yet, so long as girls remain invisible within the programs and policies of international, national, and local groups, these steps will not be taken. One, then, must ask what the future may bring to these countries when girls become women after such socialization. How will gender roles be (re)constructed given their experiences and those of boys? What will happen to their children? Very little is known about the latter except for sporadic reports of "war babies" whose development shows aberrations and whose social adjustment is poor.

In addressing these questions, within the countries of Northern Uganda and Sierra Leone much depends on empowering communities to work with these girls and their children. Because community is so central to the health and well being of people in these two countries, Western-style individualistic approaches are usually inappropriate—including psychotherapeutic approaches that diagnose children as having PTSD. A key strategy in working with these girls is to enlist the leadership of women elders to talk and listen to their stories and assist them in learning or re-learning normal behavior. These girls also need practical assistance. High priority must be given to their obtaining primary health care including, importantly, reproductive health care. They must be given opportunities to go to school or learn a skill and to participate in activities that foster healing, if they are to be empowered to become citizens in a culture of peace.

So long as these girls continue to be hidden from the world's attention and even invisible within their own communities except to be stigmatized and provoked, their strengths will not be realized. By calling their situations to attention within broader discussions of gender and terrorism, one step can be made in the right direction toward addressing the injustices they have experienced and advocating for international action on their behalf.

Notes

1. Barry Levy and Victor Sidel, *War and Public Health* (New York and Oxford: Oxford University Press, 1997); United Nations, *Women, Peace and Security* (New York: Author, 2002); UN. (2004, June 14). *Renewed commitment to decision action for protecting civilians in armed conflict needed now more than ever, security council told: Emergency relief coordinator Jan Egeland urges new Council Resolution supporting further measures to improve civilian protection.* Security Council 4990th Meeting. Press Release SC/8122. Available at (www.un.org/News/Press/docs/2004/sc8122.doc).
2. Deborah Galvin, "The female terrorist: A socio-psychological perspective," *Behavioral Sciences and the Law*, 1(2) (1983), p. 20.
3. UN, 2004.
4. See for example, Karla Cunningham, "Cross-regional trends in female terrorism," *Studies in Conflict and Terrorism, 26* (2003), pp. 171–195; Sharon Pickering and Amanda Third, "Castrating conflict: Gender(ed) terrorists and terrorism domesticated," *Social Alternatives*, 22(2) (2003), pp. 8–15.
5. For further discussion, see Dyan Mazurana and Susan McKay, *Child soldiers: What about the Girls? Bulletin of the Atomic Scientists, 57*(5), (2001, September/October), pp. 30–35; Dyan Mazurana, Susan McKay, Khristopher Carlson, and Janel Kasper, "Girls in fighting forces and groups: Their recruitment, participation, demobilization, and reintegration," *Peace and Conflict: Journal of Peace Psychology, 8*(2), (2002), pp. 97–123; Susan McKay and Dyan Mazurana, *Where are the Girls? Girls in Fighting Forces in Northern Uganda, Sierra Leone and Mozambique: Their Lives During and After War* (Montreal: Rights and Democracy, 2004).
6. For further discussion, see Myriam Denov and Richard Maclure, "Girls and armed conflict in Sierra Leone: Victimization, participation, and resistance. In V. Farr and A. Schnabel, eds., *Gender Perspectives on Small Arms and Light Weapons* (Tokyo: United Nations University Press, in press). Yvonne Keairns, October 2002, *The Voices of Girl Child Soldiers* (New York and Geneva: Quaker United Nations Office); Susan McKay and Dyan Mazurana, *Where are the Girls?*
7. McKay and Mazurana, *Where are the Girls?*
8. This study's co-investigators were Susan McKay and Dyan Mazurana. It was funded by the Canadian International Development Agency's Child Protection Research Fund and implemented in partnership with Rights and Democracy, Montreal. The study examined the presence and experiences of girls in fighting forces and groups within the context of three African armed conflicts—Mozambique, Northern Uganda, and Sierra Leone. Also, funding was contributed for Susan McKay's research by the University of Wyoming Graduate School Research Office, the Women's Studies Program, the Office of the Dean of Arts and Sciences, the International Studies Program, the Provost's Office, and School of Nursing, and the Office of International Travel.
9. Coalition to Stop the Use of Child Soldiers [Coalition]. *Child Soldiers Global Report: Global Report on Child Soldiers 2001* (London: Author); McKay & Mazurana, *Where are the Girls?*
10. Susan McKay, "The effects of armed conflict on girls and women," *Peace and Conflict: Journal of Peace Psychology, 5*

(1998), pp. 381–392; Elizabeth Rehn and Ellen Sirleaf, "Women, war and peace: The independent experts' assessment of armed conflict on women and women's role in peace-building" (New York: United Nations Development Fund for Women [UNIFEM], 2002); Indai Sajor, "Common grounds: Violence against women in war and armed conflict situations" (Quezon City, Philippines: Asian Center for Women's Human Rights, 1998); UN, *The Impact of Armed Conflict on Children: Report of the Expert of the Secretary-General, Ms. Graça Machel* (New York: Author, 1996).

11. Linda DePauw, "Battle cries and lullabies: Women in war from prehistory to the present" (Norman: University of Oklahoma Press, 1998).
12. See UN. *Beijing Platform for Action* (New York: Author, 1995).
13. United Nations Children's Fund [UNICEF]. (1997, April 30). *Cape Town Annotated Principles and Best Practices.* Adopted by the participants in the Symposium on the Prevention of Recruitment of Children into the Armed Forces and Demobilization and Social Reintegration of Child Soldiers in Africa, organized by UNICEF in cooperation with the NGO Sub-group of the NGO Working Group on the Convention on the Rights of the Child, Cape Town, South Africa, p. 1.
14. Coalition (2001); Mazurana et al., "Girls in fighting forces and groups"; McKay and Mazurana, *Where are the Girls?*
15. McKay and Mazurana, *Where are the Girls?*
16. Coalition. (2000, May). Asia report: Executive summary, Child participation in armed conflict in Asia. Available at (www.child-soldiers.org).
17. Coalition. (2000). Girls with guns: An agenda on child soldiers for "Beijing Plus Five." Available at (www.child-soldiers.org/reports/special%20reports).
18. Coalition. (2000, May). Asia report: Executive summary.
19. Keairns, October 2002, *The Voices of Girl Child Soldiers* (New York and Geneva: Quaker United Nations Office).
20. Coalition to Stop the Use of Child Soldiers [Coalition]. (2000). *Girls with Guns;* Cunningham, "Cross-regional trends in female terrorism."
21. Coalition. (2001).
22. This observation was written in an unpublished draft document (2000, July) by Ken Bush who has worked extensively on the issue of child soldiers in Sri Lanka. *Stolen Childhood: The Impact of Militarized Violence on Children in Sri Lanka.*
23. Erika Páez, *Girls in the Colombian Armed Groups, A Diagnosis: Briefing* (Germany: Terre de Hommes, 2001).
24. Angela Veale. (2003). From child soldier to ex-fighter, a political journey: Female fighters, demobililisation, and reintegration in Ethiopia. Monograph No. 85, Institute of Security Studies, South Africa.
25. Human Rights Watch, *How to Fight, How to Kill: Child Soldiers in Liberia* (New York: Author, 2004, February). Available at (http://hrw.org/reports/2004/liberia0204/6.htm#_Toc61673969).
26. Dyan Mazurana and Susan McKay, *Women and Peacebuilding* (Montreal: Rights and Democracy, 1999).
27. Human Rights Watch, *How to Fight, How to Kill.*
28. Rachel Brett and Irma Specht, *Young Soldiers: Why They Choose to Fight* (Geneva: ILO and Boulder, CO: Lynne Rienner, 2004); Páez, *Girls in the Colombian Armed Groups.*
29. Denov and Maclure, "Girls and armed conflict in Sierra Leone."
30. Brett and Specht, *Young Soldiers;* Keairns, *The Voices of Child Soldiers; Páez, Girls in the Colombian Armed Groups.*
31. McKay and Mazurana, *Where are the Girls?*
32. Ibid.
33. Coalition, 2001; Mazurana and McKay, *Child Soldiers.*
34. Mazurana and McKay, *Child Soldiers.*
35. Yvonne Keairns, *The Voices of Girl Child Soldiers: Colombia* (New York, Geneva, and London: Quaker UN Office and Coalition, 2003).
36. See for example, ibid.; Yvonne Keairns, *The Voices of Child Soldiers: Angola* (New York, Geneva, and London: Quaker UN Office and Coalition, 2003); Yvonne Keairns, *The Voices of Child Soldiers: The Philippines* (New York, Geneva, and London: Quaker UN Office and Coalition, 2003); Yvonne Keairns, *The Voices of Child Soldiers: Sri Lanka* (New York, Geneva, and London: Quaker UN Office and Coalition, 2003); Veale, *From Child Soldier to Ex-fighter.*
37. Keairns, *The Voices of Child Soldiers;* McKay and Mazurana, *Where are the Girls?;* Veale, *From Child Soldier to Ex-fighter.*
38. McKay and Mazurana, *Where are the Girls?*
39. Ibid.
40. Interview conducted in Northern Uganda by Dyan Mazurana and Susan McKay on 26 November 2001.
41. Human Rights Watch, *Stolen Children: Abduction and Recruitment in Northern Uganda, 15*(7A) (2003, March), pp. 1–24.
42. McKay and Mazurana, *Where are the Girls?*
43. Ibid.
44. Ibid.
45. Interview conducted in Sierra Leone by Susan McKay on 11 June 2002.
46. Denov and Maclure, "Girls and armed conflict in Sierra Leone."
47. Interview conducted in Sierra Leone by Susan McKay on 31 May 2002.
48. Interview conducted in Northern Uganda by Dyan Mazurana and Susan McKay on 26 November 2001.
49. Denov and Maclure, "Girls and armed coflictin Sierra Leone"; V. Sherrow, *Encyclopedia of Youth and War: Young People as Participants and Victim* (Phoenix, Arizona: The Oryx Press, 2000).
50. McKay and Mazurana, *Where are the Girls?*
51. Human Rights Watch, *Stolen Children.*
52. Ibid.
53. Interview conducted in Northern Uganda by Dyan Mazurana & Susan McKay on 27 November 27, 2001.
54. Ibid.
55. Interview conducted in Northern Uganda by Dyan Mazurana & Susan McKay on 28 November 2001.
56. McKay and Mazurana, *Where are the Girls?*
57. Denov and Maclure, "Girls and armed conflict in Sierra Leone."
58. Interview conducted in Sierra Leone by Susan McKay on 6 June 2002.
59. Ibid.
60. Interview conducted in Sierra Leone by Susan McKay on 11 June 2002.
61. Interview conducted in Sierra Leone by Susan McKay on 6 June 2002.
62. Denov & Maclure, "Girls and armed conflict in Siera Leone."
63. Ibid.
64. Ibid.; McKay and Mazurana, *Where are the Girls?*
65. Ibid.
66. Mazurana, McKay, Karlson, and Kasper, "Girls in fighting forces and groups."
67. McKay and Mazurana, *Where are the Girls?*
68. Coalition, 2000; Keairns, October 2002, *The Voices of Girl Child Soldiers* (New York and Geneva: Quaker United Nations Office); Isobel McConnan and Sara Uppard, *Children, Not Soldiers* (London: Save the Children, 2001); McKay and Mazurana, *Where are the Girls?*
69. Elise Fredrikke Barth, *Peace as Disappointment: The Reintegration of Female Soldiers in Post-Conflict Societies: A Comparative Study from Africa.* PRIO Report 3/2002. Available at (www.prio.no/publications/reports/female soldiers); Susan McKay, Mary Burman, Maria Gonsalves, and Miranda Worthen, "Known but invisible: Girl mothers returning from fighting forces," *Child Soldiers Newsletter, 11,* 10–11. McKay and Mazurana, *Where are the Girls?.*
70. Coalition, 2000; McKay and Mazurana, *Where are the Girls?*

UNIT 9

Government Response

Unit Selections

28. **Port Security Is Still a House of Cards**, Stephen E. Flynn
29. **Are We Ready Yet?**, Christopher Conte
30. **The Double-Edged Effect in South Asia**, V. R. Raghavan

Key Points to Consider

- How can security at U.S. ports be improved?
- Have ongoing federal efforts to prepare for bioterrorism undermined our public health system?
- How did 9/11 alter India's counterterrorism policy? Explain.

Student Website

www.mhcls.com/online

Internet References

Further information regarding these websites may be found in this book's preface or online.

Coalition for International Justice
http://www.cij.org/index.cfm?fuseaction=homepage

Counter-Terrorism Page
http://counterterrorism.com

ReliefWeb
http://www.reliefweb.int

The South Asian Terrorism Portal
http://www.satp.org/

Government response to terrorism is multifaceted and complex. Choices about domestic spending, the use of military force, and long-term foreign policy objectives are increasingly shaped by our commitment to a Global War on Terrorism.

While counterterrorism spending has increased significantly since September 11th, choices about how this money is to be used have become more difficult as various constituencies lobby to have their voices heard. As policymakers struggle to allay public concerns, choices between spending for security today and preparing for the threats of the future have become more difficult. The tragedy of September 11 and the subsequent anthrax attacks have fueled fears about catastrophic terrorism. This makes choices about public policy priorities even more difficult. Given limited resources, should governments focus their efforts on existing crises and the most likely threats or should they focus their resources on countless potential vulnerabilities and catastrophic threats, which many experts agree may be possible but not likely? Ideally governments should do both. Realistically, even in a resource-rich environment, governments have to make choices.

Decisions about when or how to use military force are equally complex. Should governments adopt preemptive or defensive postures? Should governments focus their resources on state-sponsors of terrorism, or should they focus their efforts on capturing or killing the leaders of existing terrorist groups? Does the long-term deployment of a military force to a foreign country increase or reduce the threat of terrorism? Should nonproliferation be a priority in the war on terrorism?

Finally, it is important to note that the U.S. commitment to a Global War on Terrorism has not only an impact on long-term U.S. foreign policy objectives but also the foreign policies of others. There is an opportunity cost to foreign policy decisions. By prioritizing a particular set of objectives, governments inevitably sacrifice others. By making terrorism a policy priority we influence and shape the policies of others, as states may act in support of U.S. policy or take advantage of the vacuums created by such policies.

As the Global War on Terrorism continues to influence public policy at home and abroad, this section examines the methods and policies governments use to respond to the threat of international terrorism. In the first article, Stephen Flynn argues that port security should be placed higher on the U.S. government's list of priorities and criticizes ongoing efforts to secure domestic ports and monitor foreign points of origin. Flynn highlights three important weaknesses of the maritime security apparatus and offers policy recommendations designed to address these issues. The second article discusses efforts by various public health agencies to prepare local communities for bioterrorism. In the article Conte contends that preparations for bioterrorism are drawing resources away from more prevalent health crises. Finally, V. R. Raghavan examines ongoing Indian and Pakistani efforts to reduce the threat of terrorism in South Asia.

Port Security Is Still A House of Cards

Stephen E. Flynn

As one of the world's busiest ports, it is fitting that Hong Kong played host to the World Trade Organization's December 2005 meeting. After all, seaports serve as the on- and off-ramps for the vast majority of traded goods. Still, the leaders of the 145 delegations that convened in Hong Kong undoubtedly did not have much more than a sightseer's interest in the host city's magnificent and frenetic harbor. For the most part, finance and trade ministers see trade liberalization as involving efforts to negotiate rules that open markets and level the playing field. They take as a given the availability of transportation infrastructures that physically link markets separated by vast distances.

But the days when policy makers could take safe transportation for granted are long past. The Sept. 11, 2001 attacks on New York and subsequent attacks on Madrid and London show that transport systems have become favored targets for terrorist organizations. It is only a matter of time before terrorists breach the superficial security measures in place to protect the ports, ships and the millions of intermodal containers that link global producers to consumers.

Should that breach involve a weapon of mass destruction, the United States and other countries will likely raise the port security alert system to its highest level, while investigators sort out what happened and establish whether or not a follow-on attack is likely. In the interim, the flow of all inbound traffic will be slowed so that the entire intermodal container system will grind to a halt. In economic terms, the costs associated with managing the attack's aftermath will substantially dwarf the actual destruction from the terrorist event itself.

Fortunately, there are pragmatic measures that governments and the private sector can pursue right now that would substantially enhance the integrity and resilience of global trade lanes. Trade security can be improved with modest upfront investments that enhance supply chain visibility and accountability, allowing companies to better manage the choreography of global logistics—and, in the process, improve their financial returns. In short, there is both a public safety imperative and a powerful economic case for advancing trade security.

A Brittle System

Though advocates for more open global markets rarely acknowledge it, when it comes to converting free trade from theory to practice the now-ubiquitous cargo container deserves a great deal of credit. On any given day, millions of containers carrying up to 32 tons of goods each are moving on trucks, trains and ships. These movements have become remarkably affordable, efficient, and reliable, resulting in increasingly complex and economically expedient global supply chains for manufacturers and retailers.

From a commercial standpoint, this has been all for the good. But there is a problem: as enterprises' dependence on the intermodal transportation system rises, they become extremely vulnerable to the consequences of a disruption in the system. To appreciate why that is so requires a brief primer on how that system has evolved.

Arguably, one of the most unheralded revolutions of the 20th century was the widespread adoption of the cargo container to move manufactured and perishable goods around the planet. In the middle of the last century, shipping most goods was labor intensive: items had to be individually moved from a loading dock at a factory to the back of a truck and then offloaded and reloaded onto a ship. Upon arrival in a foreign port, cargo had to be removed by longshoremen from the ship's holds, then moved to dock warehouses where the shipments would be examined by customs inspectors. Then they were loaded onto another transportation conveyance to be delivered to their final destination. This constant packing and repacking was inefficient and costly. It also routinely involved damage and theft. As a practical matter, this clumsy process was a barrier to trade.

The cargo container changed all that. Now goods can be placed in a container at a factory and be moved from one mode of transportation to another without being manually handled by intermediaries along the way. Larger vessels can be built to carry several thousand containers in a single voyage. In short, as global trade liberalization accelerated, the transportation system was able to accommodate the growing number of buyers and sellers.

Arguably, East Asia has been the biggest beneficiary of this transportation revolution. Despite the distance between Asia and the U.S., a container can be shipped from Hong Kong, Shanghai, or Singapore to the West Coast for roughly $4,000. This cost represents a small fraction of the $66,000 average value of goods in each container that is destined for the U.S.

However, multiple port closures in the U.S. and elsewhere would quickly throw this system into chaos. U.S.-bound container ships would be stuck in docks, unable to unload their cargo. Marine terminals would have to close their gates to all incoming containers since they would have no place to store them. Perishable cargo would spoil. Soon, factories would be idle and retailers' shelves bare.

In short, a terrorist event involving the intermodal transportation system could lead to unprecedented disruption of the global trade system, and East Asia has the most to lose.

What Has Been Done?

THE POSSIBILITY THAT terrorists could compromise the maritime and intermodal transportation system has led several U.S. agencies to pursue initiatives to manage this risk. The U.S. Coast Guard chose to take a primarily multilateral approach by working through the London-based International Maritime Organization to establish new international standards for improving security practices on vessels and within ports, known as the International Ship and Port Facility Code (ISPS). As of July 1, 2004, each member state was obliged to certify that the ships that fly their flag or the facilities under their jurisdiction are code-compliant.

The Coast Guard also requires that ships destined for the U.S. provide a notice of their arrival a minimum of 96 hours in advance and include a description of their cargoes as well as a crew and passenger list. The agency then assesses the potential risk the vessel might pose. If the available intelligence indicates a pre-arrival security check may be warranted, it arranges to intercept the ship at sea or as it enters the harbor in order to conduct an inspection.

The new U.S. Customs and Border Protection Agency (CBP), which was established within the Department of Homeland Security, mandated that ocean carriers must electronically file cargo manifests outlining the contents of U.S.-bound containers 24 hours in advance of their being loaded overseas. These manifests are then analyzed against the intelligence databases at CBP's National Targeting Center to determine if the container may pose a risk.

If so, it will likely be inspected overseas before it is loaded on a U.S.-bound ship under a new protocol called the Container Security Initiative (CSI). As of November 2005, there were 41 CSI port agreements in place where the host country permits U.S. customs inspectors to operate within its jurisdiction and agrees to pre-loading inspections of any targeted containers.

Decisions about which containers will not be subjected to an inspection are informed by an importer's willingness to participate in another post-9/11 initiative, known as the Customs-Trade Partnership Against Terrorism (C-TPAT). C-TPAT importers and transportation companies agree voluntarily to conduct self-assessments of their company operations and supply chains, and then put in place security measures to address any security vulnerabilities they find. At the multilateral level, U.S. customs authorities have worked with the Brussels-based World Customs Organization on establishing a new framework to improve trade security for all countries.

In addition to these Coast Guard and Customs initiatives, the U.S. Department of Energy and Department of Defense have developed their own programs aimed at the potential threat of weapons of mass destruction. They have been focused primarily on developing the means to detect a "dirty bomb" or a nuclear weapon.

The Energy Department has been funding and deploying radiation sensors in many of the world's largest ports as a part of a program called the Megaport Initiative. These sensors are designed to detect radioactive material within containers. The Pentagon has undertaken a counterproliferation initiative that involves obtaining permission from seafaring countries to allow specially trained U.S Navy boarding teams to conduct inspections of a flag vessel on the seas when there is intelligence that points to the possibility that nuclear material or a weapon may be part of the ship's cargo.

Finally, in September 2005, the White House weighed in with its new National Maritime Security Strategy. This purports to "present a comprehensive national effort to promote global economic stability and protect legitimate activities while preventing hostile or illegal acts within the maritime domain."

A House of Cards

OSTENSIBLY, THE FLURRY of U.S. government initiatives since 9/11 suggests substantial progress is being made in securing the global trade and transportation system. Unfortunately, all this activity should not be confused with real capability. For one thing, the approach has been piecemeal, with each agency pursuing its signature program with little regard for other initiatives. There are also vast disparities in the resources that the agencies have been allocated, ranging from an $800 million budget for the Department of Energy's Megaport initiative to no additional funding for the Coast Guard to support its congressionally mandated compliance to the ISPS Code. Even more problematic are some of the questionable assumptions about the nature of the terrorist threat that underpin these programs.

East Asia has the most to lose if a terrorist event disrupts the global trading system.

In an effort to secure funding and public support, agency heads and the White House have oversold the contributions of these new initiatives. Against a backdrop of inflated and unrealistic expectations, the public is likely to be highly skeptical of official assurances in the aftermath of a terrorist attack involving the intermodal transportation system. Scrambling for fresh alternatives to reassure anxious and angry citizens, the White House and Congress are likely to impose Draconian inspection protocols that dramatically raise costs and disrupt crossborder trade flows.

The new risk-management programs advanced by the CBP are especially vulnerable to being discredited, should terrorists succeed at turning a container into a poor man's missile. Before stepping down as commissioner in late November 2005, Robert Bonner repeatedly stated in public and before Congress that his inspectors were "inspecting 100% of the right 5% of containers." That implies the CBP's intelligence and analytical tools can be relied upon to pinpoint dangerous containers.

Former Commissioner Bonner is correct in identifying only a tiny percentage of containers as potential security risks. Unfortunately, CBP's risk-management framework is not up to the task of reliably identifying them, much less screening the low- or medium-risk cargoes that constitute the majority of containerized shipments and pass mostly uninspected into U.S. ports. There is very little counterterrorism intelligence available to support the agency's targeting system.

That leaves customs inspectors to rely primarily on their past experience in identifying criminal or regulatory misconduct to determine if a containerized shipment might potentially be compromised. This does not inspire confidence, given that the U.S. Congress's watchdog, the Government Accountability Office (GAO), and the U.S. Department of Homeland Security's own inspector general have documented glaring weaknesses with current customs targeting practices.

Prior to 9/11, the cornerstone of the risk-assessment framework used by customs inspectors was to identify "known shippers" that had an established track record of engaging in legitimate commercial activity. After 9/11, the agency expanded that model by extracting a commitment from shippers to follow the supply chain security practices outlined in C-TPAT. As long as there is no specific intelligence to tell inspectors otherwise, shipments from C-TPAT-compliant companies are viewed as low-risk.

The problem with this method is that it is designed to fight conventional crime; such an approach is not necessarily effective in combating determined terrorists. An attack involving a weapon of mass destruction differs in three important ways from organized criminal activity.

First, it is likely to be a one-time operation, and most private company security measures are not designed to prevent single-event infractions. Instead, corporate security officers try to detect infractions when they occur, conduct investigations *after* the fact, and adapt precautionary strategies accordingly.

Second, terrorists will likely target a legitimate company with a well-known brand name precisely because they can count on these shipments entering the U.S. with negligible or no inspection. It is no secret which companies are viewed by U.S. customs inspectors as "trusted" shippers; many companies enlisted in C-TPAT have advertised their participation. All a terrorist organization needs to do is find a single weak link within a "trusted" shipper's complex supply chain, such as a poorly paid truck driver taking a container from a remote factory to a port. They can then gain access to the container in one of the half-dozen ways well known to experienced smugglers.

Third, this terrorist threat is unique in terms of the severity of the economic disruption. If a weapon of mass destruction arrives in the U.S., especially if it enters via a trusted shipper, the risk-management system that customs authorities rely on will come under intense scrutiny. In the interim, it will become impossible to treat crossborder shipments by other trusted shippers as low-risk. When every container is assumed to be potentially high-risk, everything must be examined, freezing the worldwide intermodal transportation system. The credibility of the ISPS code as a risk-detection tool is not likely to survive the aftermath of such a maritime terrorist attack, and its collapse could exacerbate a climate of insecurity that could likely exist after a successful attack.

Moreover, the radiation-detection technology currently used in the world's ports by the Coast Guard and Customs and Border Protection Agency is not adequately capable of detecting a nuclear weapon or a lightly shielded dirty bomb. This is because nuclear weapons are extremely well-shielded and give off very little radioactivity. If terrorists obtained a dirty bomb and put it in a box lined with lead, it's unlikely radiation sensors would detect the bomb's low levels of radioactivity.

The flaws in detection technology require the Pentagon's counterproliferation teams to physically board container ships at sea to determine if they are carrying weapons of mass destruction. Even if there were enough trained boarding teams to perform these inspections on a regular basis—and there are not—there is still the practical problem of inspecting the contents of cargo containers at sea. Such inspections are almost impossible because containers are so closely packed on a container ship that they are often simply inaccessible. This factor, when added to the sheer number of containers on each ship—upwards of 3,000—guarantees that in the absence of very detailed intelligence, inspectors will be able to perform only the most superficial of examinations.

In the end, the U.S. government's container-security policy resembles a house of cards. In all likelihood, any terrorist attack on U.S. soil that involved a maritime container would come in contact with most, or even all, of the existing maritime security protocols. Consequently, a successful seaborne attack would implicate the entire security regime, generating tremendous political pressure to abandon it.

The Way Ahead

WE CAN DO better. The Association of Southeast Asian Nations should work with the U.S. and the European Union in authorizing third parties to conduct validation audits in accordance with the security protocols outlined in the International Ship and Port Facility Security Code and the World Customs Organization's new framework for security and trade facilitation.

A multilateral auditing organization made up of experienced inspectors should be created to periodically audit the third party auditors. This organization also should be charged with investigating major incidents and recommending appropriate changes to established security protocols.

To minimize the risk that containers will be targeted between the factory and loading port, governments should create incentives for the speedy adoption of technical standards developed by the International Standards Organization for tracking a container and monitoring its integrity. The technology now used by the U.S. Department of Defense for the global movement of military goods can provide a model for such a regime.

Asean and the EU should also endorse a pilot project being sponsored by the Container Terminal Operators Association (CTOA) of Hong Kong, in which every container that arrives passes through a gamma-ray content-scanning machine, as well as a radiation portal to record the levels of radioactivity within the container. Optical character recognition cameras then photograph the number painted on several sides of the container. These scanned images, radiation profiles, and digital photos are then stored in a database where they can be immediately retrieved if necessary.

The marine terminals in Hong Kong have invested in this system because they hope that a 100% scanning regime will deter a terrorist organization from placing a weapon of mass destruction in a container passing through their port facilities. Since each container's contents are scanned, if a terrorist tries to shield radioactive material to defeat the radiation portals, it will be relatively easy to detect the shielding material because of its density.

Another reason for making this investment is to minimize the disruption associated with targeting containers for portside inspection. The system allows the container to receive a remote preliminary inspection without the container leaving the marine terminal.

By maintaining a record of each container's contents, the port is able to provide government authorities with a forensic tool that can aid a follow-up investigation should a container with a weapon of mass destruction still slip through. This tool would allow authorities to quickly isolate the point in the supply chain where the security compromise took place, thereby minimizing the chance for a port-wide shut-down. By scanning every container, the marine terminals in Hong Kong are well-positioned to indemnify the port for security breaches. As a result, a terrorist would be unable to successfully generate enough fear and uncertainty to warrant disrupting the global trade system.

This low-cost inspection system is being carried out without impeding the operations of busy marine terminals. It could be put in place in every major container port in the world at a cost of $1.5 billion, or approximately $15 per container. Once such a system is operating globally, each nation would be in a position to monitor its exports and to check their imports against the images first collected at the loading port.

The total cost of third-party compliance inspections, deploying "smart" containers, and operating a cargo scanning system such as Hong Kong's is likely to reach $50 to $100 per container depending on the number of containers an importer has and the complexity of its supply chain. Even if the final price tag came in at $100 additional cost per container, it would raise the average price of cargo moved by, say, Wal-Mart or Target by only 0.06%. What importers and consumers are getting in return is the reduced risk of a catastrophic terrorist attack and its economic consequences.

In short, such an investment would allow container security to move from the current "trust, but don't verify" system to a more robust "trust but verify" regime. That would bring benefits to everyone but criminals and terrorists.

***Mr. Flynn** is the Jeane J. Kirkpatrick senior fellow for national security studies at the Council on Foreign Relations and author of* America the Vulnerable *(HarperCollins, 2005).*

Are We Ready Yet?

We're building our defenses against bioterrorism, but we face growing questions about costs and priorities.

CHRISTOPHER CONTE

Patrick Libbey likes to be positive. When the executive director of the National Association of County and City Health Officials is asked about how ready state and local governments are to respond to bioterrorism—it's a question he hears frequently—he'll tell you, "Significant progress has been made."

It's true. As the nation moves into its fifth year of a crash effort to build defenses against a possible bioterrorist attack, states have developed a wide range of emergency plans, conducted numerous exercises, and modernized their information and communications systems. But Libbey is not really satisfied with his answer. That's because there is no "consistent measure of where we are or where we need to be," he says. "That is probably our largest failing since the whole push began."

Libbey is not alone. George Hardy, executive director of the Association of State and Territorial Health Officers, sings a similar refrain. "It is absolutely critical that we develop some metrics," Hardy says. "We need accountability indicators for federal funds that are going out, and we need indicators for the overall level of community preparedness."

The two men, who together represent public health agencies serving the majority of Americans, no doubt have complex motives in pressing for standards to gauge preparedness efforts. On one hand, they are eager to keep federal bioterrorism funds—the first significant influx of new money into the public health system in decades—flowing. Many state and local public health agencies have launched ambitious infrastructure-building efforts on the expectation that bioterrorism readiness will continue to be a federal priority, and they clearly would like more support. Although the federal government has been sinking $1 billion annually into public health readiness, researchers at the RAND Corp. recently concluded from an assessment of California public health agencies that spending is only one-third to one-half of what local officials believe is needed.

At the same time, Libbey and Hardy also may be eager to draw some boundaries around bioterrorism-preparedness activities. The emphasis on preparing for mass-casualty incidents is diverting attention from other mounting public health problems—such as diabetes, tuberculosis and sexually-transmitted diseases—that may be relegated to the back burner for some time to come. "Our initial response to building emergency capacity was to treat the problem as if it were an emergency," Libbey says. "We thought we would drop everything we were doing, tend to the problem, and then everything would go back to normal. But there isn't the old 'normal' to go back to. Emergency preparedness is now part of our ongoing work."

All of this suggests that Americans are due for some serious soul-searching about what to expect from their public health system. What does it mean to be prepared? How do we balance the goal of making people safer with keeping them healthier? And what are the long-term implications of the growing role Washington is assuming in shaping a public health system that traditionally has been local in nature?

Mixed Results

Until recently, most health departments have been too busy trying to get a seat at the emergency-planning table to spend much time on such big-picture questions. "We spent the first year or more just building relationships," notes Darren Collins, director for the DeKalb County, Georgia, Board of Health's Center for Public Health Preparedness. That effort, at least, has paid off. Before 2001, public health departments rarely played a role in disaster planning. But now, they are integral players in most local emergency planning teams, serving on—and in some cases leading—unified incident command systems that can be activated any time there is an emergency. "We have much greater visibility in the community now," says Martin Fenstersheib, the health officer for Santa Clara County, just south of San Francisco. "With the advent of biological issues, we are now in at the top among first responders."

Along the way, public health agencies have acquired many new tools. Public health workers in Santa Clara County, for instance, now have "go-kits"—duffle bags containing personal protective gear, cameras, computers, specimen-gathering equipment and other tools to help them diagnose health problems in the field. The county also issues packs of laminated cards and reference materials so doctors can quickly recognize outbreaks of anthrax, smallpox, plague or Ricin toxin.

North Carolina's Health Department uses the Internet to review hospital admission records and check on the availability of hospital beds in real time—capabilities that could help it detect disease outbreaks more quickly and respond more effectively to mass-casualty situations. The state also has seven Public Health Regional Surveillance Teams, each consisting of a physician, an epidemiologist, an industrial hygienist and a field veterinary officer. A team can be deployed anywhere in the state in an emergency.

Montgomery County, Maryland, has been training public works and other county employees, as well as lay educators, parish nurses and other medical professionals, to serve as volunteers in emergencies, and it is helping schools, nursing homes and group homes make their own preparations to respond to emergencies. "In a crisis, we won't be able to get to people right away," says Kay Aaby, project coordinator at the county's Department of Health and Human Services. "They have to know how to shelter in place."

Should a pandemic of influenza hit, the Seattle-King County public health department has procedures for imposing a large-scale quarantine. It has identified a facility where sick people could be isolated from the general population. The department also is working with area governments and businesses to plan how they will keep operating during a pandemic, in which absenteeism could run as high as 30 percent.

Despite such progress, however, most public health officials concede the country still is not ready for a public health calamity. One reason for the gap is the sheer magnitude of the job and the dearth of trained personnel to do it. While public health laboratories have acquired a lot of new equipment in the past four years, more than half say they don't have enough scientists to run tests for anthrax or plague, and a majority lack sufficient capabilities to test for chemical terrorism, according to a report issued last year by Trust for America's Health, a public health advocacy group.

Similarly, two-thirds of the states still don't use the Internet to contribute to a national database on disease outbreaks—a capacity the federal Centers for Disease Control and Prevention believes would greatly increase the public health system's ability to identify and stamp out harmful biological agents. "A lot of gears are moving, but they aren't meshing together," says Shelley Hearne, TFAH's executive director. "The jalopy has gotten better oil and faster wheels, but it is not a race car that can get you from zero to 60 when you give it a good push on the gas."

Meanwhile, low pay and an aging workforce are complicating efforts to bolster public health staffs. Some 30 of 37 states that responded to a 2003 survey conducted by the Association of State and Territorial Health Officers reported shortages of public health nurses, 15 lacked enough epidemiologists, 11 didn't have as many laboratory workers as they need, and 11 others said they are shy of environmental specialists. And it may become difficult to stay even, let alone get ahead. The average age of public health workers is 46, compared with 40 for the overall U.S. workforce, and retirement rates are projected to run as high as 45 percent over the next five years.

Perhaps most important, there is growing concern among some public health officials about the impact preparedness efforts have had on our ability to deal with our chronic health problems.

Low pay and an aging work force are complicating much-needed efforts to bolster the public health system.

RAND researchers have noted that the increased emphasis on bioterrorism has led to retrenchments in programs to control sexually-transmitted diseases and tuberculosis, as well as teen pregnancy-prevention programs. In Madison County, North Carolina, public health nurse Jan Lounsbury had to drop work on programs to discourage smoking and promote exercise and better nutrition when she became bioterrorism planner for the county, in the state's western mountains. "Health promotion has been put on the back burner," says Lounsbury. And Russell Jones, a state epidemiologist based in Temple, Texas, warns that budget cuts have left the Lone Star State seriously vulnerable to an outbreak of pertussis (it had one several years ago). "In bioterrorism, we're adequate," Jones says. But otherwise, "if something bad happens, we are not prepared."

The New York-based Century Foundation, meanwhile, has questioned the cost-effectiveness of syndromic surveillance, a glitzy new technology that has gotten a big boost thanks to bioterrorism funding. It involves amassing a wide range of information, such as hospital admission forms, 911 calls, over-the-counter drug sales, school absenteeism figures and more, into a single database that allows analysts to look for clusters of symptoms that might indicate the beginning of a new disease outbreak. But it's expensive—New York City's system, widely considered the state of the art, costs $1.5 million annually, or about $4,000 a day—and so far largely unproven. "If it is sensitive enough to catch the few additional cases that may be the harbinger of a bioterrorist attack, it is likely to generate many false alarms as well and to command scarce resources," the foundation warns.

Defining Moments

One of the most glaring deficiencies in the public health system today is a serious gap between many health departments and the communities they serve. When RAND researchers assessed California public health departments, for instance, they noted that many local health departments have shockingly little information about their own communities. "In some jurisdictions," the analysts wrote, "representatives from police and fire departments appeared to have better knowledge of vulnerable populations than the health departments had."

Although this problem didn't begin with the national push for bioterrorism readiness, it hasn't been helped by it either. Some public health officials believe it may be a major reason why we are less safe than we need to be. That is the view of leaders of the Alameda County Public Health Department in California (it's unknown whether the department was part of the RAND study, which was conducted under a pledge of confidentiality). The department has devoted considerable time and effort to reaching out to neighborhoods. Most of its public health nurses work out of satellite offices in various communities, not out of the department's office in downtown Oakland. In some low-income communities, public health workers have gone door-to-door to survey residents, organize community meetings and otherwise help neighbors learn how to address common concerns. Public Health Director Arnold Perkins believes such "capacity building" will improve health in the targeted communities by reducing the myriad tensions and sense of isolation that significantly contribute to problems ranging from high infant mortality to heart disease, hypertension and asthma.

When federal bioterrorism funds began flowing to Alameda County, the department worked hard to integrate them into its ongoing activities, rather than let them be a diversion. One solution was to provide special "survival kits"—first aid items, food bars, water boxes, thermal blankets, ponchos, water purification tablets, dust masks, vinyl gloves and a combination radio-siren-flashlight—to people in the troubled communities in return for their participation in the capacity-building effort.

Are the kits just a clever subversion of a federal program? Not according to Anthony Iton, the county health officer. People in low-income neighborhoods have too many daily concerns—many lack health insurance or have high blood pressure; the local food stores sell liquor and junk food rather than fresh fruit and vegetables; and the neighborhood park has been taken over by drug dealers—to spend much time worrying about bioterrorism. By lacing information and self-help tools about bioterrorism into a broader effort to address residents' concerns, the department believes it has a better chance of getting its message across than if it tried to deal with preparedness separately. Moreover, by showing people that it is working to make their daily lives better, the health department is increasing the chances they will cooperate with it in an emergency—by staying home to slow the spread of a new virus or by coming to a mass inoculation site, for instance. "It's all about trust," says Iton. "If people don't trust us, they won't follow our instructions in emergencies."

Iton contends that the health department's work with poor neighborhoods effectively addresses one of the weakest links in current bioterrorism defenses. Most epidemics take root first in poor communities and then spread from there. In affluent communities, people are quick to see doctors if they get sick, increasing the chances that any new epidemic will be quickly detected. But in poor communities, people avoid going to doctors because they lack health insurance, can't afford to pay for health care or distrust mainstream institutions. That gives germs a better chance to incubate and spread. And because low-income people often commute from their homes to places such as airport concession stands, where diverse people commingle, these bugs have abundant opportunities to leap to the population at large. "Lack of health insurance, unfamiliarity with diseases and what to do about them, and the fact that the health care system turns poor people away—those are our greatest causes of vulnerability to bioterrorism in Alameda County," Iton argues.

Federal agencies aren't likely to accept as expansive a definition of emergency preparedness as Iton's views imply are necessary. To be sure, the Bush administration does view preparedness in broader terms today than it did a few years ago. When the U.S. Department of Homeland Security recently issued a "target capabilities list" for emergency planning, for instance, it identified 15 possible health calamities that would have "national significance"—including some, such as pandemic flu and major earthquakes and hurricanes, that aren't terrorism-related. But the list doesn't include lack of health insurance or community cohesion.

Nor does a list of 62 "critical tasks" and 34 performance measures recently issued by the CDC in response to calls for clearer standards to govern emergency public health preparations. Some of the proposed performance measures are highly specific (one says, for instance, that health departments should be able to bring the "initial wave" of personnel on board to conduct emergency operations within 90 minutes). Others deal with preparedness quite broadly (a knowledgeable public health professional should be available to answer a call reporting a suspicious disease within 15 minutes). But the CDC rules don't go so far as to tackle community capacity-building or require public health departments to spend more time meeting constituents in their own neighborhoods.

The current federal rules won't be the final world on preparedness; the CDC has pledged to meet with major players in the bioterrorism-preparedness arena to discuss them in detail. But the rules nevertheless represent a significant assertion of federal influence in a field that has generally been a state and local responsibility—one that nudges local health departments more in the direction of emergency preparedness. Given the absence of similar pressures to address other long-standing public health concerns, it's easy to imagine that local health departments' ability to deal with non-emergency health problems will become even more attenuated.

So here's a word of advice to NACCHO's Libbey, ASTHO's Hardy and other state and local representatives who will be discussing the issues with Washington in the months ahead: Be careful what you ask for.

***Christopher Conte** can be reached at crconte@earthlink.net*

The Double-Edged Effect in South Asia

V. R. Raghavan

The terms of engagement in the global war against terrorism waged in the aftermath of the September 11 attacks are being defined more or less exclusively by the United States. In this new environment, individual states and separate regional groups are repositioning themselves to maximize their room to maneuver. Such post–September 11 hegemonic politics ill serve the new security needs of the developing world, the most critical being economic growth coupled with social and political stability. Positing terrorism as an Islamic phenomenon—despite assertions to the contrary by the United States—has placed many developing countries in a strategic dilemma. South and Southeast Asian developing countries are either Islamic or contain significant numbers of adherents to Islam. The leadership of these countries has found it difficult to support the United States in the war on Islamic terrorism with the sullen if not hostile response by their populations.

Indeed, the sole superpower has introduced a new type of state into the realm of international law: the "harboring state," defined as one that funds, trains, or allows its territory to be used by proscribed terrorist groups. Once designated as such, the harboring state, potentially also called a rogue or outlaw state,[1] ceases to be entitled to the sovereignty guaranteed to each state under international law. India had expected and worked to have Pakistan included in such a category in light of its known support for groups practicing terrorism in Jammu and Kashmir.

Significant discontinuities in traditional alliances have also continued to emerge[2] as part of the North-South dichotomy in fighting the war on terrorism. The states of South and Southeast Asia, which see terrorism as a technique adopted to gain political ends, traditionally have taken the route of political and social assimilation of disaffected population segments that had taken to terrorism. Based in part on a long and unfortunate experience dealing with terrorism and its causes, developing states, including India, have pursued a strategy that has included both negotiations and military pressure, but not regime change or the annihilation of terrorist groups, to combat terrorism.

The Impact on South Asian Security

The events of September 11, 2001, changed South Asia's security calculus in unexpected ways. Transnational threats to U.S. security, particularly in Pakistan and Central Asia, brought a U.S. military presence nearer to India. The U.S. presence in the region after the attacks gave India an unexpected opportunity in its own war against terrorism. New Delhi expected the United States to see Pakistani-backed terrorism in Kashmir as being of the same ilk as the United States' terrorist enemies and the situation in Kashmir as a terrorist war. To India's surprise, however, its partnership with the United States did not lead Washington to seek to isolate Pakistan despite the fact that, since the nuclear tests of 1998, India had carefully developed that partnership in numerous ways. New Delhi had sought to reassure the United States, among others, about its nuclear policy by making a firm commitment to a moratorium on nuclear tests and to a no-first-use policy; the Indian government had energetically pushed economic reform and established a range of economic and trade partnerships with the United States; and India had welcomed some elements of U.S. policies pertaining to the Anti-Ballistic Missile (ABM) Treaty and missile defense.

The September 11 attacks occurred at a point when Indo-U.S. relations were on a significant upward trend and terrorism in Jammu and Kashmir was at a peak, most of it organized by groups operating from Pakistan. The Indian government thus expected that Pakistan, whose connections with the Taliban and Al Qaeda had become evident, would come under intense U.S. pressure, including the demand for a change in Pakistan's policy in Jammu and Kashmir. It soon became apparent, however, that, notwithstanding Pakistan's role in accommodating the Taliban and Al Qaeda, the United States saw Pakistan's potential to eliminate these two groups and change the political structures in Afghanistan as more important. Pakistan's president, Gen. Pervez Musharraf, had become the United States' most important ally in the war against Al Qaeda.

Two months later, the reality of the new U.S. need for Pakistani cooperation was reinforced. In response to the December 13, 2001, attacks on India's parliament and the subsequent January 2002 attack on the Kaluchak army camp, the Indian government ordered a military mobilization along the India-Pakistan border. In light of each country's possession of nuclear weapons, the Indian military mobilization raised the prospect of war between India and Pakistan spiraling toward a nuclear exchange, significantly raising the international stakes in avoiding such a war. From India's point of view, the U.S. response at the time was unsatisfactory. New Delhi believed that Pakistan knew the United States felt it had to maintain good bilateral relations with Islamabad. Given its new regional relationships, the proximate cause of India's 2002 military mobilization essentially landed the United States in the middle of the zero-sum contest between India and Pakistan, making "[o]ne of the most irreconcilable conflicts in the world ... Washington's business."[3]

India was surprised that Washington did not seek to isolate Pakistan after 9/11.

This sequence of regional crises demonstrated that nuclear deterrence, albeit fragile, was viable in the region. Previously, the 1999 Kargil conflict had demonstrated that the Pakistani leadership believed that nuclear deterrence enabled limited operations. As a result, in 1999 it occupied positions on the Indian side of the line of control (LOC) in Kashmir. The obvious military strategy for India would have been to widen the conflict by seizing Pakistani-held territories across the LOC, but India chose instead to confine its substantial military operations to its side of the LOC, a decision influenced in no small degree by the two countries' nuclear capabilities. At that time, New Delhi concluded that the risks and potential costs of a general conflict and possible nuclear exchange usurped attempts to resolve the dispute in Kashmir by force.

After India's military mobilization in 2002, Pakistan's military regime temporarily curtailed the infiltration of militants into India, in no small part as a result of immense U.S. and international pressure on Musharraf to do so. Traditionally, India had dealt with terrorism in Kashmir through defensive and reactive strategies. Security operations had been confined to searching for, arresting, and destroying militant groups in the state. Yet, this approach was insufficient for coping with an endless flow of armed terrorist groups from Pakistan into Jammu and Kashmir. New Delhi termed this terrorism "state-sponsored terrorism" because of the fact that the leadership of such groups operated openly from within Pakistan. These groups had close links with the Taliban and Al Qaeda, and many of them were trained in Afghanistan.

The 2002 military mobilization shows that, after Kargil, Indian strategy had graduated from defensive to proactive, offensive responses to terrorism. Surely, New Delhi's thinking was influenced by the response to the September 11 attacks and subsequent U.S. antiterrorist operations in Afghanistan. The Kargil conflict had already demonstrated that a limited war would not necessarily lead to a nuclear exchange. The link between the 1999 Kargil low-intensity conflict, the September 11 terrorist attacks, the December 13 attack on India's parliament, and India's mobilization of troops in 2002 was complete.

India's political and military leadership began espousing a theory of limited war in a nuclear environment by which India could retaliate directly against Pakistan and would be morally justified in doing so. Henceforth, it was up to the regime in Islamabad to decide whether it wanted further escalation. On the other hand, because India limited its response to troop mobilization in 2002, it retained some significant options to deter, and to use in the event of, Pakistan's escalation of terrorism. We may therefore see India respond in the future with punitive military actions such as air strikes against terrorist infrastructure and military forays to take out terrorist bases in Pakistani territory.

In 2002, Indian strategy graduated from defensive to proactive responses to terrorism.

Simultaneously being developed after the September 11 attacks, the U.S. counterterrorism posture bolstered the Indian government's stance. In mobilizing its own troops, India was in a better political position than it had been in a decade to pursue a strategy to compel Pakistan to stop harboring or otherwise supporting terrorists.[4] Because Pakistan's complicity in the role of the Taliban and Al Qaeda in international terrorism had become unequivocally clear, and the United States was already launching operations from Pakistani territory into Afghanistan, India expected its own military responses against terrorists and governments harboring terrorists to be accepted, if not supported internationally.

A new set of regional dynamics thus emerged as a result of tectonic changes in strategic relations after September 11. First, the prospects of a nuclear exchange were believed to be credible through an escalatory process of conventional military conflict. Second, 2002 showed that conventional war could start as a result of terrorist acts. Third, both Indian prime minister Atal Bihari Vajpayee and U.S. president George W. Bush faced similar challenges: the two elected leaders of liberal democracies had to respond to public pressure and the expectations of determined and decisive action in the face of major terrorist acts.

Changes in Pakistan

Pakistan's posture in South Asia has been significantly affected by the new international security environment and particularly its participation in the U.S.-led war on terrorism. To assure its national security, Pakistan could ill afford to remain isolated and be singled out as a rogue state. It had to disassociate itself from the Taliban and Al Qaeda, cooperate in the installation of a new government in Kabul, and rethink its policy choices on Jammu and Kashmir. Geopolitical factors made Pakistan's cooperation in the war on terrorism all the more necessary.

Traditionally, Pakistan perceived that Afghanistan provided strategic depth against threats from India and had pursued ties with the Taliban and Al Qaeda to obtain that strategic advantage. Yet, after Washington had already launched military operations into Afghanistan, it threatened to do the same in Pakistan. Under pressure from the United States and the rest of the international community,[5] Musharraf reversed some of the four fundamental pillars of Pakistan's security and foreign policy regarding Kashmir, Afghanistan, its nuclear weapons program, and military rule.

A Changed International Security Environment

Beyond South Asia, the scale and impact of the attacks on the United States have raised terrorism from a local or regional phenomenon to a global strategic threat. Post–September 11, the very nature of terrorism itself has changed. Were this period in international relations to be defined in terms of war, it might best be described as an era of asymmetric war generated by terrorist attacks. Modern technology coupled with the new nature of the terrorist threat present unprecedented challenges.

Two to three decades ago, terrorism was about local issues and conducted by small groups that sought to draw attention to their cause through terrorist acts. Terrorism was designed to kill a few and have large audiences watching it for effect. Today, it is ideologically motivated, its agenda is not limited to one country, and it is international in character. The asymmetric character of international terrorism, conducted by elusive perpetrators, has a decapacitating impact on the people and on the state it targets. It forces states and leaders to reconsider policies and respond to terrorist groups through an international effort. The world faces this new form of what some have called fourth-generation warfare, which "pits nations against nonnational organizations and networks that include not only fundamental extremists, but ethnic groups, mafias, and narcotraffickers as well. Its evolutionary roots may lie in guerrilla warfare, but it is rendered more pervasive and effective by technologies, mobilities, and miniaturized instrumentalities spawned by the age of computers and mass communication."[6]

Clearly, therefore, terrorism's increasing lethality, access to sophisticated weapons and technology, the force-multiplying effect of state-sponsored terrorism, religious motivation, the proliferation of amateurs, and operational competence all make the current threat posed by asymmetric warfare no less than horrific. In the current context, amateurs have ready and easy access to the means and methods of terrorism. Terrorism has become accessible to anyone with a grievance, an agenda, a purpose, or any idiosyncratic combination of the above. Terrorists are particularly dangerous because it is even more difficult to track and anticipate their attacks.[7] Today, modern technology in the service of terrorism provides no warning, and its perpetrators vanish with the act they have committed.[8]

In the future, India may use air strikes to take out terrorist bases in Pakistani territory.

The role of all states, democratic and not, as well as the challenges they face in this new international context of asymmetric warfare are also changing in fundamental ways. Global terrorism cannot flourish without the support of states that either overtly sympathize with or acquiesce in its actions. Cold War principles of deterrence are almost impossible to implement when a multiplicity of states are involved, some of them harboring terrorists who are in a position to wreak havoc. The transnational nature of terrorism has led governments to adopt new doctrines and develop collective regional efforts.

South Asia has dealt with conflict-generating terrorism for more than 20 years. The experience has been marked by state sponsorship of terrorism and, in other cases, controlled by elements outside the disputant countries. Even though terrorist groups are operationally separate, they share many similar dynamics and goals. To combat these threats, local responses should similarly match international efforts. The South Asian Association for Regional Cooperation (SAARC) Additional Protocol on Terrorism, signed in early 2004, is a step in this direction for regional efforts to combat terrorism by establishing and maintaining a financial intelligence unit to fight terrorism. Previously, the 1987 SAARC Agreement on the Suppression of Terrorism had fallen short of taking specific counterterrorism measures against terrorist financing.

Pakistan made a deliberate decision to join the 2004 SAARC Additional Protocol on Terrorism to show its determination to handle international terrorism, but the problem remains. Although Pakistan seeks to cooperate in the international effort to fight terrorism, it continues to encourage terrorist activity on its own soil to serve its goals in Kashmir. Even though the region resolves to fight terrorism, a bidirectional approach by a state clearly exists. Despite the fact that the Indian elections in Kashmir changed the political dynamic completely, because of interference and support from Pakistan, militant challenges remain.

When states threaten one another for incongruent reasons in a situation such as this, who deters whom and in the face of what kind of provocation?[9] In the twenty-first century, states face the arduous challenges of identifying (1) the enemy (whether a terrorist organization or a regime); (2) the terrorists' location (their territory, ideology, human resources, and financial base); and (3) the situation and the level at which military power should be used (against whom and where). To speak of a global war on terrorism distorts thinking by suggesting that there is an easily identifiable enemy and an obvious means of attack. Counterterrorism involves aggressive deterrence and prevention on several levels, but, after all, against whom should a state wage war? Should Spain be attacked because the Madrid bombers lived and plotted there?

A long list of states has directly, covertly, or even unintentionally contributed to the success of international terrorist groups. September 11 has changed international security and the international system so drastically that threat perceptions and responses have to be reexamined. The new face of terrorism is one of dozens of local groups across the world connected by a global ideology. U.S. foreign policy has changed perceptibly to deal with such threats, and military preemption has come to form the core of its policy options. Terrorism is now viewed as the principal foreign policy challenge to the United States.

Similarly, the strategic future in South Asia is vulnerable; any terrorist attack akin to those of December 13, 2001, and Kaluchak in January 2002 could bring about a new crisis. Indian policy imperatives now envisage a compellence strategy that has been bolstered by the events of September 11, 2001. Although the distinction between terrorist and military acts was apparent earlier, this is no longer the case. The distinctions between regular armies, irregular armies, and mujahideen have been blurred. This implies that Indian military forces should be kept at a high state of readiness.

Impact on India

U.S. policy substantially affected India's interests in South Asia following September 11 by requiring Pakistani-U.S. relations to fulfill U.S. strategic and military objectives in Afghanistan and in the oil- and gas-rich Central Asian region. The U.S. need for Pakistan to have a substantial role in handling Afghanistan and the Taliban places a new perspective on India's approaches to conflict resolution and dispute settlement with Pakistan on the issue of Kashmir.

If India had hoped for a constructive response to its being targeted by global terrorism, international action after the December 13 parliament attack leaves no doubt about the future course of action. India carries its burden of combating terrorism on its own. It would need to act alone to force a change of attitude and conviction in Pakistan; the September 11 attacks and international opinion can help only to a certain extent. For this reason, India's peace initiatives with Pakistan broke new ground after September 11, 2001.

India's peace initiatives with Pakistan broke new ground after September 11

Pakistan now also finds itself increasingly vulnerable to major terrorist attacks. Musharraf and some of his top military commanders repeatedly have experienced assassination attempts. Such developments have in turn led to a new understanding of the need to stabilize Indo-Pakistani relations. Vajpayee began the process in April 2002 by extending his hand in friendship to Pakistan on Kashmiri soil. At the 12th SAARC summit in Islamabad in early January 2004, Vajpayee set a conciliatory tone in his speech by focusing on strengthening the organization. India also agreed on the additional protocol updating the 1987 Convention Against Terrorism. After the 2004 elections, the new government in New Delhi is sustaining the momentum created by Vajpayee's Bharatiya Janata Party government.

Clearly, September 11 has served as a catalyst to move diplomatic relations between India and Pakistan forward. Although the immediate aftermath of India's own December 13 terrorist attacks resulted in the 2002 border confrontation and seemed to increase the risk of war, the recent dialogue process outlines just the opposite: both countries argue that nuclear weapons actually add to regional stability. In that environment, a range of discussions between the two governments to resolve all outstanding disputes has gained currency. The importance accorded to improved ties by the new government in New Delhi is evident in its efforts immediately upon assuming power to seek a close relationship with its counterpart in Islamabad. Irrespective of the outcome of the dialogue process, the intentions are clearly to build rapprochement so as to combat international terrorism together. This is the most promising and positive impact of the September 11 attacks.

Notes

1. Michael Krepon, *Cooperative Threat Reduction, Missile Defense and the Nuclear Future* (New York: Palgrave Macmillan, 2003).
2. For a detailed description of the elements of continuity and discontinuity since the September 11 attacks, see Nicholas Williams, "September 11: New Challenges and Problems for Democratic Oversight," presentation at the Geneva Centre for the Democratic Control of Armed Forces workshop on "Criteria for Success and Failure in Security Sector Reform," Geneva, September 5–7, 2002.
3. Jessica T. Matthews, "September 11, One Year Later: A World of Change," *Policy Brief*, no. 18 (August 2002): 7, http://www.ceip.org/files/pdf/Policybrief18.pdf (accessed July 3, 2004).
4. "The Delicate Strategic Balance in South Asia," *Strategic Survey 2002–2003* (Oxford, UK: Oxford University Press).
5. Ajay Darshan Behera, "On the Edge of Metamorphosis," in *Pakistan in a Changing Strategic Context*, eds. Ajay Darshan Behera and Joseph C. Matthew (New Delhi: Knowledge World, 2004).
6. Harold A. Gould and Franklin C. Spinney, "Fourth-Generation Warfare," *Hindu*, October 10, 2001.
7. Ian O. Lesser, "Coalition Dynamics in the War Against Terrorism," *International Spectator*, February 2002.
8. Henry Kissinger, "Pre-emption and the End of Westphalia," *New Perspectives Quarterly* 19, no. 4 (Fall 2002).
9. Ibid.

***V. R. Raghavan** is a retired lieutenant general and former director general of military operations of the Indian Army. He is currently president of the Centre for Security Analysis in Chennai and director of the Delhi Policy Group.*

UNIT 10

Future Threats

Unit Selections

31. **The Changing Face of Al Qaeda and the Global War on Terrorism**, Bruce Hoffman
32. **The Terrorism to Come**, Walter Laqueur

Key Points to Consider

- Could terrorists acquire a nuclear capability? What factors would influence their decision to use a nuclear device?
- How should the U.S. government respond to a more "nimble, flexible, and adaptive," al-Qaeda?
- Can the war on terrorism be won?

Student Website

www.mhcls.com/online

Internet References

Further information regarding these websites may be found in this book's preface or online.

Centers for Disease Control and Prevention—Bioterrorism
http://www.bt.cdc.gov

Nuclear Terrorism
http://www.nci.org/nci/nci-nt.htm

Terrorism will, undoubtedly, remain a major policy issue for the United States well into the 21st century. Opinions as to what future perpetrators will look like and what methods they will pursue continue to vary. While some argue that the traditional methods of terrorism, such as bombing, kidnapping, and hostage taking, will continue to dominate the new millennium, others warn that weapons of mass destruction or weapons of mass disruption, such as biological and chemical weapons, or even nuclear or radiological weapons, will be the weapons of choice for terrorists in the future.

Experts believe that there are certain trends that will characterize international terrorism in the coming years. Some scholars predict that the continuing rise of Islamic extremists will give rise to a new generation of violent, anti-American terrorists. Others warn of a rejuvenation of left-wing terrorism in Europe. Most believe that the tactics employed by terrorists will be more complex. Future terrorism will cause more casualties and will involve the use of weapons of mass destruction.

Bruce Hoffman in "The Changing Face of Al Qaeda and the Global War on Terrorism" examines the shift in strategies and tactics used by Al Qaeda. He maintains that Al Qaeda has become more "nimble, flexible, and adaptive." He suggests that that in order to cope with terrorism, the U.S. government must do the same. Finally, Walter Laqueur in "The Terrorism to Come" treats the reader to a sweeping overview of modern terrorism. He closes with a hard dose of realism arguing that "... there can be no victory, only an uphill struggle, at times successful, at others not."

The Changing Face of Al Qaeda and the Global War on Terrorism

This article assesses current trends and development in terrorism within the context of the overall progress being achieved in the global war on terrorism (GWOT). It examines first the transformation that Al Qaeda has achieved in the time since the 11 September 2001 attacks and the variety of affiliated or associated groups (e.g., what are often referred to as Al Qaeda "clones" or "franchises") that have emerged to prosecute the jihadist struggle. It then focuses on recent developments in Saudi Arabia and especially Iraq in order to shed further light on Al Qaeda's current strategy and operations. In conclusion, this article offers some broad recommendations regarding the future conduct of the GWOT.

BRUCE HOFFMAN

The RAND Corporation
Washington, DC, USA

The Al Qaida of the 9/11 period is under catastrophic stress. They are being hunted down, their days are numbered.

—Amb. Cofer Black, U.S. State Department, Counter-Terrorism Coordinator, January 2004.[1]

The Americans only have predications and old intelligence left. It will take them a long time to understand the new form of al-Qaida.

—Thabet bin Qais, Al Qaeda spokesperson, May 2003.[2]

The plots were textbook Al Qaeda, even if the would-be perpetrators were not. Hijack a jet plane loaded with ordinary travelers and deliberately crash it into the tower of a prominent local landmark. Simultaneously dispatch multiple truck bombs to destroy embassies and other diplomatic facilities. Use a small boat laden with explosives to sink a large, powerful warship. Each reprised an infamous Al Qaeda operation: 9/11, the 1998 East Africa embassy bombings, and the attack on the *U.S.S. Cole*—but in this case all the attacks were to take place in Singapore. Moreover, the plotters were not battle-hardened *mujahideen* ("holy warriors") who had cut their teeth fighting Egypt's security forces or against the Northern Alliance in Afghanistan. Nor were they the usual Al Qaeda cadre favored for such spectacular operations: young, Arab males, mostly from Saudi Arabia, Yemen, or the Gulf, whose background, family ties, and bona fides inspired the trust and confidence of that movement's senior leadership. Rather, they were an utterly unremarkable group of middle-aged Singaporeans. Some were married and some were single. Some were businessmen with university-level degrees, whereas others were cab drivers or janitors. What they did have in common was a profoundly deep devotion to their Muslim faith alongside an all-consuming hatred of the United States and the West. They had acquired both convictions not as impressionable youths either as students in *madressehs* (Islamic schools) or worshippers in radical mosques, but comparatively late in life by attending small meetings and religious sessions held in the living rooms and kitchens of the ubiquitous high-rise apartments that dot Singapore. They were therefore likely regarded by the infamous KSM—Khalid Sheik Mohammed, the bin Laden lieutenant to whom their cell circuitously reported—as the "ultimate fifth column," whose age, background, and Asian—as opposed to Arab—appearance were calculated to allay rather than arouse the suspicions of domestic and foreign security and intelligence offices alike. The fact that many of the cell's members came from the traditionally moderate, English-speaking, expatriate community of Malabari Indian Muslims long resident in Singapore was an added bonus. Indeed, KSM and the group's other controllers would likely have known that although Singapore's highly efficient Internal Security Department would have kept close watch over newer immigrants and the more radicalized communities among that city-state's large Malay-speaking Muslim population, they might have paid less attention to such groups as the well-established Malabaris.[3]

The Singaporean cell embodies a new breed of post-9/11 terrorist: men animated and inspired by Al Qaeda and bin Laden, but who neither belong specifically to Al Qaeda nor directly follow orders issued by bin Laden. The Singaporeans were members of an entirely separate group, albeit one closely associated with Al Qaeda, known as Jemaah Islamiyah (JI, or the "Islamic Organization"). The aim of this predominantly Indonesian organization, whose origins can be traced to the early 1970s, is to establish a unified Islamic nation, guided by strict interpretation of *shari'a* (Islamic law) among the countries of Southeast Asia—including Indonesia, Malaysia, the Philippines, Brunei, and Singapore. Jemaah Islamiyah and the other radical Islamic Southeast Asian movements like it (e.g., the Moro Islamic Liberation Front in the Philippines, and the Kumpulan Mujahidin Malaysia) potentially represent an even more insidious and pernicious threat than Al Qaeda.

The Singaporean JI cell also reflects how, in the time since 9/11, Al Qaeda has deliberately sought to exploit local causes and re-align mostly parochial interests with its own transnational, pan-Islamist ideology. The transformation of the Islamic Movement of Uzbekistan (IMU) from an organization once focused mainly on Central Asia into one that now champions bin Laden's ambitious international vision of a re-established Caliphate is a case in point. In other instances, moreover, local cells have been surreptitiously co-opted by Al Qaeda so that, unbeknownst to their rank and file, the group pursues Al Qaeda's broader, long-range goals in addition to (or even instead of) its own, more provincial goals. This process has been particularly evident among some Algerian terrorist cells operating in European countries.

Finally, the Singapore cell encapsulates the terrorism–counterterrorism conundrum that exists today. As counterterrorism measures improve and become stronger, Al Qaeda and its affiliates and associates must constantly scramble to adapt themselves to the less congenial operational environments in which they now have to operate. For the terrorists, this inevitably entails tapping into new and different pools of recruits, adjusting targeting and modus operandi to obviate governmental countermeasures and an enforced evolutionary process on which their survival depends. Indeed, the main challenge for the radical jihadist movement today is to promote and ensure its durability as an ideology and concept. It can only do this by staying in the news. New attacks are therefore needed to maintain their relevance as a force in international politics and to enhance their powers of coercion and intimidation.

These processes largely explain the current patterns of terrorism being seen. Indeed, as military and government targets increase their protection levels, softer targets such as comparatively more vulnerable economic and commercial sites have thereby become more attractive. The attacks staged by Al Qaeda associates or affiliates such as JI in Bali in October 2002 and in Jakarta last August; by Assiriyat al-Moustaqim in Morocco in May 2003; by the Islamic Great Eastern Raiders Front in Turkey the following November;[4] by the jihadist cell comprised mostly of Moroccan nationals responsible for the bombings of commuter trains in Madrid last March (that killed more than 200 and injured over 1,600 others);[5] the semi-autonomous operations of Al Qaeda's Arabian Peninsula unit in Saudi Arabia; and that of Abu Musab Zarqawi and his Jamaat al Tawhid and Islam and Jihad organization in Iraq, evidence this pattern.

Al Qaeda Today

Since 9/11, Al Qaeda has clearly shown itself to be a nimble, flexible, and adaptive entity. Because of its remarkable durability, the progress that the United States and allies achieved during the first phase of the GWOT—when Al Qaeda's training camps and operational bases, infrastructure and command-and-control nucleus in Afghanistan were destroyed and uprooted—has thus far proven elusive during this subsequent phase. In retrospect, the loss of Afghanistan does not appear to have affected Al Qaeda's ability to mount terrorist attacks to the extent hoped.[6] Afghanistan's main importance to Al Qaeda was as a massive base from which to prosecute a conventional civil war against the late Ahmad Shah Masoud's Northern Alliance. Arms dumps, training camps, staging areas, and networks of forward and rear headquarters were therefore required for the prosecution of this type of conflict. These accoutrements, however, are mostly irrelevant to the prosecution of an international terrorism campaign—as events since 9/11 have repeatedly demonstrated.[7] Indeed, Al Qaeda had rebounded from its Afghanistan setbacks within months of the last set-piece battles that were fought in the White Mountains along the Pakistani border at Shoh-e-Kot, Tora Bora, and elsewhere between December 2001 and March 2002. The attacks in Tunisia in April 2002 and in Pakistan the next month provided the first signs of this movement's resiliency. These were followed in turn by the attacks in Bali, Yemen, and Kuwait the following October, and then by the coordinated, near-simultaneous incidents in Kenya that November.[8]

Perhaps Al Qaeda's greatest achievement, though, has been the makeover it has given itself since 2001.[9] On the eve of 9/11, Al Qaeda was a unitary organization, assuming the dimensions of a lumbering bureaucracy. The troves of documents and voluminous data from computer hard disks captured in Afghanistan, for example, revealed as much mundane bumf as grandiose plots: complaints about expensive cellphone bills and expenditures for superfluous office equipment[10] as well as crude designs for dreamt-about nuclear weapons.[11] Because of its logistical bases and infrastructure, that now-anachronistic version of Al Qaeda had a clear, distinct center of gravity. As seen in the systematic and rapid destruction inflicted during the military operations in Afghanistan during the GWOT's first phase, that structure was not only extremely vulnerable to the application of conventional military power, but played precisely to the American military's vast technological strengths. But in the time since 9/11, bin Laden and his lieutenants have engineered nothing short of a stunning transformation of Al Qaeda from the more or less unitary, near bureaucratic entity it once had been to something more akin to an ideology. Al Qaeda today, as other analysts have noted, has become more an idea or a concept than an organization;[12] an amorphous movement tenuously held together by a loosely networked transnational constituency rather than a monolithic, international terrorist organization with either a defined or identifiable command and control apparatus. Al Qaeda in essence has transformed itself from a bureaucratic entity that could be destroyed or an irregular army that could be defeated on the battlefield to a less-tangible transnational movement true to its name—the "base of operation" or "foundation" or, as other translations have it, the "precept" or "method."[13] The result is that today there are many Al Qaedas rather than the single Al Qaeda of the past. It has become a vast enterprise—an international movement or franchise operation with like-minded local representatives, loosely connected to a central ideological or motivational base, but advancing their common goal independently of one another.

Amazingly, Al Qaeda also claims that it is stronger and more capable today than it was on 9/11.[14] Al Qaeda propagandists on websites and other forums point repeatedly to a newfound vitality that has facilitated a capability to carry out at least two major attacks per year since 9/11 compared to the one attack every two years that it could implement before 9/11. "We are still chasing the Americans and their allies everywhere," Ayman al Zawahiri crowed in December 2003, "even in their homeland."[15] Irrespective of whether the U.S. definition of a major attack and Al Qaeda's are the same, propaganda does not have to be true to be believed: all that matters is that it is communicated effectively and persuasively—precisely the two essential components of information operations that Al Qaeda has arguably mastered.

That Al Qaeda can continue to prosecute this struggle is a reflection not only of its transformative qualities and communications skills, but also of the deep well of trained jihadists from which it can still draw. According to the authoritative annual *Strategic Survey*, published by the London-based International Institute for Strategic Studies, a cadre of at least 18,000 individuals who trained in Al Qaeda's Afghanistan camps between 1996 and 2001 are today theoretically positioned in some 60 countries throughout the world.[16] Moreover, Al Qaeda's management reserves seem to be similarly robust—at least to an extent perhaps not previously appreciated. A "corporate succession" plan of sorts has seemed to function even during a time when Al Qaeda has been relentlessly tracked, harassed, and weakened. Al Qaeda thus appears to retain at least some depth in managerial personnel as evidenced by its abilities to produce successor echelons for the mid-level operational commanders who have been killed or captured. It also still retains some form of a centralized command and control structure responsible for gathering intelligence, planning, and perhaps even overseeing more spectacular attacks against what are deemed the movement's most important, high-value targets in the United States. The computer records, e-mail traffic, and other documents seized by Pakistani authorities when a computer-savvy Al Qaeda operative named Mohammed Naeem Noor Khan was apprehended in August 2004 point to the existence of a more robust, centralized entity than had previously been assumed.[17]

Moreover, despite the vast inroads made in reducing terrorist finances and especially financial contributions, Al Qaeda doubtless still has sufficient funds with which to continue to prosecute its struggle. According to one open source estimate, some $120 million of *identifiable* Al Qaeda assets has been seized or frozen.[18] Given that bin Laden reputedly amassed a war chest in the billions of dollars, ample funds may still be at the disposal of his minions. At one point, for example, bin Laden was reputed to own or control some 80 companies around the world. In the Sudan alone, according to Peter Bergen, he owned all of that country's most profitable businesses, including construction, manufacturing, currency trading, import-export, and agricultural enterprises.[19] Not only were many of these well managed to the extent that they regularly turned a profit, but this largesse in turn was funneled to local Al Qaeda cells that in essence became entirely self-sufficient, self-reliant terrorist entities in the countries within which they operated.[20] The previously cited recent IISS report focuses on this issue as well. "While the organization and its affiliates and friends do not enjoy the financial fluidity that they did before the post-11 September counterterrorism mobilization," the report notes, "neither do they appear shorn of resources." The analysis disquietingly also explains that

> terrorist operations are asymmetrically inexpensive. The Bali bombing cost under $35,000, the *USS Cole* operations about $50,000 and the 11 September attacks less than $500,000. Moving large amounts of cash therefore is not an operational necessity. Furthermore, since the Afghanistan intervention forced al-Qaeda to decentralize and eliminated the financial burden of maintaining a large physical base, al-Qaeda has needed less money to operate.[21]

Finally, and above all, despite the damage and destruction, personnel and key leadership losses that it has sustained over the past three years, Al Qaeda stubbornly adheres to its fundamental strategy and objectives thereby continuing to inspire the broader radical jihadist movement. Bin Laden years ago defined this strategy as a two-pronged assault on both the "far enemy" (the U.S. and the West) and the "near enemy" (those reprobate, authoritarian, anti-Islamic regimes in the Middle East, Central Asia, South Asia, and South East Asia against whom the global jihadist movement is implacably opposed).[22] During the past year, for example, terrorist strikes have rocked Madrid and Istanbul (representing the "far enemy") as well as Riyadh, Baghdad, Islamabad and Jakarta (the "near enemies"). The periodic release of fresh targeting guidance and operational instructions has helped to give renewed focus and sustain this strategy. A particularly important guidance document, titled the "Camp al Battar [the sword] Magazine," was released around 31 March 2004. Reportedly written by the late Abdul Azziz al-Moqrin,[23] the reputed commander of Al Qaeda's operations on the Arabian Peninsula, it sheds considerable light on the current pattern of jihadist attacks in Saudi Arabia and Iraq. For example, it identifies as high attack priorities economic targets in the Middle East—and especially those connected with the region's oil industry. "The purpose of these targets," Moqrin wrote,

> is to destabilize the situation and not allow the economic recovery such as hitting oil wells and pipelines that will scare foreign companies from working there and stealing Muslim treasures. Another purpose is to have foreign investment withdrawn from local markets. Some of the benefits of those operations are the effect it has on the economic powers like the one that had happened recently in Madrid where the whole European economy was affected.[24]

To some extent, this strategy has already begun to bear fruit. The U.S. State Department, for example, has advised American workers and their families to leave Saudi Arabia. Following the murder in April of five expatriate employees of a petrochemical complex in the Saudi industrial city of Yanbu, foreign companies were reported to have evacuated personnel from the country.[25] Whatever optimism remained that the situation might quickly improve were dashed with the May attack on a housing complex in Khobar, where 22 foreigners were killed, and in the past weeks by the beheading of an American defense contractor, Paul M. Johnson, Jr.[26]

This same targeting guidance also explains the spate of kidnappings and execution of foreign workers in Iraq. Within a week of this targeting instruction's dissemination, for instance, the current wave of kidnapping of foreign nationals commenced in Iraq. The first victim was Mohammed Rifat, a Canadian, who was seized on 8 April and is still missing. Since then more than 60 others have been kidnapped. Although the majority has been released, to date five have been murdered and six are still missing. Among the dead is Nicholas Berg, who was kidnapped on 9 April and whose decapitated body was found a month later as well as a Korean national, Kim Sun II, who was kidnapped on 17 June and whose headless body was discovered five days later.[27] Moqrin deemed "[a]ssassinating Jewish businessmen" a special priority in order to "teach lessons to those who cooperate with them":[28] thus explaining why Berg, a Jewish-American, was doubtless symbolically so valuable a victim to Zarqawi and his followers.

Moqrin elaborated at some length on the "practical examples" of his targeting guidance. Following Jews—"American and Israeli Jews first, the British Jews and then French Jews and so on"— are "Christians: Their importance is as follows: Americans, British, Spanish, Australians, Canadians, Italians." Within these categories there are further distinctions: "Businessmen, bankers, and economists, because money is very important in this age"; followed by "Diplomats, politicians, scholars, analysts, and diplomatic missions"; "Scientists, associates and experts"; "Military commander and soldiers"; and, lastly, "Tourists and entertainment missions and anybody that was warned by mujahideen not to go to step in the lands of Moslems."[29] Jews and businessmen, as is seen, head the list.

The clearest explication of Al Qaeda's broad strategy in Iraq is perhaps the one provided by Zawahiri himself on the occasion of the second anniversary of the 9/11 attacks. "We thank God," he declared, "for appeasing us with the dilemmas in Iraq and Afghanistan. The Americans are facing a delicate situation in both countries. If they withdraw they will lose everything and if they stay, they will continue to bleed to death."[30] Indeed, what U.S. military commanders optimistically described last year as the jihadist "magnet" or terrorist "flytrap"[31] orchestrated by the U.S. invasion of Iraq, is viewed very differently by Al Qaeda. "Two years after Tora Bora," Zawahiri observed in December 2003, "the American bloodshed [has] started to increase in Iraq and the Americans are unable to defend themselves."[32] For Al Qaeda, therefore, Iraq's preeminent utility has been as a useful side-show: an effective means to preoccupy American military forces and distract U.S. attention while Al Qaeda and its confederates make new inroads and strike elsewhere. On a personal level, it may have also provided bin Laden with the breathing space that he desperately needed to further obfuscate his trail.

But most significant perhaps is that bin Laden and his fellow jihadists did not drive the Soviets out of Afghanistan by taking the fight to an organized enemy on a battlefield of its choosing. In fact, the idea that Al Qaeda wanted to make Iraq the central battlefield of jihad was first suggested by Al Qaeda itself. In February 2003, before the coalition invasion of Iraq even began, the group's information department produced a series of articles titled *In the Shadow of the Lances* that gave practical advice to Iraqis and foreign jihadists on how guerrilla warfare tactics could be used against the American and British troops. The call to arms by Al Qaeda only intensified after the fall of Baghdad, when its intermittent website, al Neda, similarly extolled the virtues of guerrilla warfare: invoking prominent lessons of history—including America's defeat in Vietnam and the Red Army's in Afghanistan. Under the caption "Guerrilla Warfare Is the Most Powerful Weapon Muslims Have, and It is The Best Method to Continue the Conflict with the Crusader Enemy," these lessons of history were cited to rally the jihadists for renewed battle. "With guerilla warfare," the statement explained,

> the Americans were defeated in Vietnam and the Soviets were defeated in Afghanistan. This is the method that expelled the direct Crusader colonialism from most of the Muslim lands, with Algeria the most well known. We still see how this method stopped Jewish immigration to Palestine, and caused reverse immigration of Jews from Palestine. The successful attempts of dealing defeat to invaders using guerilla warfare were many, and we will not expound on them. However, these attempts have proven that the most effective method for the materially weak against the strong is guerrilla warfare.[33]

But, as useful as Iraq undoubtedly has been as a rallying cry for jihad, it has been a conspicuously less prominent rallying point, at least in terms of men and money. The Coalition Provisional Authority (CPA) may be right that hundreds, perhaps even a few thousand, of foreign fighters have converged on Iraq. But few who have been captured have any demonstrable *direct* ties to Al Qaeda. Nor is there evidence of any direct command-and-control relationship between the Al Qaeda central leadership and the insurgents. If there are Al Qaeda warriors in Iraq, they are likely cannon fodder, possibly recruited through Al Qaeda networks and routed to Iraq via jihadist "rat lines," rather than battle-hardened, veteran mujahideen.

Al Qaeda's interest in Iraq, therefore, appears partly to exploit the occupation as a rousing propaganda and recruitment tool for the global jihadist cause.[34] Its primary intention may in fact have been to preoccupy the U.S with Iraq, thus enabling Al Qaeda and its affiliates to strike elsewhere. For them, Iraq is but one of many battlefields scattered throughout the world. Bin Laden and Al Qaeda's propagandists have long and repeatedly said as much. Indeed, while America has been tied down in Iraq, the international terrorism network has been active on other fronts. Various attacks undertaken by Al Qaeda and its affiliates since the occupation began have taken place in countries that are longstanding sources of bin Laden's enmity (e.g., Saudi Arabia)[35] or where an opportunity has presented itself (e.g., the suicide bombings by associated groups in Morocco last May,[36] Indonesia last August, and Turkey in November). In fact, although Saif al-Adel, the senior Al Qaeda operational commander who wrote the aforementioned *In the Shadow of the Lances* installments, is believed to have been behind the series of five suicide attacks that rocked Riyadh last May,[37] he has yet to be linked to any incidents in Iraq. Thus, it may not be a coincidence that, within weeks of President Bush's May 2003 declaration of the end of the war in Iraq, the jihadists struck in quick succession in Saudi Arabia and Morocco.

Finally, as America bears down on Iraq, Al Qaeda is doubtless bearing down on the United States. Chatter on Al Qaeda-linked web sites has revealed that the jihadists are constantly monitoring America, studying and gauging reactions to intelligence gathered on them and adapting their own plans accordingly. "If we know the importance of the information for the enemy, even if it is a small piece of information, then we can understand how important are [sic] the information that we know," was the admonition that appeared on a jihadist web site in June 2003.[38]

Conclusion: The Future Conduct of the GWOT

In summary, three main themes have framed this article:

- First, terrorism and the terrorist threat are changing. This is as much a reflection of success in the war against terrorism, as of the terrorists' own determination, adaptation, adjustment, and resiliency. There is no doubt, therefore, that the United States faces a formidable, capable, and implacable enemy.
- Second, the war on terrorism as it has unfolded over the past two plus years has come to bear little resemblance to any past conventional wars that the United States has fought. Because of the American military's overwhelming technological and doctrinal superiority, U.S. adversaries are deliberately using asymmetric warfare (e.g., terrorism and insurgency) and hopefully to lock the United States into a war of attrition. Indeed, whereas since the end of the Cold War the United States has become used to wars that last months if not weeks (e.g., the 1991 Gulf War and the 2002 invasion of Iraq), its adversaries have defined this conflict as a war of attrition: designed to wear down resolve, undermine confidence in U.S. leaders, and erode public support for the GWOT. In this respect, they have envisioned a struggle lasting years, if not decades. How this war is effectively fought will therefore require new approaches and perspectives and a different mindset from that which the United States brought to previous such conflicts.
- Third, terrorism itself is becoming a more diffuse and amorphous phenomenon: less centralized and with more opaque command and control relationships. The traditional way of understanding terrorism and looking at terrorists based on organizational definitions and attributes given Al Qaeda's evolution and development is no longer relevant. This inevitably will necessitate changes in how this evolving form of terrorism is studied and countered.

Given these observations, what can the United States do given these changed circumstances and this highly dynamic threat? Eight broad imperatives or policy options appear most relevant.

1. The preeminent lesson of 9/11 is not to be lulled into a false sense of complacency or to rest on past laurels: especially in a struggle that U.S. adversaries have defined as a war of attrition. In these circumstances, the main challenge faced is to retain focus and maintain vigilance and keep up pressure on terrorists by adapting and adjusting—rapidly and efficiently—to the changes unfolding with respect to terrorism. To do so, the United States needs to better understand Al Qaeda's operations and evolution and thus more effectively anticipate changes in radical international jihadism and better assess the implications of those changes.
2. It must be ensured that the new Iraq succeeds. The stakes are enormous. Regardless of whether or not it is Al Qaeda's central front, Iraq has become a critical arena and test of America's strength and resolve. That a democratic, stable government takes root in Iraq, that the Iraqi people are united in having a stake in that outcome, that security is achieved throughout the country; and, that the January 2005 democratic elections are held have indisputably become among the most important metrics not only for assessing success in Iraq, but inevitably now in the war on terrorism. Ensuring that Iraq succeeds, however, is an international imperative, not only an American one. To that end, increased and active international involvement and assistance (including military commitments involving troop deployments along with civilian expertise) will be needed to strengthen the new Iraqi government and facilitate its ability to stabilize the country and promote the longevity of the democratic values to which it aspires.
3. America must systematically and thoroughly overhaul communications with, and create a more positive image of, the United States in the Muslim world. These communications were already fractured and efforts both stillborn and maladroit before the invasion of Iraq and the revelations about the treatment of Iraqi detainees at Abu Ghraib surfaced. Fixing these efforts and repairing the damage done has accordingly become critical. The United States today is increasingly viewed as a malignant force among Muslims throughout the world: thus furnishing Al Qaeda propagandists with fresh ammunition and alienating precisely that community that must be America's closest allies in the struggle against terrorism. Greater resources and more sustained focused efforts need to be committed to improving public diplomacy in the Muslim world as well as to develop more effective initiatives to counter the messages of radicalism and hate promulgated with greater fervor by the jihadists.
4. Part and parcel of this, the United States should recognize that it cannot compete with Al-Jazeera and Al-Arabiya and other Arab media simply by creating rival outlets such as the Arabic-language television station, Al-Hura, and radio station, Radio Sawa. In addition to those American-backed stations, which will inevitably take time to win their own significant audience share, America must meanwhile find ways to communicate more effectively using precisely those often hostile media like Al-Jazeera and Al-Arabiya to get its message across and directly challenge and counter the misperceptions that they foster.
5. America must address and conclusively resolve the open-ended legal status of the Guantanamo detainees and others held elsewhere. This is already a growing source of worldwide anger and opprobrium directed at the United States, especially in the aftermath of the Abu Ghraib revelations. Failure to arrive at an acceptable international legal determination regarding the detainees' status and ultimate disposition will remain an open sore in how the United States is perceived abroad and especially in the Muslim world.

6. America must embark on a renewed and concerted effort to resolve the Palestinian–Israeli conflict. Neither Americans nor anyone else should be under any illusion that resolving this conflict will magically end global terrorism. Bin Laden and Al Qaeda in fact took root and flowered in the late-1990s-precisely at a time when Palestinian–Israeli relations were at their zenith as a result of the Oslo Accords. But, it is nonetheless indisputable that being seen to play a more active and equitable role in resolving this conflict will have an enormously salutary effect on Middle Eastern stability, global Muslim attitudes toward the United States, and America's image abroad. The active involvement and assistance of its European allies and Russia as well as the United Nations will be a vital part of any reinvigorated peace process.
7. America must more instinctively regard its relations with friends and allies in the war on terrorism as a perishable commodity: not taken for granted and regularly repaired, replenished, and strengthened. Notwithstanding the sometimes profound policy differences that surfaced between the United States and even some of its closest allies over the war in Iraq, working-level intelligence and law enforcement cooperation in the war on terrorism has remained remarkably strong. However, these critically important relationships should neither be taken for granted nor be allowed to weaken. This will entail repeated and ongoing sharing of intelligence, consultation and consensus, and continued unity of effort if America and its allies are to prevail against the international jihadist threat. Moreover, for the war on terrorism to succeed, enhanced multilateral efforts will need to be strengthened to accompany the already existent, strong bilateral relations.
8. Finally, the United States must enunciate a clear policy for countering terrorism and from that policy develop a comprehensive strategy. Nearly three years into this global war on terrorism there is no clear policy and there is, in turn, a too vague and ill-formed strategy. In the confrontation with communism following World War II, the United States did not declare a "war on communism." Rather, it articulated the policy of containment and within that intellectual framework developed a clever, comprehensive, multifaceted strategy—that did not rely exclusively on the military option—to serve that policy. Similar clarity of thought and focus is urgently needed today with respect to the global war on terrorism to guide and shape thinking and direct efforts through the subsequent phases of what will likely be a long struggle.

Notes

1. "U.S.: Al Qaida is 70 percent gone, their 'days are numbered,'" *World Tribune. Com* 23 January 2004.
2. Sarah el Deeb, "Al-Qaida Reportedly Plans Big New Attack," *Associated Press*, 8 May 2003.
3. Information provided by Singaporean authorities, June 2002 and September 2004. See also Ministry of Home Affairs, *White Paper: The Jemaah lslamiyah Arrests And The Threat of Terrorism: Presented to Parliament by Command of The President of the Republic of Singapore*, Cmd. 2 of 2003 (Singapore: Ministry of Home Affairs, 7 January 2003), pp. 10–17.
4. See, for example, Craig S. Smith, "Turkey Links Synagogue Bombers to Al Qaeda," New York Times, 18 November 2003; Louis Meixler, "Suicide bombers kill 27 in attacks on British consulate," *Associated Press*, 20 November 2003; Michael Isikoff and Mark Hosenball, "Al Qaeda's New Strategy: In Turkey, the terror group adopted a new strategy of directing homegrown militants," *Newsweek*, 10 December 2003.
5. See Associated Press, "Videotape Claims Responsibility for Madrid Attacks on Behalf of Al-Qaeda," New York Times, 13 March 2004; idem., "Transcript of Purported Al Qaeda Videotape," *New York Times*, 14 March 2004; CNN.com, "Video claims Al Qaeda to blame," 15 March 2004 available at (`http://www.cnn.worldnews/printthis.htm`); Glenn Frankel, Peter Finn and Keith B. Richburg, "Al Qaeda Implicated In Madrid Bombings," *Washington Post*, 15 March 2004; Craig S. Smith, "A Long Fuse Links Tangier to Bombings in Madrid," *New York Times*, 28 March 2004; Mark Huband, "Tangier's unlikely hothouse of radicalism" and Joshua Levitt, "Hunt for Madrid bomb suspects ends as four blow themselves up," *Financial Times* (London), 5 April 2004.
6. See, for example, Associated Press, "Expert Warns of al-Qaida-Linked Groups," 7 January 2004; Ellen Nakashima, "Thai Officials Probe Tie To Al Qaeda in Attacks," *Washington Post*, 9 January 2004; Associated Press, "Saudis Discover al-Qaida Training Camps," 15 January 2004.
7. Indeed, previous "high-end" Al Qaeda plots predating its comfortable relationship with the Taleban in Afghanistan demonstrate that the movement's strength is not in geographical possession or occupation of a defined geographical territory, but in its fluidity and impermanence. The activities of the peripatetic Rarnzi Ahmad Yousef, reputed mastermind of the first World Trade Center bombing, during his sojourn in the Philippines during 1994 and 1995 is a case in point. Yousefs grand scheme to bomb simultaneously 12 American commercial aircraft in midflight over the Pacific Ocean (the infamous "Bojinka" plot), for example, did not require extensive operational bases and command and control headquarters in an existing country to facilitate the planning and execution of those attacks.
8. See William Wallis, "Kenya terror attacks 'planned from Somalia,'" *Financial Times*, 5 November 2003.
9. This point is also made in International Institute for Strategic Studies, *Strategic Survey 2003/4* (Oxford; Oxford University Press, 2004), p. 6, where the authors note: "The Afghanistan intervention offensively hobbled, but defensively benefited, al-Qaeda. While al-Qaeda lost a recruiting magnet and a training, command and operations base, it was compelled to disperse and become even more decentralized, 'virtual' and invisible."
10. See Alan Cullison, "Inside Al-Qaeda's Hard Drive: A fortuitous discovery reveals budget squabbles, baby pictures, office rivalries—and the path to 9/11," *The Atlantic Monthly* 294(2), no. 2 (September 2004), pp. 63–64.
11. Presentation by CNN correspondent Mike Boetcher, at the "Centre for the Study of Terrorism and Political Violence Symposium on Islamic Extremism and Terrorism in the Greater Middle East," University of St Andrews, St Andrews, Scotland, 7–8 June 2002.
12. See, especially, Jason Burke, "Think Again: Al Qaeda," *Foreign Policy* (May/June 2004), available at (`http://www.foreignpolicy.com`).
13. Ibid.
14. See Dana Priest and Walter Pincus, "New Target and Tone: Message Shows Al Qaeda's Adaptability," *Washington Post*, 16 April 2004; and, Geoffrey Nunberg, "Bin Laden's Low-Tech Weapon," *New York Times*, 18 April 2004.

15. Associated Press, "Purported al-Qaida Tape Warns of Attacks," 19 December 2003.
16. International Institute for Strategic Studies, *Strategic Survey 2003/4* (Oxford; Oxford University Press, 2004), p. 6. As one commentary explained, the "IISS's figure of 18,000 potential operatives is calculated by deducting the 2,000 suspects killed or captured since the September 11, 2001 attacks from the estimated 20,000 recruits thought to have passed through al-Qaeda training camps in Afghanistan between 1996 and 2001." Mark Huband and David Buchan, "Al Qaeda may have access to 18,000 'potential operatives', says think-tank," *Financial Times* (London), 26 May 2004. A figure of 20,000 is similarly cited in *Staff Report No. 15,* the National Commission on Terrorist Attacks Upon the United States ("9/11 Commission") on p. 10.

    Note that according to the report issued by the Joint Inquiry of the Senate and House Intelligence Committees, an estimated 70–120,000 persons trained in Afghanistan between 1979–2001, thus suggesting that the IISS estimate could be a conservative one. See Joint Inquiry of the Senate and House Intelligence Committees, p. 38; and "Al Qaeda Trained at Least 70,000 in Terrorist Camps, Senator Says," *Los Angeles Times*, 14 July 2003.
17. See, for example, David Johnston and David E. Sanger, "New Generation of Leaders is Emerging for Al Qaeda," *New York Times*, 10 August 2004; Josh Meyer and Greg Miller, "Fresh Details Back Threats," *Los Angeles Times*, 3 August 2004; Walter Pincus and John Mintz, "Pakistani-U.S. Raid Uncovered terrorist Cell's Surveillance Data," *Washington Post*, 2 August 2004; and, Glen Kessler, "Old Data, New Credibility Issues," *Washington Post*, 4 August 2004.
18. Electronic newsletter of the Orion Group, 24 April 2003.
19. Peter Bergen, *Holy Terror, Inc.: Inside the Secret World of Osama bin Laden* (New York: Free Press, 2001), pp. 47–49.
20. Anonymous, *Through Our Enemies' Eyes: Osama bin Laden, Radical Islam, and the Future of America* (Dulles, VA: Brassey's, 2002), p. 34. This book, written by a 20-year veteran of the CIA's operations directorate, is without doubt the preeminent work on bin Laden and Al Qaeda.
21. IISS, *Strategic Survey*, p. 8. In its *Staff Report No. 16*, the National Commission on Terrorist Attacks Upon the United States ("9/11 Commission") notes on p. 16 that the September 11th operation cost between $400–500,000 to mount.
22. Burke, "Think Again: Al Qaeda."
23. Moqrin was reportedly killed by Saudi security forces in Riyadh on 18 June 2004.
24. IntelCenter, al-Qaeda Targeting Guidance—v.1.0, Thursday, 1 April 2004 (Alexandria, VA: IntelCenter/Tempest Publishing, 2004), pp. 6–9.
25. See Neal MacFarquhar, "As Terrorists Strike Arab Targets, Escalation Fears Arise," *New York Times*, 30 April 2004; idem., "Firm Pulls 100 From Saudi Arabia After 5 Deaths," 2 May 2004; idem., "After Attack, Company's Staff Plans to Leave Saudi Arabia," 3 May 2004; and, Kim Ghattas and Roula Khalaf, "Shooting spree in Saudi city spreads jitters among western companies," *Financial Times* (London), 3 May 2004.
26. Craig Whitlock, "Islamic Radicals Behead American in Saudi Arabia," *Washington Post*, 19 June 2004.
27. Fabrizio Quattrocchi, an Italian and Henrik Frandsen, a Dane, and Hussein Ali Alyan a Leanese are the other foreign kidnap victims who have killed. See "Kidnapped in Iraq," *Washington Post*, 23 June 2004.
28. IntelCenter, al-Qaeda Targeting Guidance—v.1.0, Thursday, 1 April 2004 (Alexandria, VA: IntelCenter/Tempest Publishing, 2004), pp. 6–9.
29. Ibid.
30. Quoted in Anonymous, *Imperial Hubris: Why America is Losing the War Against Terrorism* (Alexandria, VA: Brassey's, 2004), p. xxi.
31. Quoted in Bruce Hoffman, "Saddam Is Ours. Does Al Qaeda Care?" *New York Times*, 17 December 2003.
32. Quoted in Associated Press, "Purported al-Qaida Tape Warns of Attacks."
33. MEMRI Special Dispatch-Jihad and Terrorism Studies, 11 April 2003, no. 493 quoting (`http://www.cubezero.nt/vhsvideo/imagis/?subject=2&rec=1043`).
34. See, for example, Daniel Williams, "Italy Targeted By Recruiters For Terrorists," *Washington Post*, 17 December 2003 where it notes how a suicide bomber recruited in Italy is believed to have been responsible for the August 2003 bombing of the UN Headquarters in Baghdad and the rocket attack on the al-Rashid Hotel there in October.
35. See Simon Henderson and Matthew Levitt, "U.S.-Saudi Counterterrorism Cooperation In The Wake Of The Riyadh Bombing," Policywatch, No. 759, *Washington Institute of Near East Policy*, 23 May 2003; and, Craig Whitlock, "For Saudi Arabia, Al Qaeda Threat Is Now Hitting Home," *Washington Post*, 8 June 2004.
36. See, for example, Agence France Presse, "Qaeda 'paid' $50,000 to Morocco Bombers," 25 May 2003; Eliane Sciolino, "Moroccans Say Al Qaeda Was Behind Casablanca Bombings," *New York Times*, 23 May 2003; Jonathan Schanzer, "Intensify the hunt," *The Baltimore Sun*, 28 May 2003.
37. See Dana Priest and Susan Schmidt, "Al Qaeda Figure Tied to Riyadh Blasts," *Washington Post*, 18 May 2003.
38. "Do Not Be With the Enemy Against Us," article posted on *Islamic Studies and Research* (Al Qaeda-affiliated website), June 2003.

This article was presented at the RAND Center for Middle East Public Policy and Geneva Center for Security Policy 5th Annual Conference, "The U.S., Europe and the Wider Middle East," Geneva, Switzerland, 27–29 June 2004. This is the third such article on this subject that the author has presented at this series of conferences, each building on, and therefore incorporating some material from its predecessors. The two previous articles were subsequently published as "Rethinking Terrorism and Counterterrorism Since 9/11," *Studies in Conflict and Terrorism* 25(5) (September–October 2002); and, "Al Qaeda, Trends in Terrorism, and Future Potentialities: An Assessment," *Studies in Conflict and Terrorism* 26(6) (November–December 2003).

Address correspondence to Bruce Hoffman, the RAND Corporation, Washington, DC, USA. E-mail: hoffman@rand.org

The Terrorism to Come

By WALTER LAQUEUR

TERRORISM HAS BECOME over a number of years the topic of ceaseless comment, debate, controversy, and search for roots and motives, and it figures on top of the national and international agenda. It is also at present one of the most highly emotionally charged topics of public debate, though quite why this should be the case is not entirely clear, because the overwhelming majority of participants do not sympathize with terrorism.

Confusion prevails, but confusion alone does not explain the emotions. There is always confusion when a new international phenomenon appears on the scene. This was the case, for instance, when communism first appeared (it was thought to be aiming largely at the nationalization of women and the burning of priests) and also fascism. But terrorism is not an unprecedented phenomenon; it is as old as the hills.

Thirty years ago, when the terrorism debate got underway, it was widely asserted that terrorism was basically a left-wing revolutionary movement caused by oppression and exploitation. Hence the conclusion: Find a political and social solution, remedy the underlying evil—no oppression, no terrorism. The argument about the left-wing character of terrorism is no longer frequently heard, but the belief in a fatal link between poverty and violence has persisted. Whenever a major terrorist attack has taken place, one has heard appeals from high and low to provide credits and loans, to deal at long last with the deeper, true causes of terrorism, the roots rather than the symptoms and outward manifestations. And these roots are believed to be poverty, unemployment, backwardness, and inequality.

It is not too difficult to examine whether there is such a correlation between poverty and terrorism, and all the investigations have shown that this is not the case. The experts have maintained for a long time that poverty does not cause terrorism and prosperity does not cure it. In the world's 50 poorest countries there is little or no terrorism. A study by scholars Alan Krueger and Jitka Maleckova reached the conclusion that the terrorists are not poor people and do not come from poor societies. A Harvard economist has shown that economic growth is closely related to a society's ability to manage conflicts. More recently, a study of India has demonstrated that terrorism in the subcontinent has occurred in the most prosperous (Punjab) and most egalitarian (Kashmir, with a poverty ratio of 3.5 compared with the national average of 26 percent) regions and that, on the other hand, the poorest regions such as North Bihar have been free of terrorism. In the Arab countries (such as Egypt and Saudi Arabia, but also in North Africa), the terrorists originated not in the poorest and most neglected districts but hailed from places with concentrations of radical preachers. The backwardness, if any, was intellectual and cultural—not economic and social.

It is no secret that terrorists operating in Europe and America are usually of middle-class origin.

These findings, however, have had little impact on public opinion (or on many politicians), and it is not difficult to see why. There is the general feeling that poverty and backwardness with all their concomitants are bad—and that there is an urgent need to do much more about these problems. Hence the inclination to couple the two issues and the belief that if the (comparatively) wealthy Western nations would contribute much more to the development and welfare of the less fortunate, in cooperation with their governments, this would be in a long-term perspective the best, perhaps the only, effective way to solve the terrorist problem.

Reducing poverty in the Third World is a moral as well as a political and economic imperative, but to expect from it a decisive change in the foreseeable future as far as terrorism is concerned is unrealistic, to say the least. It ignores both the causes of backwardness and poverty and the motives for terrorism.

Poverty combined with youth unemployment does create a social and psychological climate in which Islamism and various populist and religious sects flourish, which in turn provide some of the footfolk for violent groups in internal conflicts. According to some projections, the number of young unemployed in the Arab world and North Africa could reach 50 million in two decades. Such a situation will not be conducive to political stability; it will increase the demographic pressure on Europe, since according to polls a majority of these young people want to emigrate. Politically, the populist discontent will be directed against the rulers—Islamist in Iran, moderate in countries such as Egypt, Jordan, or Morocco. But how to help the failed economies of the Middle East and North Africa? What are the reasons for backwardness and stagnation in this part of the world? The countries that have made economic progress—such as China and India, Korea and Taiwan, Malaysia and Turkey—did so without massive foreign help.

All this points to a deep malaise and impending danger, but not to a direct link between the economic situation and international terrorism. There is of course a negative link: Terrorists will not hesitate to bring about a further aggravation in the situation; they certainly did great harm to the tourist industries in Bali and Egypt, in Palestine, Jordan, and Morocco. One of the main targets of terrorism in Iraq was the oil industry. It is no longer a secret that the carriers of international terrorism operating in Europe and America hail not from the poor, downtrodden, and unemployed but are usually of middle-class origin.

The local element

THE LINK BETWEEN terrorism and nationalist, ethnic, religious, and tribal conflict is far more tangible. These instances of terrorism are many and need not be enumerated in detail. Solving these conflicts would probably bring about a certain reduction in the incidence of terrorism. But the conflicts are many, and if some of them have been defused in recent years, other, new ones have emerged. Nor are the issues usually clear-cut or the bones of contention easy to define—let alone to solve.

If the issue at stake is a certain territory or the demand for autonomy, a compromise through negotiations might be achieved. But it ought to be recalled that al Qaeda was founded and September 11 occurred not because of a territorial dispute or the feeling of national oppression but because of a religious commandment—jihad and the establishment of *shari'ah.* Terrorist attacks in Central Asia and Morocco, in Saudi Arabia, Algeria, and partly in Iraq were directed against fellow Muslims, not against infidels. Appeasement may work in individual cases, but terrorist groups with global ambitions cannot be appeased by territorial concessions.

As in the war against poverty, the initiatives to solve local conflicts are overdue and should be welcomed. In an ideal world, the United Nations would be the main conflict resolver, but so far the record of the U.N. has been more than modest, and it is unlikely that this will change in the foreseeable future. Making peace is not an easy option; it involves funds and in some cases the stationing of armed forces. There is no great international crush to join the ranks of the volunteers: China, Russia, and Europe do not want to be bothered, and the United States is overstretched. In brief, as is so often the case, a fresh impetus is likely to occur only if the situation gets considerably worse and if the interests of some of the powers in restoring order happen to coincide.

Lastly, there should be no illusions with regard to the wider effect of a peaceful solution of one conflict or another. To give but one obvious example: Peace (or at least the absence of war) between Israel and the Palestinians would be a blessing for those concerned. It may be necessary to impose a solution since the chances of making any progress in this direction are nil but for some outside intervention. However, the assumption that a solution of a local conflict (even one of great symbolic importance) would have a dramatic effect in other parts of the world is unfounded. Osama bin Laden did not go to war because of Gaza and Nablus; he did not send his warriors to fight in Palestine. Even the disappearance of the "Zionist entity" would not have a significant impact on his supporters, except perhaps to provide encouragement for further action.

Osama bin Laden did not go to war because of Gaza and Nablus.

Such a warning against illusions is called for because there is a great deal of wishful thinking and naïveté in this respect—a belief in quick fixes and miracle solutions: If only there would be peace between Israelis and Palestinians, all the other conflicts would become manageable. But the problems are as much in Europe, Asia, and Africa as in the Middle East; there is a great deal of free-floating aggression which could (and probably would) easily turn in other directions once one conflict has been defused.

It seems likely, for instance, that in the years to come the struggle against the "near enemy" (the governments of the Arab and some non-Arab Muslim countries) will again feature prominently. There has been for some time a truce on the part of al Qaeda and related groups, partly for strategic reasons (to concentrate on the fight against America and the West) and partly because attacks against fellow Muslims, even if they are considered apostates, are bound to be less popular than fighting the infidels. But this truce, as events in Saudi Arabia and elsewhere show, may be coming to an end.

Tackling these supposed sources of terrorism, even for the wrong reasons, will do no harm and may bring some good. But it does not bring us any nearer to an understanding of the real sources of terrorism, a field that has become something akin to a circus ground for riding hobbyhorses and peddling preconceived notions.

How to explain the fact that in an inordinate number of instances where there has been a great deal of explosive material, there has been no terrorism? The gypsies of Europe certainly had many grievances and the Dalets (untouchables) of India and other Asian countries even more. But there has been no terrorism on their part—just as the Chechens have been up in arms but not the Tartars of Russia, the Basque but not the Catalans of Spain. The list could easily be lengthened.

Accident may play a role (the absence or presence of a militant leadership), but there could also be a cultural-psychological predisposition. How to explain that out of 100 militants believing with equal intensity in the justice of their cause, only a very few will actually engage in terrorist actions? And out of this small minority even fewer will be willing to sacrifice their lives as suicide bombers? Imponderable factors might be involved: indoctrination but also psychological motives. Neither economic nor political analysis will be of much help in gaining an understanding, and it may not be sheer accident that there has been great reluctance to explore this political-intellectual minefield.

The focus on Islamist terrorism

TO MAKE PREDICTIONS about the future course of terrorism is even more risky than political predictions in general. We are dealing here not with mass movements but small—sometimes very small—groups of people, and there is no known way at present to account for the movement of small particles either in the physical world or in human societies.

It is certain that terrorism will continue to operate. At the present time almost all attention is focused on Islamist terrorism, but it is useful to remember from time to time that this was not always the case—even less than 30 years ago—and that there are a great many conflicts, perceived oppressions, and other causes calling for radical action in the world which may come to the fore in the years to come. These need not even be major conflicts in an age in which small groups will have access to weapons of mass destruction.

At present, Islamist terrorism all but monopolizes our attention, and it certainly has not yet run its course. But it is unlikely that its present fanaticism will last forever; religious-nationalist fervor does not constantly burn with the same intensity. There is a phenomenon known in Egypt as "Salafi burnout," the mellowing of radical young people, the weakening of the original fanatical impetus. Like all other movements in history, messianic groups are subject to routinization, to the circulation of generations, to changing political circumstances, and to sudden or gradual changes in the intensity of religious belief. This could happen as a result of either victories or defeats. One day, it might be possible to appease militant Islamism—though hardly in a period of burning aggression when confidence and faith in global victory have not yet been broken.

More likely the terrorist impetus will decline as a result of setbacks. Fanaticism, as history shows, is not easy to transfer from one generation to the next; attacks will continue, and some will be crowned with success (perhaps spectacular success), but many will not. When Alfred Nobel invented dynamite, many terrorists thought that this was the answer to their prayers, but theirs was a false hope. The trust put today in that new invincible weapon, namely suicide terrorism, may in the end be equally misplaced. Even the use of weapons of mass destruction might not be the terrorist panacea some believe it will be. Perhaps their effect will be less deadly than anticipated; perhaps it will be so destructive as to be considered counterproductive. Statistics show that in the terrorist attacks over the past decade, considerably more Muslims were killed than infidels. Since terrorists do not operate in a vacuum, this is bound to lead to dissent among their followers and even among the fanatical preachers.

Over the past decade, more Muslims were killed in terrorist attacks than infidels.

There are likely to be splits among the terrorist groups even though their structure is not highly centralized. In brief, there is a probability that a united terrorist front will not last. It is unlikely that Osama and his close followers will be challenged on theological grounds, but there has been criticism for tactical reasons: Assuming that America and the West in general are in a state of decline, why did he not have more patience? Why did he have to launch a big attack while the infidels were still in a position to retaliate massively?

Some leading students of Islam have argued for a long time that radical Islamism passed its peak years ago and that its downfall and disappearance are only a question of time, perhaps not much time. It is true that societies that were exposed to the rule of fundamentalist fanatics (such as Iran) or to radical Islamist attack (such as Algeria) have been immunized to a certain extent. However, in a country of 60 million, some fanatics can always be found; as these lines are written, volunteers for suicide missions are being enlisted in Teheran and other cities of Iran. In any case, many countries have not yet undergone such first-hand experience; for them the rule of the *shari'ah* and the restoration of the caliphate are still brilliant dreams. By and large, therefore, the predictions about the impending demise of Islamism have been premature, while no doubt correct in the long run. Nor do we know what will follow. An interesting study on what happens "when prophecy fails" (by Leon Festinger) was published not long after World War II. We now need a similar study on the likely circumstances and consequences of the failure of fanaticism. The history of religions (and political religions) offers some clues, as does the history of terrorism.

These, then, are the likely perspectives for the more distant future. But in a shorter-term perspective the danger remains acute and may, in fact, grow. Where and when are terrorist attacks most likely to occur? They will not necessarily be directed against the greatest and most dangerous enemy as perceived by the terrorist gurus. Much depends on where terrorists are strong and believe the enemy to be weak. That terrorist attacks are likely to continue in the Middle East goes without saying; other main danger zones are Central Asia and, above all, Pakistan.

The founders of Pakistan were secular politicians. The religious establishment and in particular the extremists among the Indian Muslims had opposed the emergence of the state. But once Pakistan came into being, they began to try with considerable success to dominate it. Their alternative educational system, the many thousand madrassas, became the breeding ground for jihad fighters. Ayub Khan, the first military ruler, tried to break their stranglehold but failed. Subsequent rulers, military and civilian, have not even tried. It is more than doubtful whether Pervez Musharraf will have any success in limiting their power. The tens of thousands of graduates they annually produce formed the backbone of the Taliban. Their leaders will find employment for them at home and in Central Asia, even if there is a deescalation in tensions with India over Kashmir. Their most radical leaders aim at the destruction of India. Given Pakistan's internal weakness this may appear more than a little fanciful, but their destructive power is still considerable, and they can count on certain sympathies in the army and the intelligence service. A failed Pakistan with nuclear weapons at its disposal would be a major nightmare. Still, Pakistani terrorism—like Palestinian and Middle Eastern in general—remains territorial, likely to be limited to the subcontinent and Central Asia.

Battlefield Europe

EUROPE IS PROBABLY the most vulnerable battlefield. To carry out operations in Europe and America, talents are needed that are not normally found among those who have no direct personal experience of life in the West. The Pakistani diaspora has not been very active in the terrorist field, except for a few militants in the United Kingdom.

Western Europe has become over a number of years the main base of terrorist support groups. This process has been facilitated by the growth of Muslim communities, the growing tensions with the native population, and the relative freedom with which radicals could organize in certain mosques and cultural organizations. Indoctrination was provided by militants who came to these countries as religious dignitaries. This freedom of action was considerably greater than that enjoyed in the Arab and Muslim world; not a few terrorists convicted of capital crimes in countries such as Egypt, Jordan, Morocco, and Algeria were given political asylum in Europe. True, there were some arrests and closer controls after September 11, but given the legal and political restrictions under which the European security services were laboring, effective counteraction was still exceedingly difficult.

West European governments have been frequently criticized for not having done enough to integrate Muslim newcomers into their societies, but cultural and social integration was certainly not what the newcomers wanted. They wanted to preserve their religious and ethnic identity and their way of life, and they resented intervention by secular authorities. In its great majority, the first generation of immigrants wanted to live in peace and quiet and to make a living for their families. But today they no longer have much control over their offspring.

Non-Muslims began to feel threatened in streets they could once walk without fear.

This is a common phenomenon all over the world: the radicalization of the second generation of immigrants. This generation has been superficially acculturated (speaking fluently the language of the host country) yet at the same time feels resentment and hostility more acutely. It is not necessarily the power of the fundamentalist message (the young are not the most pious believers when it comes to carrying out all the religious commandments) which inspires many of the younger radical activists or sympathizers. It is the feeling of deep resentment because, unlike immigrants from other parts of the world, they could not successfully compete in the educational field, nor quite often make it at the work place. Feelings of being excluded, sexual repression (a taboo subject in this context), and other factors led to free-floating aggression and crime directed against the authorities and their neighbors.

As a result, non-Muslims began to feel threatened in streets they could once walk without fear. They came to regard the new immigrants as antisocial elements who wanted to change the traditional character of their homeland and their way of life, and consequently tensions continued to increase. Pressure on European governments is growing from all sides, right and left, to stop immigration and to restore law and order.

This, in briefest outline, is the milieu in which Islamist terrorism and terrorist support groups in Western Europe developed. There is little reason to assume that this trend will fundamentally change in the near future. On the contrary, the more the young generation of immigrants asserts itself, the more violence occurs in the streets, and the more terrorist attacks take place, the greater the anti-Muslim resentment on the part of the rest of the population. The rapid demographic growth of the Muslim communities further strengthens the impression among the old residents that they are swamped and deprived of their rights in their own homeland, not even entitled to speak the truth about the prevailing situation (such as, for instance, to reveal the statistics of prison inmates with Muslim backgrounds). Hence the violent reaction in even the most liberal European countries such as the Netherlands, Belgium, and Denmark. The fear of the veil turns into the fear that in the foreseeable future they too, having become a minority, will be compelled to conform to the commandments of another religion and culture.

True, the number of extremists is still very small. Among British Muslims, for instance, only 13 percent have expressed sympathy and support for terrorist attacks. But this still amounts to several hundred thousands, far more than needed for staging a terrorist campaign. The figure is suspect in any case because not all of those sharing radical views will openly express them to strangers, for reasons that hardly need be elaborated. Lastly, such a minority will not feel isolated in their own community as long as the majority remains silent—which has been the case in France and most other European countries.

Extremists may be repelled by the decadence of the society facing them, but they are also attracted by it.

The prospects for terrorism based on a substantial Islamist periphery could hardly appear to be more promising, but there are certain circumstances that make the picture appear somewhat less threatening. The tensions are not equally strong in all countries. They are less palpably felt in Germany and Britain than in France and the Netherlands. Muslims in Germany are predominantly of Turkish origin and have (always with some exceptions) shown less inclination to take violent action than communities mainly composed of Arab and North African immigrants.

If acculturation and integration has been a failure in the short run, prospects are less hopeless in a longer perspective. The temptations of Western civilization are corrosive; young Muslims cannot be kept in a hermetically sealed ghetto (even though a strong attempt is made). They are disgusted and repelled by alcohol, loose morals, general decadence, and all the other wickedness of the society facing them, but they are at the same time fascinated and attracted by them. This is bound to affect their activist fervor, and they will be exposed not only to the negative aspects of the world surrounding them but also its values. Other religions had to face these temptations over the ages and by and large have been fighting a losing battle.

It is often forgotten that only a relatively short period passed from the primitive beginnings of Islam in the Arabian desert to the splendor and luxury (and learning and poetry) of Harun al Rashid's Baghdad—from the austerity of the Koran to the not-so-austere Arabian Nights. The pulse of contemporary history is beating much faster, but is it beating fast enough? For it is a race against time. The advent of megaterrorism and the access to weapons of mass destruction is dangerous enough, but coupled with fanaticism it generates scenarios too unpleasant even to contemplate.

Enduring asymmetry

THERE CAN BE no final victory in the fight against terrorism, for terrorism (rather than full-scale war) is the contemporary manifestation of conflict, and conflict will not disappear from earth as far as one can look ahead and human nature has not undergone a basic change. But it will be in our power to make life for terrorists and potential terrorists much more difficult.

Who ought to conduct the struggle against terrorism? Obviously, the military should play only a limited role in this context, and not only because it has not been trained for this purpose. The military may have to be called in for restoring order in countries that have failed to function and have become terrorist havens. It may have to intervene to prevent or stop massacres. It may be needed to deliver blows against terrorist concentrations. But these are not the most typical or frequent terrorist situations.

The key role in asymmetric warfare (a redundant new term for something that has been known for many centuries) should be played by intelligence and security services that may need a military arm.

As far as terrorism and also guerrilla warfare are concerned, there can be no general, overall doctrine in the way that Clausewitz or Jomini and others developed a regular warfare philosophy. An airplane or a battleship do not change their character wherever they operate, but the character of terrorism and guerrilla warfare depends largely on the motivations of those engaging in it and the conditions under which it takes place. Over the past centuries rules and laws of war have developed, and even earlier on there were certain rules that were by and large adhered to.

But terrorists cannot possibly accept these rules. It would be suicidal from their point of view if, to give but one example, they were to wear uniforms or other distinguishing marks. The essence of their operations rests on hiding their identities. On the other hand, they and their well-wishers insist that when captured, they should enjoy all the rights and benefits accorded to belligerents, that they be humanely treated, even paid some money and released after the end of hostilities. When regular soldiers do not stick to the rules of warfare, killing or maiming prisoners, carrying out massacres, taking hostages or committing crimes against the civilian population, they will be treated as war criminals.

If terrorists behaved according to these norms they would have little if any chance of success; the essence of terrorist operations now is indiscriminate attacks against civilians. But governments defending themselves against terrorism are widely expected not to behave in a similar way but to adhere to international law as it developed in conditions quite different from those prevailing today.

Terrorism does not accept laws and rules, whereas governments are bound by them; this, in briefest outline, is asymmetric warfare. If governments were to behave in a similar way, not feeling bound by existing rules and laws such as those against the killing of prisoners, this would be bitterly denounced. When the late Syrian President Hafez Assad faced an insurgency (and an attempted assassination) on the part of the Muslim Brotherhood in the city of Hama in 1980, his soldiers massacred some 20,000 inhabitants. This put an end to all ideas of terrorism and guerrilla warfare.

Such behavior on the part of democratic governments would be denounced as barbaric, a relapse into the practices of long-gone pre-civilized days. But if governments accept the principle of asymmetric warfare they will be severely, possibly fatally, handicapped. They cannot accept that terrorists are protected by the Geneva Conventions, which would mean, among other things, that they should be paid a salary while in captivity. Should they be regarded like the pirates of a bygone age as *hostes generis humani*, enemies of humankind, and be treated according to the principle of *a un corsaire, un corsaire et demi*—"to catch a thief, it takes a thief," to quote one of Karl Marx's favorite sayings?

Should terrorists be regarded, like pirates of a bygone age, as enemies of humankind?

The problem will not arise if the terrorist group is small and not very dangerous. In this case normal legal procedures will be sufficient to deal with the problem (but even this is not quite certain once weapons of mass destruction become more readily accessible). Nor will the issue of shedding legal restraint arise if the issues at stake are of marginal importance, if in other words no core interests of the governments involved are concerned. If, on the other hand, the very survival of a society is at stake, it is most unlikely that governments will be impeded in their defense by laws and norms belonging to a bygone (and more humane) age.

It is often argued that such action is counterproductive because terrorism cannot be defeated by weapons alone, but is a struggle for the hearts and minds of people, a confrontation of ideas (or ideologies). If it were only that easy. It is not the terrorist ideas which cause the damage, but their weapons. Each case is different, but many terrorist groups do not have any specific idea or ideology, but a fervent belief, be it of a religious character or of a political religion. They fight for demands, territorial or otherwise, that seem to them self-evident, and they want to defeat their enemies. They are not open to dialogue or rational debate. When Mussolini was asked about his program by the socialists during the early days of fascism, he said that his program was to smash the skulls of the socialists.

Experience teaches that a little force is indeed counterproductive except in instances where small groups are involved. The use of massive, overwhelming force, on the other hand, is usually effective. But the use of massive force is almost always unpopular at home and abroad, and it will be applied only if core interests of the state are involved. To give but one example: The Russian government could deport the Chechens (or a significant portion), thus solving the problem according to the Stalinist pattern. If the Chechens were to threaten Moscow or St. Petersburg or the functioning of the Russian state or its fuel supply, there is but little doubt that such measures would be taken by the Russian or indeed any other government. But as long as the threat is only a marginal and peripheral one, the price to be paid for the application of massive force will be considered too high.

Two lessons follow: First, governments should launch an anti-terrorist campaign only if they are able and willing to apply massive force if need be. Second, terrorists have to ask themselves whether it is in their own best interest to cross the line between nuisance operations and attacks that threaten the vital interests of their enemies and will inevitably lead to massive counterblows.

Terrorists want total war—not in the sense that they will (or could) mobilize unlimited resources; in this respect their possibilities are limited. But they want their attacks to be unfettered by laws, norms, regulations, and conventions. In the terrorist conception of warfare there is no room for the Red Cross.

Love or respect?

THE WHY-DO-THEY-HATE-US question is raised in this context, along with the question of what could be done about it—that is, the use of soft power in combating terrorism. Disturbing figures have been published about the low (and decreasing) popularity of America in foreign parts. Yet it is too often forgotten that international relations is not a popularity contest and that big and powerful countries have always been feared, resented, and envied; in short, they have not been loved. This has been the case since the days of the Assyrians and the Roman Empire. Neither the Ottoman nor the Spanish Empire, the Chinese, the Russian, nor the Japanese was ever popular. British sports were emulated in the colonies and French culture impressed the local elites in North Africa and Indochina, but this did not lead to political support, let alone identification with the rulers. Had there been public opinion polls in the days of Alexander the Great (let alone Ghengis Khan), the results, one suspects, would have been quite negative.

Big powers have been respected and feared but not loved for good reasons—even if benevolent, tactful, and on their best behavior, they were threatening simply because of their very existence. Smaller nations could not feel comfortable, especially if they were located close to them. This was the case even in times when there was more than one big power (which allowed for the possibility of playing one against the other). It is all the more so at a time when only one superpower is left and the perceived threat looms even larger.

There is no known way for a big power to reduce this feeling on the part of other, smaller countries—short of committing suicide or, at the very least, by somehow becoming weaker and less threatening. A moderate and intelligent policy on the part of the great power, concessions, and good deeds may mitigate somewhat the perceived threat, but it cannot remove it, because potentially the big power remains dangerous. It could always change its policy and become nasty, arrogant, and aggressive. These are the unfortunate facts of international life.

Soft power is important but has its limitations. Joseph S. Nye has described it as based on culture and political ideas, as influenced by the seductiveness of democracy, human rights, and individual opportunity. This is a powerful argument, and it is true that Washington has seldom used all its opportunities, the public diplomacy budget being about one-quarter of one percentage point of the defense budget. But the question is always to be asked: Who is to be influenced by our values and ideas? They could be quite effective in Europe, less so in a country like Russia, and not at all among the radical Islamists who abhor democracy (for all sovereignty rests with Allah rather than the people), who believe that human rights and tolerance are imperialist inventions, and who want to have nothing to do with deeper Western values which are not those of the Koran as they interpret it.

Big, powerful countries have always been feared, resented, and envied.

The work of the American radio stations during the Cold War ought to be recalled. They operated against much resistance at home but certainly had an impact on public opinion in Eastern Europe; according to evidence later received, even the Beatles had an influence on the younger generation in the Soviet Union. But, at present, radio and television has to be beamed to an audience 70 percent of which firmly believes that the operations of September 11 were staged by the Mossad. Such an audience will not be impressed by exposure to Western pop culture or a truthful, matter-of-fact coverage of the news. These societies may be vulnerable to covert manipulation of the kind conducted by the British government during World War II: black (or at least gray) propaganda, rumors, half-truths, and outright lies. Societies steeped in belief in conspiracy theories will give credence to even the wildest rumors. But it is easy to imagine how an attempt to generate such propaganda would be received at home: It would be utterly rejected. Democratic countries are not able to engage in such practices except in a case of a major emergency, which at the present time has not yet arisen.

Big powers will never be loved, but in the terrorist context it is essential that they should be respected. As bin Laden's declarations prior to September 11 show, it was lack of respect for America that made him launch his attacks; he felt certain that the risk he was running was small, for the United States was a paper tiger, lacking both the will and the capability to strike back. After all, the Americans ran from Beirut in the 1980s and from Mogadishu in 1993 after only a few attacks, and there was every reason to believe that they would do so again.

Response in proportion to threat

LIFE COULD BE made more difficult for terrorists by imposing more controls and restrictions wherever useful. But neither the rules of national nor those of international law are adequate to deal with terrorism. Many terrorists or suspected terrorists have been detained in America and in Europe, but only a handful have been put on trial and convicted, because inadmissible evidence was submitted or the authorities were reluctant to reveal the sources of their information—and thus lose those sources. As a result, many who were almost certainly involved in terrorist operations were never arrested, while others were acquitted or released from detention.

As for those who are still detained, there have been loud protests against a violation of elementary human rights. Activists have argued that the real danger is not terrorism (the extent and the consequences of which have been greatly exaggerated) but the war against terrorism. Is it not true that American society could survive a disaster on the scale of September 11 even if it occurred once a year? Should free societies so easily give up their freedoms, which have been fought for and achieved over many centuries?

Some have foretold the coming of fascism in America (and to a lesser extent in Europe); others have predicted an authoritarian regime gradually introduced by governments cleverly exploiting the present situation for their own anti-democratic purposes. And it is quite likely indeed that among those detained there have been and are innocent people and that some of the controls introduced have interfered with human rights. However, there is much reason to think that to combat terrorism effectively, considerably more stringent measures will be needed than those presently in force.

But these measures can be adopted only if there is overwhelming public support, and it would be unwise even to try to push them through until the learning process about the danger of terrorism in an age of weapons of mass destruction has made further progress. Time will tell. If devastating attacks do not occur, stringent anti-terrorist measures will not be necessary. But if they do happen, the demand for effective countermeasures will be overwhelming. One could perhaps argue that further limitations of freedom are bound to be ineffective because terrorist groups are likely to be small or very small in the future and therefore likely to slip through safety nets. This is indeed a danger—but the advice to abstain from safety measures is a counsel of despair unlikely to be accepted.

There are political reasons to use these restrictions with caution, because Muslim groups are bound to be under special scrutiny and every precaution should be taken not to antagonize moderate elements in this community. Muslim organizations in Britain have complained that a young Pakistani or Arab is 10 times more likely to be stopped and interrogated by the police than other youths. The same is true for France and other countries. But the police, after all, have some reasons to be particularly interested in these young people rather than those from other groups. It will not be easy to find a just and easy way out of the dilemma, and those who have to deal with it are not to be envied.

It could well be that, as far as the recent past is concerned, the danger of terrorism has been overstated. In the two world wars, more people were sometimes killed and more material damage caused in a few hours than through all the terrorist attacks in a recent year. True, our societies have since become more vulnerable and also far more sensitive regarding the loss of life, but the real issue at stake is not the attacks of the past few years but the coming dangers. Megaterrorism has not yet arrived; even 9-11 was a stage in between old-fashioned terrorism and the shape of things to come: the use of weapons of mass destruction.

The real issue at stake is not the attacks of the past few years but the coming dangers.

The idea that such weapons should be used goes back at least 150 years. It was first enunciated by Karl Heinzen, a German radical—later a resident of Louisville, Kentucky and Boston, Massachusetts—soon after some Irish militants considered the use of poison gas in the British Parliament. But these were fantasies by a few eccentrics, too farfetched even for the science fiction writers of the day.

Today these have become real possibilities. For the first time in human history very small groups have, or will have, the potential to cause immense destruction. In a situation such as the present one there is always the danger of focusing entirely on the situation at hand—radical nationalist or religious groups with whom political solutions may be found. There is a danger of concentrating on Islamism and forgetting that the problem is a far wider one. Political solutions to deal with their grievances may sometimes be possible, but frequently they are not. Today's terrorists, in their majority, are not diplomats eager to negotiate or to find compromises. And even if some of them would be satisfied with less than total victory and the annihilation of the enemy, there will always be a more radical group eager to continue the struggle.

This was always the case, but in the past it mattered little: If some Irish radicals wanted to continue the struggle against the British in 1921-22, even after the mainstream rebels had signed a treaty with the British government which gave them a free state, they were quickly defeated. Today even small groups matter a great deal precisely because of their enormous potential destructive power, their relative independence, the fact that they are not rational actors, and the possibility that their motivation may not be political in the first place.

Perhaps the scenario is too pessimistic; perhaps the weapons of mass destruction, for whatever reason, will never be used. But it would be the first time in human history that such arms, once invented, had not been used. In the last resort, the problem is, of course, the human condition.

In 1932, when Einstein attempted to induce Freud to support pacifism, Freud replied that there was no likelihood of suppressing humanity's aggressive tendencies. If there was any reason for hope, it was that people would turn away on rational grounds—that war had become too destructive, that there was no scope anymore in war for acts of heroism according to the old ideals.

Freud was partly correct: War (at least between great powers) has become far less likely for rational reasons. But his argument does not apply to terrorism motivated mainly not by political or economic interests, based not just on aggression but also on fanaticism with an admixture of madness.

Terrorism, therefore, will continue—not perhaps with the same intensity at all times, and some parts of the globe may be spared altogether. But there can be no victory, only an uphill struggle, at times successful, at others not.

Walter Laqueur is co-chair of the International Research Council at the Center for Strategic and International Studies. He is the author of some of the basic texts on terrorism, most recently Voices of Terror *(Reed Publishing, 2004).* *The present article is part of a larger project; the author wishes to thank the Earhart Foundation for its support.*

Index

Test Your Knowledge Form

We encourage you to photocopy and use this page as a tool to assess how the articles in *Annual Editions* expand on the information in your textbook. By reflecting on the articles you will gain enhanced text information. You can also access this useful form on a product's book support Web site at *http://www.mhcls.com/online/*.

NAME: DATE:

TITLE AND NUMBER OF ARTICLE:

BRIEFLY STATE THE MAIN IDEA OF THIS ARTICLE:

LIST THREE IMPORTANT FACTS THAT THE AUTHOR USES TO SUPPORT THE MAIN IDEA:

WHAT INFORMATION OR IDEAS DISCUSSED IN THIS ARTICLE ARE ALSO DISCUSSED IN YOUR TEXTBOOK OR OTHER READINGS THAT YOU HAVE DONE? LIST THE TEXTBOOK CHAPTERS AND PAGE NUMBERS:

LIST ANY EXAMPLES OF BIAS OR FAULTY REASONING THAT YOU FOUND IN THE ARTICLE:

LIST ANY NEW TERMS/CONCEPTS THAT WERE DISCUSSED IN THE ARTICLE, AND WRITE A SHORT DEFINITION:

We Want Your Advice

ANNUAL EDITIONS revisions depend on two major opinion sources: one is our Advisory Board, listed in the front of this volume, which works with us in scanning the thousands of articles published in the public press each year; the other is you—the person actually using the book. Please help us and the users of the next edition by completing the prepaid article rating form on this page and returning it to us. Thank you for your help!

ANNUAL EDITIONS: Violence and Terrorism 07/08

ARTICLE RATING FORM

Here is an opportunity for you to have direct input into the next revision of this volume.
We would like you to rate each of the articles listed below, using the following scale:

1. **Excellent: should definitely be retained**
2. **Above average: should probably be retained**
3. **Below average: should probably be deleted**
4. **Poor: should definitely be deleted**

Your ratings will play a vital part in the next revision.
Please mail this prepaid form to us as soon as possible.
Thanks for your help!

RATING	ARTICLE
____________	1. Ghosts of Our Past
____________	2. An Essay on Terrorism
____________	3. The Origins of the New Terrorism
____________	4. Terrorists' New Tactic: Assassination
____________	5. Paying for Terror
____________	6. The Moral Logic and Growth of Suicide Terrorism
____________	7. Iran: Confronting Terrorism
____________	8. The Growing Syrian Missile Threat: Syria after Lebanon
____________	9. Terrorists Don't Need States
____________	10. Guerrilla Nation
____________	11. Extremist Groups in Egypt
____________	12. Colombia and the United States: From Counternarcotics to Counterterrorism
____________	13. Root Causes of Chechen Terror
____________	14. End of Terrorism?
____________	15. Homegrown Terror
____________	16. Speaking for the Animals, or the Terrorists?
____________	17. Women and Organized Racial Terrorism in the United States
____________	18. José Padilla and the War on Rights
____________	19. Terrorism as Breaking News: Attack on America
____________	20. A Violent Episode in the Virtual World
____________	21. Terror's Server
____________	22. High Anxiety
____________	23. Holy Orders: Religious Opposition to Modern States
____________	24. The Madrassa Scapegoat
____________	25. Cross-Regional Trends in Female Terrorism
____________	26. Explosive Baggage: Female Palestinian Suicide Bombers and the Rhetoric of Emotion
____________	27. Girls as "Weapons of Terror" in Northern Uganda and Sierra Leonean Rebel Fighting Forces
____________	28. Port Security Is Still a House of Cards
____________	29. Are We Ready Yet?
____________	30. The Double-Edged Effect in South Asia
____________	31. The Changing Face of Al Qaeda and the Global War on Terrorism
____________	32. The Terrorism to Come

(Continued on next page)

NO POSTAGE NECESSARY IF MAILED IN THE UNITED STATES

BUSINESS REPLY MAIL
FIRST CLASS MAIL PERMIT NO. 551 DUBUQUE IA

POSTAGE WILL BE PAID BY ADDRESEE

McGraw-Hill Contemporary Learning Series
2460 KERPER BLVD
DUBUQUE, IA 52001-9902

ABOUT YOU

Name Date

Are you a teacher? ❒ A student? ❒
Your school's name

Department

Address City State Zip

School telephone #

YOUR COMMENTS ARE IMPORTANT TO US!

Please fill in the following information:
For which course did you use this book?

Did you use a text with this ANNUAL EDITION? ❒ yes ❒ no
What was the title of the text?

What are your general reactions to the *Annual Editions* concept?

Have you read any pertinent articles recently that you think should be included in the next edition? Explain.

Are there any articles that you feel should be replaced in the next edition? Why?

Are there any World Wide Web sites that you feel should be included in the next edition? Please annotate.

May we contact you for editorial input? ❒ yes ❒ no
May we quote your comments? ❒ yes ❒ no